DATE			

Technion

The Story of Israel's Institute of Technology

Carl Alpert

American Technion Society • Technion–Israel Institute of Technology

New York • Haifa

Library of Congress Cataloguing in Publication Data

Alpert, Carl, 1913–
  Technion: the story of Israel's Institute of Technology.

  Bibliography: p.
  Includes index.
  1. Tekhniyon, Makhon tekhnologi le-Yisrael. I. Title.
T173.H139A46 1982   607'.115694   82-11556
ISBN 0-87203-102-0

Manufactured in Israel by Keterpress Enterprises
New York Office: Keter Inc., 440 Park Avenue South, N.Y. 10016

to Nechama
with love and gratitude

"So necessary for our survival...."

The Technion, Israel Institute of Technology, is still one of the corner-stones of Israel's development. Its laboratories and its workshops, its scientific and its technical staff are constantly at the disposal of industry, Government and the Defense Forces of the nation. Even more important to the State is the steady supply of highly trained engineers and men and women of science who are educated at the Technion.... The measure of a nation's security may today be gauged in terms of its ability to keep abreast of the progress of science as well as its mastery of the skills of technology. It is for that reason that we value so highly the Technion where the youth of Israel obtain their education in engineering and the fields of applied science.... As the needs of the State expand, we shall continue to demand of the Technion that more and still more of our young people be provided with the technological training so necessary for our survival....

David Ben Gurion

Preface

Few public institutions have had as exciting, as dramatic a history as Technion, Israel Institute of Technology. It was created against the background of the intense European rivalries which eventually led to the First World War; it was closely associated with the rebirth of Jewish nationalism and the revival of the Hebrew language. Its buildings were occupied successively by the armies of three nations, which used them for purposes quite remote from higher education.

It went bankrupt and was sold at public auction to the highest bidder. It became a bitter bone of contention between realists who "proved" that the training of engineers in Palestine was a waste of money, and visionaries who foresaw a burgeoning industrial economy which would require highly skilled technical manpower.

The Technion withered as a result of strikes and empty coffers. It thrived on the devotion of students, teachers, staff and friends, many of whom cheerfully bore unusual sacrifices to assure its development.

It served as a supply and training center for an illegal underground movement, helped create a national air force, was the center of a famous espionage case and was locale of a Hollywood movie. All this and much more in the period of almost 80 years since someone first proposed establishment of an engineering university in Palestine.

It emerged as Israel's great center of technology, engineering and applied science, serving the general economy and the defense needs of the country alike. Hardly any aspect of the industrial, agricultural, scientific or military complex of life in Israel has not been affected and influenced by Technion's graduates or faculty members, or by the fruits of the research and consultation services which it has made available. Little wonder that David Ben Gurion more than once referred to the Technion as "the cornerstone of the State of Israel".

This is the story of the birth, development and flowering of a major institution of higher learning. In a larger sense it also reflects the story of an epoch in Jewish history which included two world wars, the Holocaust, and the establishment of the State of Israel.

The history of the Technion was created by the men and women who brought the institution into existence, nursed it through difficult days, taught its students, carried out research in its laboratories, and manned its various offices and institutions, academic, administrative and technical. The names of many are mentioned in these pages, but there are countless other individuals who also played their roles faithfully as well. Their names may not appear here but they are not forgotten. For them I cite the text of the inscription written in mosaic on the floor of an ancient synagogue uncovered in Jericho. The plaque gives credit to all who took part in the construction of the holy community, its elders and its youth, who gave of their efforts "... may they be well remembered, may their memory be for good...." Then we expect the list of names to follow. But there are no names at all. The text concludes: "He who knows their names and the names of their children and the names of the people of their households — He shall write them in the Book of Life together with the Just."

This is the story of the Technion, told not only in dates, statistics and resolutions, but also in anecdotes which reflect the human side of the institution. It is not a history of the various Technion Societies, which have played an important role in the growth and development of the Institute. Each has its own story of dedication and achievement. References are made to specific Societies and special activities to the extent that they impinge on matters central to the history of the Technion.

In narrating the story of the Technion's birth, growth and development, this volume can not go into the details of the academic program and achievements. For a listing of the curricula of the various Faculties and Departments, the reader must be referred to the annual Catalogue, which in its latest English edition ran to 475 pages. For details on the richly creative research program, the reader must turn to the 500-page Research Report, published periodically in English. Both of these volumes are necessary supplements for anyone who wishes to obtain the full picture of Technion's academic activity. Further valuable documentation will be found in the pages of the many hundreds of international scientific and professional journals which publish the studies and the findings of Technion's staff members. Such published papers have in no small measure added to the prestige which the Institute enjoys in the academic world.

No attempt has been made to gloss over those aspects of the history which may not always make pleasant reading. A product of human creation, the Technion may at times also have been subject to human weaknesses, and the story is told as it happened and was recorded.

I must make special acknowledgment to General Amos Horev, President of the Technion (1973–1982), who gave me the initial assignment and followed it up with encouragement and support. I am grateful to the American Technion Society, which undertook to have the book published. I had the opportunity of consulting and interviewing dozens of the principals who played a role in the making of this history. Many of them consented to read portions of the book, and their comments were most helpful, but the final responsibility for what appears here is mine alone.

In transliterating Hebrew words and phrases I have not followed any academic determination which properly seeks to introduce uniformity into a field in which there has long been linguistic anarchy. Instead I have chosen to use the spellings which were commonly and popularly accepted in the respective periods. Words and phrases that may be new to the reader will be found listed and explained in the Glossary.

The major sources from which I have drawn my information are listed separately. I am grateful for access given to me by the Zionist Archives, the National and University Library, and the American Jewish Archives, all in Jerusalem, as well as the Weizmann Archives in Rehovot. My principal source, of course, has been Technion's own archives, established by the late Yehoshua Nessyahu, and maintained and expanded with fanatical devotion by Miriam Shomroni, who is able to produce necessary information and photographs with promptness and efficiency.

Most of the photographs from the early period are from Technion's own Nessyahu Archives. Pictures since 1952 are for the most part by Gedalia Anoushi, who served for many years as staff photographer. Other pictures are from the estate of the late Moshe Eldad, courtesy of Yosef Eldad, the Collection of Menahem Haklai, Herbert S. Sonnenfeld, the Zionist Archives in Jerusalem, the American Zionist Archives in New York, Yair Nahor, David Maestro, Keren-Or, Oskar Tauber, Walter Langenback (Metropolitan Photo Service Inc.), Foto-Wald, Leni Sonnenfeld, Daniel Franck, Charlotte & Gerda Meyer, David Harris, Photo Mordhai and Roy Brody.

I owe a special debt to David C. Gross, who helped prepare the manuscript for press; Bill Phillips and David Friedlander, who painstakingly went through the manuscript; Miriam Golzman, secretary of the Senate; and my secretary, Johanna Neumann, who typed the manuscript and kept a close and devoted eye on its progress.

For practical and technical reasons the book makes its appearance first in English. It is to be hoped that the edition in Hebrew will follow without too much delay.

Carl Alpert

Haifa, May, 1982

Table of Contents

"So necessary for our survival...." v

Preface vii

Chapter

I   An Idea is Born 1
II   The War of the Languages 36
III   Under Zionist Administration 76
IV   The Institute Opens 105
V   Stormy Years 133
VI   The Long, Hard Pull 163
VII   The War Years and After 200
VIII   Statehood 235
IX   From Adolescence to Maturity 255
X   Dynamic Expansion 307
XI   Consolidation 336
XII   Taking Stock 367

Epilogue 388

Bibliography 405

Glossary of Words and Terms 408

Appendices:

A   Major Engineering and Science Buildings 411
B   Major Research Laboratories and Wings 411
C   Major Faculty and Student Facilities 412
D   Student Hostels 412
E   Roads, Parks and Other Facilities 413
F   Buildings Under Construction or Being Planned 413
G   Academic Chairs and Lectureships 414
H   Special Endowment Funds 415
I   Honorary Degrees Awarded 417
J   Harvey Prize Laureates 418
K   Tenured Academic Staff 418
L   Board of Governors 423

Index 427

An Idea is Born CHAPTER ONE

Prologue

The Technion, which has played a major role in the shaping of an era, was itself, much earlier, the product of an era. The ages-old yearning of the Jewish people for a return to their ancestral homeland was given political crystallization in 1895 when the Viennese journalist, Theodor Herzl, wrote his *The Jewish State.* This was followed two years later by the convening of the First Zionist Congress in Basle, a gathering of Jewish delegates from various parts of the world, who banded together to establish the organizational framework necessary to create that State.

Herzl was the unquestioned leader and prophet of the movement, but was soon confronted with differences of opinion and emphasis among his followers. The problems of Moses, the Lawgiver, when the Children of Israel turned against him in rebellion in the desert, may have come to Herzl's mind when he faced the Fifth Zionist Congress at Basle in December, 1901. Before him was an organized opposition movement led by a group of young men who clamored, among other things, for the establishment of cultural institutions — this in a movement which had hitherto concentrated almost exclusively on political matters. In response to their needling, the Congress adopted a resolution calling for a "fundamental survey of the question of founding a Jewish University."

The survey took the form of a detailed pamphlet published in July of the following year by three young men, 24, 27 and 28 years of age respectively.

The three men were Martin Buber, of Vienna, student of philosophy, Zionist, political writer and editor; Berthold Feiwel, of Berlin, newspaperman and editor; Chaim Weizmann, student of chemistry and at the time a lecturer at Geneva University. None of them had any special affinity for engineering, yet it is significant that even as they made their case for a Jewish University, they emphasized the primacy of technology.

In their carefully prepared document they pointed out the difficulties faced by Jewish youth who sought admittance to universities in the lands in which they dwelt. They described in some detail the operation of the *numerus clausus* in Russia. The lack of opportunities for technical studies, they wrote, was much more serious for Jewish students than the obstacles placed in their way for a general university program. The problem was not only spiritual; it was also "eminently economic and social". It meant that in Russia the Jews were practically excluded from technical professions, with the result that they were pushed into urban commercial occupations, into *Luftmensch* life.

They unfolded their plan: First step should be to set up a preparatory Technikum (in Hebrew they wrote Technion), in part to train students for the University, and in part to serve as an independent institution for the training of young people in technical, agricultural and similar professions.

Graduates of the school would provide the basis for establishing and maintaining a Jewish industry. However, they were unable to present detailed plans, among other reasons, because "special study must determine the organizational form, unique to this Technikum". It was clear that they realized their own lack of experience or familiarity in this area. The absence of Jewish engineers and technologists in the early ideological planning for the Technion, as well as in later years when the Zionist Organization officially adopted the Technion, retarded the proper development of the institution and at certain stages almost led to its closing down.

Buber, Feiwel and Weizmann proceeded to develop the idea of a Jewish University as part of the national renaissance of the Jewish people. There was no doubt that it could be completely successful only in Palestine, both because it belonged there, and because it would provide the moral, scientific, cultural and economic base for the Jewish national home. Faced with reality, however, they were prepared to give consideration to the possibility of first establishing the university in another country, England or Switzerland, on the assumption that it would eventually be transferred to Palestine.

They took their proposal seriously. Dr. Weizmann opened an office of the Jewish University in Geneva, and conducted a public relations campaign from there. The role of the Technion was not forgotten, and in his meticulously kept expense account Weizmann recorded that he spent 45 francs for a trip to Zurich to study the operations of the technical secondary school at Winterthur.

The Jewish University remained very much on the Zionist agenda, but when the subject was next brought before a public forum, it was the Technion which was projected to the forefront. This was in August, 1903, and the place was the Carmel town of Zichron Yaakov. The 60,000 Jews of Palestine had held elections for their first national democratic assembly, the grandfather of the present Knesset. Indeed it was called the *Knessiah Rishonah*, the First Assembly. Product of the initiative of Dr. Menahem Ussishkin, Zionist leader, it was composed of delegates chosen to represent each of the cities and towns of the country. Because it met at almost the same time as the Sixth Zionist Congress, it has been given little attention, but it was unique in many respects. It was an abortive attempt to set up the first organs of Jewish self-government. It was also a trail blazer in that it gave women both the right to vote and to hold office, this despite noisy but ineffective protest of the orthodox elements. At that time New Zealand was the only country in the world to have given women such rights.

The Hebrew teachers of Palestine utilized this historic occasion to organize professionally, and out of their meeting in Zichron Yaakov grew the present Israel Teachers' Association.

The teachers got off to an inauspicious start. The *Knessiah* droned on with its parade of rhetoric and did not finish as scheduled. The teachers' conference began two days later than the date set. Its first session was affected by poor planning and disorganization and the 59 representatives of the teaching profession argued irritably over personalities as well as issues. On the agenda were such subjects as the need for Hebrew kindergartens; which pronunciation, Ashkenazi or Sephardi, should be used in the schools; the place of religion and of physical education in the curriculum; the need for professional teaching standards, etc.

Dr. Ussishkin was to have delivered the keynote address on Thursday, August 27, but it was not until Friday afternoon that he spoke, and he dealt with the need for an institution of higher education in Palestine. What should be the nature of such an institution?

He referred to the Buber-Feiwel-Weizmann paper and made it clear that in his opinion the institution could be established nowhere but in Palestine. He devoted considerable attention to the nature of the curriculum in the proposed university. His analysis was astute and prophetic. He identified two basic trends in university curricula, one which he called *positiviut*, science and technology, and the other *humanitariut*, humanities and social sciences. Which should get priority? There is little doubt that Ussishkin's views were colored by his own background. He had graduated from a technical high school in Moscow in 1883, had enrolled in the Moscow Technical Institute, and in 1889, at the age of 26, had received his degree of Technical Engineer, with distinction. Though he had never practiced engineering, his outlook on the realities of life was clear. He made his recommendations:

If decision had to be made on the basis of the material and cultural resources then available, then the choice must be for the humanities, he said. The reasons: first, the Hebrew language is rich in scope in the spiritual and cultural fields, but inadequate for the exact sciences. Secondly, the funds required to teach the humanities are far less, whereas the "positive" studies require expensive tools and equipment.

But if decision had to be made on the basis of what would be most useful and practical, then Ussishkin opted for the technical curriculum because the graduates would find it much easier to earn a livelihood. Since practical considerations were paramount, the choice must be for the technical field, no matter what the linguistic or financial difficulties.

On the matter of language, which was in a few years to become an agonizing bone of contention, Ussishkin was realistic: "I believe that until such happy day, which certainly will come, when we have Hebrew profes-

3

sors who will be able to teach in Hebrew, we can permit the teaching of science in a foreign language as was done, for example, in Russia and in Japan when they first set up institutions of higher learning there. They studied science in a foreign tongue (like German) and not in their own language. If there is anything to be feared from the teaching of a foreign tongue to our children it is in the lower grades, and not at university level."

He proposed reaching agreement with the Alliance Israélite Universelle to transform their Mikveh Israel Agricultural School into a university. This would not only provide existing buildings, but would also obviate the need to seek formal permission from the Turkish government to put up a new institution, as required by law. Mikveh Israel already had such permit. A few years later the problem of the permit was to endanger the new Technion at its very birth.

He called upon the convention to adopt a resolution calling for the establishment of a polytechnical institute. And to those who felt that it had been a waste of time to devote so much attention to a hypothetical proposal, fulfillment of which seemed doubtful at best, he quoted an old Hebrew aphorism which can be loosely translated as: "Time sometimes has a way of taking care of matters better than intelligence".

The conference followed his advice to the letter, and the following resolution was adopted:

"1. The Teachers' Conference voices its opinion that an institution of higher learning for the children of Israel must be established only in Palestine, and nowhere else.

"2. The Teachers' Conference voices its opinion that because of the current needs, such institution of higher learning should be a polytechnical institute, like schools overseas.

"3. The Teachers' Conference is of the opinion that students seeking to enrol at such institution of higher learning should have a scholastic level equivalent to that of students at a European Middle School, and must be masters of the Hebrew language and literature.

"4. The Conference expresses the opinion that until such time as we have suitable professors who can teach in the Hebrew tongue, the sciences should be taught in some European tongue.

"5. The Conference proposes to those individuals abroad who are working for establishment of an institute of higher learning that they seek to realize their goal by approaching the Alliance Israélite Universelle, with a view to changing its Mikveh Israel school to a polytechnical institute."

The call was issued. Nothing further happened. The year 1903 passed, and another four years thereafter. The Zionist Organization became involved in the controversy over the Uganda project. Herzl died, and the movement was embroiled in the conflict between the so-called political Zionists and the practical Zionists.

The Technion remained on paper. A fresh start had to be made, but some of the dramatis personae in those early, premature days were yet to play significant and at times even dramatic roles in the eventual development of the Institute.

The German Connection

The tensions and rivalries between the leading European powers at the turn of the century were also felt in the Near East in general and in the Ottoman Empire in particular. These were also reflected in French, German and English Jewish organizations, which frequently served as willing or unconscious agents of the national interests of their respective countries. The oldest and best known of these organizations was the Alliance Israélite Universelle, which had since 1860 carried on a far-flung program concentrated on education and defense of Jewish rights. The agricultural school at Mikveh Israel, Palestine, to which Ussishkin had referred, was established in 1870. The Alliance also operated a network of Jewish schools throughout the Balkans, North Africa and the Near East. The language of instruction was usually French, and the schools obviously provided an effective spearhead for French cultural, political and commercial penetration into these areas.

In Palestine there were Alliance schools in Jerusalem, Jaffa, Haifa, Safed, Tiberias and elsewhere, providing a modern education to some 2,000 children. The famous Laemel School, founded in Jerusalem in 1856 with Austrian money, was later taken over by the Alliance. Language of instruction in all these schools was French. The background was Jewish, but without undue emphasis.

The English got into the picture too. A school for girls founded in 1864, and teaching in French, was later taken over by the Anglo-Jewish Association, and became the Evelina de Rothschild School. The language was changed to English, though Hebrew was gradually introduced and became language of instruction for about half the subjects.

The Germans were therefore late arrivals on the scene when the Hilfsverein der deutschen Juden (German Jews' Aid Society) was established in Berlin in 1901. Its avowed purpose was to defend Jewish rights, and it extended the program to encompass philanthropic, educational and cultural activities among Jewish communities overseas. The founders and principal personalities were James Simon and Paul Nathan. Simon, a philanthropist and cotton merchant, was prominent in commerce and bank-

Menahem M. Ussishkin (1863–1941).

ing. He had important connections high in the German government and was a patron of archeology and the arts. He served as President.

Paul Nathan, who had earned a reputation as a journalist and fighter against anti-Semitism, was closer to Jewish life than Simon. As Vice President and Director of the Hilfsverein he provided most of the initiative. As early as 1898 Nathan had sought support from the German Foreign Office for his plan to establish a German Association for schools for Jews in the Near East. He failed to receive approval, possibly because he had been active as a political liberal, in opposition to the government. The Foreign Office did finally agree to give the official protection required for activity in Turkey when the Hilfsverein was formed in 1901, but active cooperation was lacking for some years.

Soon after its organization, the Hilfsverein intervened in a number of cases involving anti-Semitism and sought to enlist governmental support and public opinion against the perpetrators of the Kishinev and Homel pogroms of 1903 and the Romanian discrimination against Jews. In 1905 and 1906 Nathan went to Russia and argued with leading Russian officials about lack of effective action in preventing pogroms. Later (1913), it should be added, the Hilfsverein helped finance the defense of Mendel Beilis.

The educational program of the Hilfsverein manifested itself in the establishment of Jewish schools. Of the 50 or more such schools, 28 were in Palestine, and the remainder in Bulgaria, Romania and Galicia. At the outset Nathan sought to reassure his friends in France that the Hilfsverein was not seeking to compete with or supplant the Alliance. He went to Paris to explain that the German motives were purely philanthropic, and for the best interests of the Jewish people, without a political base. The Chief Rabbi of France, Zadoc Kahn, approved the new organization as being in the best spirit of Jewish tradition. There is little doubt, however, that the spreading influence of the Alliance schools was an important stimulus for the Hilfsverein. Again and again Nathan emphasized that one of his purposes was to educate the Jewish youth in Palestine for life in that country on a viable basis, hence the Hilfsverein schools taught practical as well as general subjects. The youth were to be educated to remain in the country, and not have to emigrate to seek education and training abroad. It could well be said that this was the first serious attempt to stem *yerida*, emigration from Israel. Of the 28 schools in Palestine, it should be added, 22 were almost exclusively Hebraic.

Judaism, as distinct from Jewish nationalism, was of great importance to Nathan. His attitude to religion is shown by a letter he wrote to Israel Zangwill in 1905, endorsing the Uganda project. He warned Zangwill that success there would depend on fostering Jewish religious consciousness. "Deep religious feelings help the settlers in the new Homeland to overcome the problems of the early days", he wrote. Though Nathan himself did not observe tradition, the spirit and curriculum of the Hilfsverein schools was orthodox. This did not necessarily win the support of the orthodox elements in Palestine, most of whom operated their own yeshivas and Talmud Torahs. They objected to any other school system, no matter how religious, which placed secular subjects on the same plane as the religious.

It is significant that from the very outset the Hilfsverein schools emphasized the use of the Hebrew language, and especially in the kindergartens. The children came from polyglot backgrounds, speaking Yiddish, Ladino, Arabic, Bokharan, Persian and Georgian, among other languages, at home. Nathan was being very practical, therefore, when he stimulated and encouraged the use of Hebrew in his schools. This policy, plus his encouragement of immigration to Palestine from Eastern Europe, earned him the support of the Zionist movement, which was beginning to find fault with the policies of the Alliance. The latter had stepped up its program with what one Zionist called "excessive attention" to French in its schools. The reason given was to secure to its pupils the capacity of making a living abroad. Any education leading to emigration was seen by the Zionists as detrimental, and they were driven even closer to the Hilfsverein schools. Dr. Jacob Thon, prominent Zionist leader and one of the founders of Tel Aviv, called their Hebrew program "a real blessing to the population of Palestine."

At the same time, the Hilfsverein saw itself as performing a valuable function in advancing German imperial interests in an important part of the world. German influence on Turkey had become a matter of hautpolitic. Nathan and Simon therefore continued to seek the closest possible relationship with the German Foreign Office, despite the cool response they got at first. Throughout, Nathan sincerely felt he was serving the best interests of the Jewish people and of the German government simultaneously.

Encouragement from German officialdom came at about the time the Hilfsverein first mooted the idea of setting up a technical school in Palestine as the crowning jewel of an educational network which already ran from kindergartens through grade schools up to a teachers' training institute and a commercial training school. Perhaps Nathan and Simon were aware that there had already been some early thoughts in this direction by the German government.

In 1907 Nathan visited the Balkan countries and went on to inspect the Hilfsverein schools in Palestine. As the head of a large philanthropic organization he was of course warmly welcomed, and was also flooded with requests and suggestions for projects. One proposal was to have the

Bas-relief of Paul Nathan, located in the entryway of the old building. The inscription pays tribute to him for initiating the establishment of the Technion in 1907.

Hilfsverein fund a local Hebrew-German theater. Another was to set up a central information office. One businessman asked for support in opening a sugar factory. The head of the Hilfsverein office in Jerusalem had previously talked to the Zionist, Schmaryahu Levin, about the feasibility of opening an Institute of Archeology in Jerusalem. The idea of setting up an institution for advanced technical training came to Nathan on the basis of his observations, and as a natural outgrowth of the times. The Turkish government was beginning to embark on large public works programs, including construction of the Hedjaz railroad, and there would be need for domestically trained personnel. In the entire Ottoman Empire there was not a single training school, and foreign contractors usually brought in their staffs from abroad.

Upon his return to Germany, Nathan submitted a memorandum to the Hilfsverein board, setting forth his ideas. In his original paper he described the proposed institution as a Manual Training School, with a three or four year program. Pupils would be between the ages of 15 and 18. Later he expanded the concept, and the intention was that pupils would begin their studies at the age of 17, and could be accepted up to the age of 24–25, thus considerably raising the proposed standard. Gradually this evolved into a two-stage program embracing a secondary school which would prepare its pupils for advanced education, and the Technikum* proper, on a college level. The differences of opinion which shortly developed with respect to the precise level for the Technikum were to plague the institution for more than 25 years.

Nathan expressed the hope that Jewish engineers and technicians, trained at the school he envisioned, would find their places in government service and in the construction industry. This would open the way for further employment of Jewish artisans and other Jewish workers with technical training. The economic standards of Jews in the Turkish Empire would be raised, and emigration would cease. Further, better conditions would attract immigration to Palestine by those fleeing from persecution in eastern Europe. At the time, Germany was being inundated by Jewish refugees from the East. Wherever possible, they were being shipped on to the United States, but if conditions were made desirable, perhaps tens of thousands of them could be diverted to the thinly populated areas of the Ottoman Empire.

The kind of school he had in mind was also intended to provide study

---

* The German name for the institution was Technikum, sometimes spelled Technicum. This name will be used in the text throughout the German period, except when reference is to the institution in its later years. As will be seen, the name went through a number of changes.

opportunities for Jewish youth in the West, who could not always gain admittance to universities there.

In press interviews Nathan emphasized the ability of Palestine to absorb more immigrants. He commented on the importance of Hebrew in the educational institutions of the country. The Zionists found his attitude encouraging, and the Hilfsverein activities were frequently and warmly reported in the Zionist press. Yet here and there some Zionist commentators warned that the Hilfsverein's motives were not pure, and were certainly not activated by Jewish nationalism.

The Hilfsverein made no secret of its purpose. James Simon submitted a formal report to the German Foreign Ministry and pointed out that the proposed technical school would promote German interests in the Near East through educational rather than political means. It would be open to non-Jews as well as Jews, and would obviously serve Turkish interests as well.

During his visit Nathan had sought support for the Hilfsverein school network in Palestine from the government in Constantinople but to no avail. Berlin's reaction was much more positive. Since the school would be set up along the lines of similar institutions in Germany, there would be need for machinery and equipment for the laboratories and workshops. These would come from Germany. Textbooks and reference books would also be in German since that would be the language of instruction. All this was taken for granted. In the long run this would lead to further penetration by German industry. Berlin's attitude to the Hilfsverein gradually warmed up, and the official support and approval which had previously been withheld began to manifest itself.

All that was now needed was financing.

First Funds

Nathan discussed his idea with his Zionist friend, Dr. Schmaryahu Levin, whom he had met on an earlier visit to Russia, where Levin had been a member of the Duma. Levin was enthusiastic and offered his full cooperation. An opportunity for help materialized earlier than they had perhaps expected. Visiting in Berlin in January, 1908, was David Wissotzky, son of the late Kalonymous Zeev Wissotzky, Russian tea merchant, who had died in 1904. In his will he had provided for a large fund, the income of which his executors were obliged to use every five years in support of a public institution. The first five years had almost elapsed, and a sum of about 100,000 rubles had already accumulated. Among the executors was Ahad Ha-Am, the distinguished Zionist philosopher, who had won the friendship, admiration and respect of the old man. Wissotzky had served as his

Kalonymous Zeev Wissotzky
(1824–1904).

9

Schmaryahu Levin (1867–1935).

Jacob H. Schiff (1847–1920).

patron in a real sense and had provided him employment with the tea company in London under conditions which enabled him to pursue his literary career in comfort. Ahad Ha-Am had sought to influence the other executors to have the money used in Palestine; after all, Wissotzky had been one of the leaders of the early *Hovevei Zion* (Lovers of Zion) movement. However, he had been unable to move them.

The meeting with the philanthropist's son in Berlin was therefore most fortuitous. Nathan explained his plan, and Levin added his own recommendation. The Wissotzky heir was impressed, more by the charitable aspects of the proposed institution than by any Jewish nationalist reasons. Indeed, he made it quite clear that he was personally not a Zionist. He took the message back to Russia, discussed it with his relatives, and reported interest, but further information was required. Would other funds be available? Would the Hilfsverein be responsible for the establishment and management of the school? Reply quickly, he wrote, "there is pressure upon us to found a school in Russia instead of in Palestine."

Little time was lost, and on March 29, 1908, the formal agreement was signed, setting up a Wissotzky Family Endowment to the Hilfsverein, for the establishment of the Technikum. The contribution was 100,000 rubles, then equivalent to about $52,000, and there was implied suggestion that when the estate made its next distribution, five years later, the income would again go to the same cause. It was further agreed that this was a philanthropic, not a Zionist enterprise. Ahad Ha-Am, who helped draft the text of the agreement, was at least able to introduce a clause assuring the school would be Jewish in character. It may be assumed that leaders of the infant Zionist movement, pleased at the prospects of having such a school in Palestine, were equally pleased they would not be saddled with its financing.

There were certain strings attached. For one thing, the institution must be "independent", and governed by its own Board. No outside organizations, neither the Hilfsverein nor the Zionist bodies, could be represented as such. The first Board was therefore composed of individuals all serving *ad personam*, though their interests and connections were known. The members of the first Board, or Kuratorium, as it was called, were: five representatives of the Wissotzky family — D. Wissotzky, J. Zeitlin, R. Gutz, Dr. Boris Gawronsky, all of Moscow, and Ahad Ha-Am, from London; five individuals all of whom happened to be closely associated with the Hilfsverein — James Simon, Paul Nathan, Prof. M. Philippson, Eugen Landau and B. Timendorfer. The odd member was Schmaryahu Levin. As a matter of fact the Hilfsverein retained full operating control; Simon was elected Chairman and Nathan Executive Director and Vice Chairman. Headquarters were established in Berlin.

There was another condition, specified in the will. Any institution established with this money must bear the name of Kalonymous Zeev Wissotzky. The Kuratorium agreed. In effect this meant that the school could be known as the Wissotzky Institute of Technology.

The immediate joy at the big windfall gift was soon tempered by realization that the contribution, while princely in amount, was not really sufficient to construct the proposed school and put it into operation. More money would be needed, and once again fate intervened.

The American philanthropist, Jacob Schiff, visited the Holy Land in 1908. He had been in Egypt previously and there had met Baron Rothschild, who invited Schiff to join him on a cruise down the Nile on his private yacht. "Even if he were to give me a million I would not give up my visit to Palestine," Schiff told his hosts in Jerusalem later. He was deeply moved by Palestine, but was most affected by the poverty and destitution which he found among many Jews. At one stage he proposed, whether in earnest or in jest, moving all the poor Jews of Jerusalem to New York. There, he said, the philanthropic agencies would quickly see to it that they would become self-supporting. In his desire to help he gave liberal contributions to every single institution or agency which made a request of him, but he still felt that the answer was in helping the poor to achieve economic self-sufficiency. He had been struck by the lack of crafts and skills among the Jewish population. En route back to New York he passed through Berlin, and there met with Paul Nathan. The idea of an institution to provide technical training struck his fancy and he pledged $100,000 for the new Technikum. But he had conditions as well, and some of them were not easy to fill.

Ahad Ha-Am (1856–1927).

For one thing, no "isms" except Judaism should be permitted in the school. This readily met with the approval of the Hilfsverein and the Wissotzky family, both of whom were anxious to keep out Zionist influence. Ahad Ha-Am objected, claiming this would constitute restriction on freedom of speech and thought in an educational institution. To avoid a clash, Schiff's condition was accepted orally.

In the second place, Schiff objected to having the school bear the name of Wissotzky. He felt it was not fitting that an institution of this nature should be named for an individual, no matter how generous he had been. Levin succeeded in obtaining the consent of the Wissotzky family, with the understanding that a suitable plaque bearing Wissotzky's name would be placed in the building when it went up. This pledge was fulfilled in 1926.

Schiff's third condition was even more difficult. He went further than the Russians in seeking independence for the institution. He wanted it taken out of the hands of the Hilfsverein so that it would be free to seek and obtain support from all groups and all factions in Jewish life. He wanted Kuratorium members from other countries as well, to give the school an international complexion.

Ahad Ha-Am and Schmaryahu Levin objected on the grounds this might introduce assimilationist elements. The Hilfsverein, intent on projecting the German coloration, also raised objections but couched them in practical terms. Under the system of capitulations existing at the time in the Ottoman Empire, the school would require the official backing and sponsorship of some powerful state, like Germany. An international character would leave it impotent and defenseless before the vagaries of Turkish officialdom. Nathan was willing to add to the Board prominent Jews from other countries, but he insisted that the actual management of the school should be in the hands of a small executive committee with headquarters in Berlin.

The compromise was the setting up of a separate body called *Jüdisches Institut für technische Erziehung in Palästina* (Jewish Institute for Technical Education in Palestine), registered in Berlin in 1909. New members added to the Kuratorium were Ludwig Schiff of Frankfurt and M. Warburg, Hamburg; Prof. Solomon Schechter, Louis Marshall, Mortimer Schiff and Samuel Strauss, all of New York, and Dr. Yechiel Tschlenow, of Moscow. The latter was unofficially recognized as the spokesman for the Zionist point of view. Actually, the new body was a legal fiction and the expanded Kuratorium continued to function through the Hilfsverein as had the original committee. The Executive Committee was composed of Simon, Nathan, Levin, Ahad Ha-Am and Ludwig Schiff.

Paragraph four of the constitution of the new society read: "The object of this Society shall be to enable the Jewish youth of the Orient to acquire a technical education. In addition, an opportunity is to be given to the students to acquire a knowledge of the Hebrew language and Jewish history to the end that they may become self-respecting Jews, reverencing their ancestry and their religion. Towards the accomplishment of this purpose a Jewish Technical Institute, together with a Preparatory School, shall be erected at Haifa, Palestine." Neither the Board nor the Executive met very often, and so Nathan remained in full charge. Headquarters were in the Hilfsverein offices. In its annual reports that organization referred to the Technikum as if it were one of its projects. It did indeed operate as an adjunct of the Hilfsverein, and the members in Russia or the United States were too far away to note or to object.

Schiff was one of the few Americans who kept up an active interest — an interest which was to manifest itself in dramatic manner at a most critical stage following the war. Though not a member of the Kuratorium, he followed developments constantly. The architect's plans, when drawn up, were sent to him for inspection. He was responsible for securing the support of others in the U.S. When there were delays in Haifa, he prodded Nathan.

Many credited Levin with having obtained Schiff's backing. In 1914 Schiff felt called upon to set the record straight. That credit belonged to Paul Nathan, he wrote in a letter to *The American Hebrew*. His interest became stimulated "because of the prospect which here opened itself for the German, Russian and American Jew, for the orthodox, the reformer, the Zionist and the anti-Zionist to cooperate harmoniously in the cause of cultural elevation and progress in Palestine."

Choice of Haifa

With the funds apparently in hand, Nathan set to work in earnest. Together with Levin he visited several technical institutions in Germany to get an idea of what was involved. In May, 1908, he returned to Palestine to fix the site and acquire the land for the Technikum. Even on his first visit he had already given serious consideration to Haifa, but he had not reached final decision. There were many factors to be considered.

For one thing, the school was intended to serve Jewish communities throughout the Ottoman Empire. The Jews in Jerusalem mounted a powerful campaign to influence Nathan and the Wissotzky trustees to construct the Technikum there. A special committee was set up and five reasons were advanced why Jerusalem, and only Jerusalem, was suitable for the purpose:

1. Jerusalem was the largest center of Jewish population, and the poorer residents would not be able to afford to send their sons to study in distant Haifa.
2. The new school would provide employment opportunities, and thus serve to bolster the local economy. The poor community of Jerusalem needed this kind of help, whereas Haifa would in any event benefit from the industrial development projected for that city.
3. Economic and industrial growth in Jerusalem had hitherto been hampered by lack of technical skills, which the Technikum would make available.
4. Jerusalem already possessed a network of educational institutions, from which the Technikum would undoubtedly be able to draw its best candidates.
5. Students from overseas would find in Jerusalem the cultural, religious and other Jewish influences so necessary for the background of Jewish student life. These were lacking in Haifa, and students there might be led astray, politically and culturally.

13

At the same time, the Jerusalem committee expressed the hope that religious circles in the Holy City would not create any difficulties in the setting up of the new school.

The Haifa Jewish community pleaded its case eloquently, and a committee headed by Nahum Wilbushewitz and Shmuel Pewsner* submitted its own memorandum to Nathan. The latter did favor Haifa, and he listed three reasons for his choice:

1. Haifa was destined to be the city of the future, a great port center of industry and shipping. With the building of the Hedjaz railroad it would be linked to Damascus and Baghdad, and would become an important crossroads for land transport as well. It was therefore an ideal location for a school of technical education.

2. The local Jewish community was not yet rigid in its organization and character, in contrast to Jerusalem, which was the center of fanatical orthodoxy, or Jaffa, which was a hotbed of Jewish nationalism. The neutrality of Haifa would serve to minimize conflicts.

3. The local Jewish community was small, and its influence hardly felt in the city. Establishment of the Technikum would give impetus to the expansion and development of the Jewish population.

The decision was courageous. Haifa was a quiet provincial town with a total population of only about 20,000, of whom close to 2,000 were Jews. Of these about two thirds were Sephardim, most of whom attended the large Alliance school. The first Hilfsverein Hebrew kindergarten was opened there in 1907. Haifa had no political importance; the seat of local government was at Acre, across the bay. In view of the unsettled conditions on the roads, the most secure means of transport between Jaffa and Haifa was sometimes by sea.

Despite the obvious problems, Nathan shared the rosy views about Haifa expressed by Theodor Herzl in his *Altneuland*. His decision was firm, but he refrained from announcing it. His representative had already begun exploring the availability of suitable sites, and if word were to get out that the "rich" Hilfsverein was in the market for land, prices of the plots would soar sky high. Nathan played his cards skillfully. In his public talk about the Technikum he spoke vaguely about building it in "Turkey". He dropped hints that Beirut, Smyrna, or perhaps even Constantinople, were under consideration.

Ahad Ha-Am was furious, and he rallied the Zionists to support siting of the school in Palestine. He, too, favored Haifa for much the same reasons advanced by Nathan. He foresaw development of the country along industrial, rather than just agricultural lines, and Haifa was the natural center

---

* In the earlier spellings the name appears as Pewsner. Later it became Pevsner.

for this development. Perhaps he had a personal interest as well. His eldest daughter, Leah, was married to Shmuel Pewsner, and lived in Haifa. Pewsner bombarded his father-in-law with letters and arguments in favor of the Carmel city. Ahad Ha-Am was particularly critical of the Hilfsverein for its apparent lack of any emotional feelings for the Holy Land.

Nathan could not defend himself publicly, but quietly proceeded with the purchase of land. The official representative of the Hilfsverein in Palestine and director of its school system, Ephraim Cohn, took Nathan to inspect various possible sites. One suggestion was an area in the German colony, at sea level. This was rejected because of the fear of dampness which would affect machinery and equipment. Another proposal was a stretch of land once occupied by Oliphant on the upper stretches of the Carmel. There was no road to the place, and when their wagon could go no further they proceeded on foot over the rocky terrain. The party finally made it to the site. The view was impressive, but the area remote and desolate. Nathan turned to the Arab wagon-driver and asked what he, the Arab, would do if he were to receive this land as a gift. Replied the Arab quickly: "Give it away immediately as a gift to someone else!"

The choice finally fell on an area part-way up the Carmel slope. Some of it was open land and part was an olive grove. The major section was owned by Arabs, principally Alexander Kassab, and by local Germans.

Cohn favored this site, but pointed out some of the problems and difficulties, to test Nathan's decision. The latter responded that if they had wanted a place without any problems they would have chosen to build the school in Potsdam or Spandau. The negotiations began. There was no meeting of minds, and on Cohn's advice the would-be purchasers departed for Jerusalem. The owners quickly came around and the sale was consummated at a good price. Rafael Hakim, a prominent Jewish merchant in Haifa, helped execute the deal. Eliyahu Sapir, of Petach Tikva, a man with many connections, helped conclude the sale and have the property properly registered. That required special Levantine arrangements. Cohn and Sapir met privately in a downtown Arab café with the local Turkish official. They explained to him that some "crazy guy" in Germany wanted to buy some worthless land on "the holy Carmel". Money quietly changed hands. The official explained that he would have to split with his superior in Acre. The registration went through without a hitch.

The first purchases were effected in May and June of 1908. At first the land was registered in the names of local people because foreign nationals were not permitted to acquire land without a formal decree of the Turkish government. A local Haifa committee to assist in the project, composed of Nathan Kaiserman and Shabbetai Levi, made great efforts to obtain the

In the beginning: guarding the site. 1909.

permission, but at first without avail. A change in the regulations later made it possible, and title to the land already acquired was transferred to the name of a German citizen, thus assuring the legal protection which would follow upon such registration. As subsequent plots were purchased they too were registered in the name of James Simon, of Berlin, a name which meant little to anyone in Haifa. The total area was 46 metric dunams, and the cost came to about 100,000 francs.

There were some who had doubts about the location. It was too far out of town, in a desolate area, they said. That situation changed with time.

The Zionist Organization quickly recognized the importance of the project and sought to become involved, despite the expressed detachment of the Technikum from any organization. In August, 1908, the Jewish National Fund decided to put up the money for the land on which the school would be built. The 100,000 francs were equivalent to about $25,000. As a matter of fact, Nathan had by that time already bought the land, but the J.N.F. money was to be earmarked as if for that purpose, thus satisfying constitutional requirements of the J.N.F. Nathan was willing to accept the money but he refused to have the land registered in the name of the Fund. Under Turkish law there was no distinction between title to land and to the buildings erected on it, and he did not want the institution to become the property of the Zionists. The compromise was that the 100,000 francs be regarded as a loan, not repayable so long as the land was used for the purpose intended. This was agreed to and final payment was made in 1910, two years after the land had been bought. Designation of the payment as a loan was to have important legal implications five years later.

Dr. Ussishkin, now one of the leaders of the Zionist movement, was keen on the project. He recognized it as a fulfillment of his own dream of

1903, even though it was being created by a non-Zionist body. He confided to Ahad Ha-Am in 1908: "It is not a pleasant feeling for the Zionist Organization to push itself into this matter when those responsible for it do not want our participation." The philosopher reassured him that anyone with the means to do so should inject himself and help. The adherence of Tschlenow to the Kuratorium was one way of assuring a degree of Zionist influence.

Construction Begins

Kaiserman, Director of the Anglo-Palestine Bank in Haifa, was asked to serve as the resident representative of the Hilfsverein. He received instructions and funds as necessary from Cohn in Jerusalem, who was in turn responsible to Berlin. The extended chain of command, and the slow communications in those days led to many delays, because the Berlin committee, namely Nathan, insisted on making decisions with respect to almost every little step.

The years 1908–1910 were busy ones for Kaiserman, and it sometimes seemed he was giving more time to Technikum affairs than to his bank. He scouted out fresh properties which could be added to the parcel being assembled, conducted negotiations for the sale, prodded the law firm

First fence around the site before beginning of construction. 1910?

which was under fire from Jerusalem and Berlin for its apparently excessive bills, and saw to it that applications were made to the local authorities for the permit, *Ruhsa*, which would enable them to proceed with actual construction.

In the meantime Cohn instructed him to obtain bids from a contractor for construction of a cistern on the property. A basic requirement was that only Jewish labor be used.

Work was to begin on a stone wall to surround the entire parcel, and it had to be extended as new plots were added. Berlin was beginning to plan the proposed buildings but nothing could be done without the *Ruhsa*. Kassab and Hakim were supposed to help obtain the permit, but nothing was moving. Cohn complained to Kaiserman: "We gave them an advance payment, even though not obliged to do so, and they haven't even started. Is some official demanding *bakshish* which Hakim does not want to pay? Kaiserman, use your influence to get on with the job, even if it costs money."

For some two years this was to be the general tenor of the activity. Great importance was attached to preserving the accumulated parcel of land as a single unbroken bloc. Again and again it was emphasized that no street or road must ever be built across the bloc, and the necessary consent documents, or *Mazbata*, had to be obtained not only from the sellers of the land, but from adjacent property owners as well. Years later this scrupulous preservation of Technion's territorial integrity was to be the cause of major traffic problems in the heart of the busy, congested Hadar Hacarmel area.

All bills were sent to Jerusalem, and from there to Berlin for payment. Every invoice had to be accompanied by the fullest possible details such as precise measurements of the fence, depth of the well being dug, etc. Cohn was apologetic. Everything — plans, purchases, buildings, planting of trees — all had to be decided in Berlin, and there were differences of opinion even in that remote control committee.

As the property began to take on form and substance there was need for a watchman. In this respect Cohn was willing to compromise with his usual requirement that all employees be Jewish. If a good Jewish guard could not be found, he instructed Kaiserman to hire an Arab. It was customary in the colonies to hire Moslem guards, thus affording better protection. "There are no Jewish thieves, thank God," he said, "and one is better protected from Arab thieves by Arab watchmen." Indeed, it used to be the practice to hire the most widely feared thief, pay him well, and give him the responsibility of safeguarding property. It was usually effective, Cohn wrote. On the records Katriel Shevelov was listed as the first watchman.

Matters became even more complicated, and still the various permits and licenses were not forthcoming from the Turkish authorities. Kaiserman wanted quick decisions. Cohn went to Europe to report to Berlin in person, but his reply to Kaiserman was not helpful. Dr. Nathan was ill, needed complete rest, and could be approached only with good news (!)

Back in Europe, interest in the new institution was kept high on two fronts. Addressing a Zionist Conference in Manchester in 1909, Dr. Chaim Weizmann reported that things were going quite well with the "National Polytechnikum". From a material point of view, it would be possible to open the school at once, because wealthy Jews from many lands had promised their generous support, and undoubtedly would keep their promises. The hitch was lack of top level Jewish professors. To be sure, Weizmann added, there were many Jewish professors and scholars of repute, but the Polytechnikum needed the very best, and there were very few who possessed both superiority in their field and Jewish nationalist inclinations. He deplored the fact that, in his opinion, only two Jews had made major contributions to human thought, Baruch Spinoza and Karl Marx. Both of these were in the fields of philosophical and speculative learning. In the world of practical science there had not yet been one truly outstanding Jew. For this reason, perhaps, he looked to the new institution in Haifa to be a source of Jewish contribution to the world of science.

There is some historical justice in the fact that from the very Manchester where Weizmann spoke, forty years later there came one of the world's most distinguished applied mathematicians, the Zionist, Dr. Sydney Goldstein, who established the Department of Aeronautical Engineering at the Technion and thereby helped change the course of history in Israel. But of that, more in its chronological place.

The Technikum item on the agenda in Berlin dealt with more immediate considerations. First was the matter of curriculum and scope of the school, and simultaneously, the physical planning of the structure. Nathan sought expert advice and assistance. In September, 1909, he called a meeting of teachers from various German technical institutions. Chief among them was Prof. George Schlesinger, the only Jew on the faculty at the Technische Hochschule at Charlottenburg, who at once assumed a leading role in the planning.

None of the participants knew anything about the problems or economy of Palestine. They came to their conclusions without careful study, and certainly without the vision which the project required. Some of their errors of judgment were quickly corrected. Others remained to haunt the institution for years to come.

The German experts set forth the functions of the Technikum as follows:

1. The institution would serve primarily as a Middle Technical School, and would be known by the German term for such school, Technikum. The purpose was to train technicians, assistants to engineers and work foremen in the two fields of construction and industry. They proposed setting up two departments: Mechanical Engineering, a two year course, preceded by three years of practical training in workshops; and a Department of Construction and Public Works, a three year program, preceded by two years of workshop or apprentice training.
2. The Technikum workshops would also be used to provide advanced training for craftsmen and artisans. The emphasis was to be on practical training but time would also be set aside for some theoretical studies.
3. A technical (vocational) school would be attached to the institute to furnish the practical training called for in the above paragraphs, and also to train technicians who would go on further with their education. This would be a seven year program. The committee also suggested adding a Reali School with two additional years, to serve as a preparatory stage for higher education.
4. The laboratories for chemistry, physics, machinery and testing of materials should not only be used for training, but should also be exploited for actual services to industry and for scientific research. They should be examples to industry.

In general outline the proposal was not too far different from the skeleton program Nathan had himself presented a year earlier. It was clear that their concept was not of a university, but of something much more limited. The dream of a technical university, granting academic degrees, was farthest from their thoughts, and was advanced at that stage only by some of the Zionist leaders.

Hindsight later revealed that the placing of mechanical studies on a par with construction was a mistake. Very little was required in the Ottoman Empire at the time in the way of mechanical or electrical trades. The need was mostly in construction, as Nathan had himself noted. The German experts conceived of the Technikum as they thought it ought to be. As a result, the expensive mechanical equipment acquired at the outset was a heavy financial burden and in the end much of it turned out to be an almost complete waste.

The experts were no less impulsive in their consideration of the physical facilities. Without ever seeing the proposed site they recommended the construction of seven buildings: a teaching edifice, workshops, secondary school, gatehouse, well-house, water tower and dormitory for overseas students. Their suggestion was cancelled when the difficult site was taken into consideration. It was long and narrow, running up the side of the hill — about 400 to 450 meters long and about 60 meters wide, or up to 140 at its widest. There were problems of clearing and grading. Water supply was another problem, and the very idea of a cistern for accumulation of rain waters was foreign to the Germans. Consideration had to be given to an independent power supply.

The physical aspects of the planning were corrected in good time. The Kuratorium took up the matter of selection of an architect. Was it necessary that he be Jewish? The question had first arisen when a German engineer in Haifa, Gottlieb Schumacher, had been given the task of building the wall around the site. Ephraim Cohn had given him the contract at the time of the purchase of the land as one way of obtaining his good will. Schumacher, a non-Jew, had been one of the owners of the plots that had been bought. However, Jewish architects had seen this as discrimination against them and had voiced their protests. Schmaryahu Levin, Ahad Ha-Am and others presented their views to the Kuratorium and as a result that body decided that since this was a wholly Jewish project, both the architect and the general contractor should be Jewish. Insofar as possible, Jewish labor should be engaged in the construction.

In October of 1909 Alexander Baerwald, an architect with the Public Works Department of the Government of Prussia, was asked to prepare a preliminary plan for the building. Before the year was up he and Levin went to Haifa to study the area.

Baerwald drew up plans in the spirit of what he saw in the country. The major building was monumental in design, European in its general lines, but topped by Eastern elements, from the dome to the crenellated roof. The Kuratorium approved the plans as well as Baerwald's estimate of about 800,000 francs for construction. In those days even architects were

Model prepared by Architect Baerwald in Berlin. 1910.

Alexander Baerwald (1877–1930).

felt to be competent to judge anticipated costs. In August, 1910, he was awarded the assignment to draw up the detailed plans and to supervise the execution.

He opened an office in Haifa and engaged as his aide Yosef Barsky, a staff member of the Bezalel School in Jerusalem, who had already shown his competence in supervising the construction of the Herzlia Gymnasium (High School) in Tel Aviv.

As the initial negotiations with contractors and sub-contractors soon revealed, the first estimate of costs was too low by about one half. The Kuratorium insisted on sticking with the original figure and ordered that the scope of the project be cut down to the funds available. It was the first, but certainly not the last of many such decisions in the history of the Technion, or for that matter, of other institutions in Palestine.

Nathan was pressed to procure more funds, and in the first stage found them close to home. His cousin, the Baroness Julie von Cohn-Oppenheim, had not only left him enough money to give him financial independence, but had also set up a foundation which he administered. He drew upon the foundation for $75,000. James Simon was recorded as adding $25,000. The German department store magnate, Oskar Tietz, gave $12,500. It was not until affairs reached a crisis about five years later that it became clear that much of this German money was actually only a loan.

Back in Haifa many on-the-site problems remained. The Kuratorium appointed a local building committee, comprised of Kaiserman, Dr. Eliyahu Auerbach and Pewsner. The latter, it will be recalled, was Ahad Ha-Am's son-in-law. He had come to Palestine in 1906 from Russia, and together with Shmuel Itzkowitz and Tuvia Dunnia had set up the first soap and oils factory, Atid, which later became the well known firm, Shemen. Even in the absence of government permits the committee went ahead with as much of the preparation as possible, relying on the approval given by the local town officials to erect two buildings, no mention being made of a school. Building materials were accumulated and stacked on the land. Suddenly local land owners raised objection to the long, unbroken stretch of the site, claiming the lack of a thoroughfare in the middle adversely affected them. The encompassing wall was already up, and the committee insisted it had the written consent (*Mazbata*) of all concerned. The objections were ignored.

From Berlin Nathan sent instructions. He indicated on the plans what area was to be levelled and what should be done with the earth, to be used later in construction. He pinpointed location of two cisterns and suggested that separate firms be engaged to build each. He insisted on receiving samples of the stone to be used, and for a while even toyed with the idea of making artificial stone for the building.

Stones cut and ready for construction to begin. 1911.

The Haifa committee printed letterheads and envelopes bearing the name of the Technikum, but quickly received instructions from Berlin to discontinue their use. Nothing should be done which might draw the attention of the Turkish authorities to the fact that they were going ahead with the construction of a school, without full permission from Constantinople.

The aggravating efforts to obtain the various permits continued for almost two years. Work on the site was stopped on January 12, 1911, when policemen appeared and declared the activity illegal. Local Arabs had complained on the basis that since this land was classified as *Miri*, authorization (*Irada*) was required from the central government. Under the circumstances, it was decided to reveal all and to make the full application for establishment of a school. The request went to the area office in Acre. It was forwarded to the regional headquarters in Beirut. From there it was sent on to Constantinople, and there it became buried in government bureaucracy. It appears that discreet payments of *bakshish* had been helpful along the way, but something more was needed now.

Ephraim Cohn had thought originally that he could bypass all the usual channels and make direct application to Constantinople. As early as April, 1908, the German Foreign Ministry had written to its Consul in Jaffa asking that all assistance be given to Nathan, who was coming to set up a technical institute in Haifa. In mid-1909 the influence of the German government

A stone for the building. 1912.

was again sought and the Secretary of State, von Schoen, asked the Embassy in Constantinople to help the Hilfsverein obtain the necessary construction permits.

The timing was not good. The Young Turks who had come to power as a result of the Turkish Revolution of 1908 were not friendly to Germany, and the connection that was supposed to be an asset turned out to be a handicap. Indeed, German settlers in Palestine had sent an appeal to the Kaiser in 1910 asking for his protection against the persecution and injustice which they claimed they were suffering at the hands of the Arabs and the Turkish officials. They itemized their complaints ranging from outright murder to theft and bureaucratic harassment. Under the circumstances it was not surprising that the German government could do little to help the Haifa project. Still, Germany was pouring generous financial help into Turkey, and pressures in the Turkish capital were increased. The German Vice Consul in Haifa took the initiative in inviting Kaiserman to call upon him to discuss the project.

Still nothing moved. Shabbetai Levi and Nathan Kaiserman in Haifa were pulling every string they knew. Their lawyer in Beirut was pushing at that end. Everyone agreed that the real action was in Constantinople. Who had influence there? They found the right man. Asher Mallah, a Salonica Jew and fervent Zionist, had received his doctorate from the University of Constantinople and in 1911 was practicing law there. He had been politically active in behalf of the Young Turks and was personally acquainted with those who later came to power and influence in Turkey.

The task was not simple. This was not an ordinary building permit that was being requested. The construction of a technical school under Jewish auspices had political connotations, and the Turks were not sympathetic with the aspirations of the Zionist movement. If they had been, the whole early history of the Zionist movement might have been quite different.

There were very specific, stringent laws governing the operation of a school of this sort. The authorities requested certification from the Governor of the Haifa District that in the immediate area of the proposed school there was no mosque, no cemetery, no church, no hospital, no military camp, no storehouse, no public buildings of similar nature. The Governor so certified, and added that he saw no local inconvenience which would be caused by construction of "the School of Arts and Trades". The application stated that students would be accepted without distinction of race or religion. There would be external students and boarding students.

Mallah was familiar with the intricacies of government operations, and he began to unravel the procedures. He established that on August 11, 1911, the petition to the Acre authorities had been transferred to Constantinople via Beirut. Late in November Mallah's contacts informed him unofficially that the permission had been granted, and he so notified Haifa. However, the precious document had to make the reverse journey, through the same channels. The lawyer in Beirut cabled on December 11 "Mallah Constantinople Cable Irada Technikum Haifa Promulgue Stop Affaire Tout A Fait Terminee." Four days later the word had been passed on to Acre, and on January 10, 1912, Shabbetai Levi could inform the office in Berlin that the text was in hand, and the case was closed.

"I congratulate you my dear friend for your success and hope that the representatives of the Technikum in Berlin appreciate your services." Thus wrote Levi to Mallah. But as a result of simple neglect the name of Mallah disappeared from the records. When he died in Paris in 1969 no one at the Technion knew of his role in the history of the Institute. It was not until 1977 that a plaque was set up on the Technion campus noting that the man who had been for many years President of the Zionist Federation of Greece and of the Union of Jewish Communities of Greece, had also played a crucial role in making possible the building of the Technion.

Work on the site resumed on March 6, 1912, and public arrangements could now be made for the festive laying of the cornerstone, scheduled for April.

Despite the conscientious efforts of the members of the Haifa committee, the authorities in Berlin felt the need for more direct control over what was going on at the construction site. In December, 1911, Schmaryahu Levin signed a contract with the Kuratorium providing for utilization of his services as of January 1, 1912. At first he would be supervisor of construction, resident in Haifa. However, Levin saw himself as playing a role in the institution transcending the initial stages. The agreement provided for his joining the staff of the school, when it opened, in an educational capacity spelled out as director of Jewish studies and as a teacher. Friends of Levin from overseas undertook to defray part of his salary.

Levin moved to Haifa and sought to expedite the activity, but quickly discovered that he was subject to the same problems which had faced the local committee. Writing to Ahad Ha-Am on March 26, 1912, Levin reported that it was difficult to move because the office in Berlin had no understanding, displayed no leadership, made no decisions, but insisted on retaining tight control.

Because of the long interruption in construction, the contractors demanded extra compensation. Levin was at a loss. More and more he found himself relying on Kaiserman, the only member remaining of the old committee. "I spend all day on the Technion site, keeping an eye open, and meeting problems as they arise," he wrote.

The problems were many, and he gave Ahad Ha-Am a running, almost

Gedalyahu Wilbushewitz (1865–1943).

Shmuel Pewsner (1879–1930).

fortnightly account of what was happening. Baerwald had instructed that they remove the topsoil until they struck bedrock, and then place the foundations on the rock. They had already gone down seven meters, Levin complained, and still no bedrock. Baerwald had estimated the foundations would cost about 7,000 francs. The newest estimate was 40,000 francs. And in all this he got no reaction from Berlin to his frantic inquiries.

Not an engineer, Levin found it necessary to seek expert advice, and attached himself to one Gedalyahu Wilbushewitz, an engineer from Minsk, who was visiting. Levin urged the Kuratorium to engage this expert who had so fortuitously appeared. Besides, Levin confided to Ahad Ha-Am, he's one of "our" people. We should get him involved so that control is not exclusively in German hands.

It was Wilbushewitz who solved the foundation problem. The excavation was getting deeper and deeper. Finally, in one corner, at 16 meters depth, they struck rock. When Schumacher had been asked what to do under the circumstances, he had advised broadening the excavation to a 16-meter depth all around. The cost would have been enormous, possibly prohibitive.

The Minsk engineer suggested sinking a beam of reinforced concrete where they had reached rock. He ordered the placing of a reinforced block and each day came to test the rate of settling. When he had determined the maximum, they proceeded with similar beams at various selected spots and could set the foundation without major excavation, from an overall depth of only two meters.

It was at this point that Wilbushewitz took over the management of construction. Barsky, who had proved to be a disappointment, was put in charge of the drafting. Hayim Hari was the construction foreman. Wilbushewitz found fault with Baerwald's plans, principally on the grounds they failed to take into consideration that the building would not only serve as a school, but as a public institution would be called upon to provide communal facilities. His recommendations included provisions for adequate public meeting halls, a large library, exhibition space, exploitation of the basement for laboratories, and finally creation of a large courtyard facing the building, for public assemblies. Baerwald accepted all these recommendations in good grace.

Interviewed years later, Hari had clear memories of how things were managed in those days. He recalled the frequent interference from "Professor" Schumacher, who had hoped to get the planning contract for himself. It was Baerwald who had brought Hari into the picture, inducing him to move to Haifa from Jerusalem. The first steps were to gather the building materials. Limestone, for interior walls, was available in Haifa.

Sandstone, for the exterior walls, was brought from Tantura and Atlit by an Arab contractor, on the backs of camels. Most of the stone work on the site was done by Yemenite masons, brought in from Jerusalem. They slept in shacks erected on the west side of the plot, but went back to Jerusalem every week for the Sabbath. They were paid by piece work.

The standstone blocks were not always perfect, and so the Yemenites used to plug up the little holes with sand. One day, during one of Baerwald's visits, a lashing rainstorm washed out much of the sand, leaving the holes exposed. The workmen hastened to fill them again with lime and sand, so Baerwald would not notice, but he did. He laughed, and told them to leave the imperfections as they were. They conformed to ancient architectural style, and were quite suitable for this building.

Porters deliver crated machinery received from Germany. 1912.

Employment of labor was always a problem. Levin would have wanted Jewish workmen exclusively, but they were not always available. When work began on the excavations Yitzhak Rosenberg, a member of the Hashomer (Jewish Watchmen) organization, was engaged to supervise the Arab laborers. He thought it a disgrace that Jews should not do the work, this at a time when Zionist ideology was preaching the dignity of labor. He offered to have the job done by a Jewish crew at the same price as Arabs would charge. He did it; the Jews were paid 4 bishlik a day, the same rate as the Arabs. When the construction got under way Wilbushewitz noted that the laborers were all Arabs; no Jewish labor was available in Haifa for this purpose. He therefore turned to the employment office of the Poale Zion in Jaffa, then headed by Shlomo Kaplansky and David Bloch-Blumenfeld, and urged that a team of skilled Jewish workmen be sent to Haifa as an opening wedge. The Jews came, and did a good job, but this time their salary demands were higher. Wilbushewitz offset this factor by teaching them advanced work methods, which made their production more economical than that of the Arabs, who still used primitive methods. The contractors complained to Berlin that all this was in violation of their contract and they demanded that Wilbushewitz's "interference" be submitted to arbitration. With fierce intensity the Russian director accelerated the pace of construction, and by the time the slow mails brought from Berlin assent to arbitration, most of the work had been completed. By mid-April 1912 Levin could report that the labor force was 60% Jewish, 40% Arab.

Arab extremists in Haifa charged that Arabs were being frozen out of the work entirely, but they were invited to check with the men on the site, and were convinced otherwise.

Yaish Nadaf was one of the Yemenites who had been brought from Jerusalem. He recalled the fastidious care with which Wilbushewitz used to check their work. If he found a stone which had not been shaped properly he would reject it. "It's my job to see to it that this building stands forever," he used to growl at the workers. He was so fussy that what didn't measure up to his standards had to be torn down and done over. Nadaf's colleague, Haim Garcy, was another who had come from Jerusalem. Garcy saved everything he earned and soon had enough money to marry a girl he had known in Yemen. He brought his bride to Haifa.

The two men recalled that the paymasters on the site were Israel and Manya Shochat, and they used to dole out the money from big bags which, rumor had it, someone had brought from overseas. Shochat was at the time one of the heads of the Hashomer society, the group of Jewish watchmen which, in a sense, was a precursor of the Haganah. They provided the armed guards to protect the outlying Jewish settlements, and Shochat's home in Haifa was the headquarters. His defense work was entirely voluntary; he earned his livelihood, together with his wife, a personality in her own right, through their duties at the Technion. Officially he was what would today be called director of personnel. Food was served in a shack on the grounds run by "Sabta" Sturman, another name prominent in the annals of settlement history in Palestine.

Construction proceeded on two buildings. The secondary school building, to house what became known as the Reali School, was first. This was completed in 1913, and much of the main building went up in the same year. Baerwald wanted the domes of the two buildings made of fitted stone. This was done in the Reali School, but the cost was great and when they got to the big dome of the main Technion building it was made of zinc plates.

Aside from the stone, most of the other material in the building came from abroad. The lime was from France, the cement from Germany. Plumbing installations and various fixtures also came from Europe, and to this day the visitor can read the German manufacturers' inscriptions on floor plates and elsewhere in the building.

The third building was to house the workshops. On Wilbushewitz's advice the structure took advantage of the slope and the basement was created on the lower side of the building.

Digging of the well also created problems. First borings went down to about 40 meters, and had to be suspended for lack of skilled labor. A special permit to import the required dynamite was obtained only with great difficulty. It was not until a foreign expert came in that work was resumed and water was finally struck at 93 meters. The well was deepened to 100 meters, and for many years thereafter became a major source of the water supply, not only for the Technion, but for the entire immediate neighborhood. It was also an important source of income for the Technion.

Perhaps the only German actually engaged in the work on the spot was Dr. Eliyahu Auerbach. He had studied medicine in Berlin and practiced there for five years. A Zionist, he convinced Paul Nathan to send him to Haifa to look after the health needs of the workers and staff. Baerwald had met him in Berlin, and when he heard that Auerbach was actually going to live in Palestine he told him he was crazy, the doctor recalled. He arrived in Haifa in 1909, at the age of 27, working on a retainer fee from the Hilfsverein. The danger of malaria was great because of the mosquitoes from swamps in the bay area. He took quinine pills daily.

There were three Arab doctors in Haifa, of limited training. Auerbach was the first Jew. Indeed, he claimed to have been the first German Zionist to settle in Palestine. He was the third member of the local Technikum management committee, together with Kaiserman and Pewsner.

It should be noted that when the war broke out Auerbach was recalled to Germany, conscripted, and sent to the front. Some years later he returned to Haifa and became physician to the Reali School, looking after Technion students and teachers as well.

The cornerstone ceremony, held on April 11, 1912, was a major event in the Palestine of those days. The Zionist, Yechiel Tschlenow, conducted the ceremony as the only member of the Kuratorium present. Though there were no Turkish officials present, he spoke carefully, discreetly, and avoided Zionist references. He referred to the loyalty of the Jewish community to the Ottoman Government and the Jewish efforts to help revive Palestine. The Technikum was cited as one step in such efforts. The German Consulate was represented, and the notes taken on that occasion were duly sent back to Berlin, as will be seen in a later context.

The entire Jewish community of Haifa turned out. One of the little boys in the crowd was Yaakov Dostrovsky, later Yaakov Dori, and a future Technion president. The authorities in Berlin did not even acknowledge the cable which Levin sent them from the ceremony. Absentee direction did not lead to good relations.

Most furious of all with Berlin was Ahad Ha-Am. As a trustee of the Wissotzky Foundation, which had made the first major contribution, and as a member of the Kuratorium he felt that he was being brushed aside when decisions were made. He was not even kept informed of what was going on. In letters to Levin, over a period of several years, he poured his heart out. Paul Nathan was singled out for particular criticism. In one letter Ahad Ha-Am referred sarcastically to Nathan as "The Creator". In another letter he attacked the "Pope" who sat in Berlin and basked in self-confidence and the justice of his ideas, even if he had appropriated them from someone else. Ahad Ha-Am wrote he was sick and tired of it all and wanted to quit, but was afraid that would hurt the cause.

Nathan Kaiserman (1863–1945).

Dr. Eliyahu Auerbach (1882–1971).

Laying of the foundation stone for the first building, April 11, 1912.

Ahad Ha-Am did not lose interest. The Wissotzky Foundation was required to make periodic distribution of its assets, and in 1912 Ahad Ha-Am proposed to the heirs and his fellow trustees that they appropriate another 100,000 rubles for the Technikum. Legally the gift could not yet be paid at the time but he procured the signatures of the four heirs in advance, to assure the gift. In the minutes it was duly recorded that Ahad Ha-Am had made the proposal. But, he wrote to Levin, when the news would be issued from Berlin it would probably be announced that the contribution had come as a result of the intercession and persuasion of Dr. Nathan.

The differences were not only personal. The two sides were diametrically opposed in their philosophy of Jewish life. Whereas the Hilfsverein and its people felt that the only bond which united Jews was religion, the Zionists were Jewish nationalists. The petty sniping in these early years was a preliminary symptom of the dramatic confrontation yet to take place.

The official group portrait at the laying of the foundation stone. Left to right (standing): Nachum Armon, Nachum Karol, not known, Kurdish well-digger, sister of Mrs. Barsky, Ing. Emsallem, Mrs. Barsky, Mr. Wexler, Mr. Assiskin, not known, Shmuel Pewsner, Nathan Kaiserman, Yehoshua Hankin, Mrs. Tschlenow, Shmuel Itzkovich, Dr. Yechiel Tschlenow, German Consul Julius Loytved Hardegg, Dr. Schmaryahu Levin, Yosef Barsky, Dr. Hillel Jaffe, Yaacov Neumann, Ephraim Cohn, Israel Shochat, Hayim Hari and Bezalel Shklarsky. (Seated): Zeigerman boy, Yemenite stone cutter, Shimon Levin, Ing. Tuvia Dunnia, Mr. Belkind, not known, Avraham Medina, Mr. Rivnikar, not known. 1912.

The tasks of management in Haifa became even more demanding. Wilbushewitz did yeoman service, but Levin, a publicist by profession, was not equipped to solve the technical and administrative problems that arose. He was at a loss when it came to handling the frequent disputes between the contractors and the Technikum, but really threw up his hands when disputes between the contractors and their hired hands affected the progress of the work. He was not happy with his work and was increasingly absent on Zionist propaganda missions overseas but did not want to give up the post because the Technikum was to provide his opportunity for permanent settlement in Palestine when the school opened. He realized, however, that an experienced manager was needed, and he asked the Kuratorium to appoint one.

Even before the walls were up inquiries were received from potential students and teachers. In the very week of the cornerstone ceremony Levin received a letter from Zvi Ben Yaakov Bruck, writing from Moscow. It was typical. Bruck said that many young people had already turned to

Main building under construction, rear view. December, 1912.

Excavating for foundations of the workshops. 1913.

Main building under construction, front view. January, 1913.

Construction material brought to the site by convenient transportation methods. 1913?

him for information, and he asked for an immediate reply to questions dealing with the level of the Technikum, courses, tuition fees, entrance requirements, degrees offered, etc. When would classes begin?

There was no one yet qualified to answer any of these questions, but the need for a managing director became all the more apparent. The committee in Berlin began to cast about for a suitable person in Germany. The Zionists had other ideas. They wanted someone who shared their views. Chaim Weizmann, then 38, was considered. Judah L. Magnes, close to the Americans who were actively interested in the Technikum, felt they would support Weizmann because he was known as a good organizer. Weizmann went to Berlin and met with the committee. He was not altogether unknown to them because in 1903, when a *Privatdozent* at the University of Geneva, he had applied for a post on the teaching staff of the Hilfsverein's Teachers' Institute in Jerusalem. He had not been accepted. This time he was not impressed with them because of their negative attitude to nationalism.

They were not impressed with him because of his Zionism. Ahad Ha-Am was prepared to conduct independent negotiations with any suitable

The walls of the workshop rise. 1913.

34

candidate, since the Berlin committee was not keeping him informed, but there was no one to interview.

One Hilfsverein member suggested that it would be good for public relations if the new Director were to be a distinguished non-Jew. The idea was summarily rejected and never raised again.

In the meantime Paul Nathan offered a contract to Alphonse Finkelstein, a German with a technical background, and the Kuratorium formally engaged him. He had studied engineering under Schlesinger, and came with the Professor's warm recommendation. He began his administrative duties from Berlin, visited Haifa to get the lay of the land, and finally in October, 1913, settled in Haifa.

Levin had hoped that financing would be forthcoming from his American friends (principally Julius Rosenwald) to make it possible for him to remain with the Technikum in some professional capacity. It had been suggested that the management be split: Finkelstein to be technical director, and Levin in charge of Jewish content. The proposal never went through, and Finkelstein took over from Levin, who left Haifa. With the departure of Wilbushewitz, his functions too were taken over. Other members of the team remained. Bezalel Shklarsky, as chief bookkeeper, was efficient and well liked. Barsky, no longer saddled with major responsibility, was a competent help. Finkelstein was technically equipped for his new administrative post. The spring of 1914 was tentatively set as the date for the beginning of classes.

The first preliminary indications of a clash came in mid-1910. As plans for the Technikum proceeded at full pace, the Hilfsverein school in Jaffa announced that its graduates would qualify for automatic acceptance into the Technikum, as part of the Hilfsverein network. Ahad Ha-Am was the first to react angrily. He denied that the Technikum was part of that network, and charged that the public was being given a false impression. He made it clear that the institute was completely independent and in due course would set its own admission criteria. In reality the Berlin committee ignored the organic separation, and continued to act as if the Technikum were another Hilfsverein project, rather than an institution governed by an independent Board of Governors with Russian and American members, as well as Germans.

The dark clouds had been gathering. The clash of philosophies resulted in a long and bitter conflict which has become known in history as the War of the Languages.

CHAPTER TWO

The War of the Languages

Primacy of German

Despite their intense devotion to Hebrew culture and the Hebrew language as part of the national revival of the Jewish people, the early Zionists did very little to create in Palestine an educational system to inculcate these values in the young. Organized education was largely left in the hands of either the various orthodox groups or the foreign organizations, of which the Alliance Israélite Universelle was the most important. The latter, as we have seen, placed no emphasis on the Hebrew language.

The Hilfsverein schools, on the other hand, were pioneers in the teaching of modern Hebrew. Their success in this field became all the more obvious when many of the pupils of those very schools carried the torch for Hebrew in the struggles that were about to break out.

During Technion's early formative years, even Paul Nathan saw Hebrew as playing an important role in the program. In his original memorandum he wrote that one of the goals of the Haifa secondary school would be to give its pupils full mastery of the Hebrew language, adequate knowledge of Arabic, and use of German as a foreign language as a means of communicating with the world of European culture. At the Technikum he thought there should also be courses in Turkish as preparation for government service, and in English for those who would seek to establish themselves in Egypt. German would be necessary for those who would hold posts with the railroads to be built with German capital. But Hebrew, he emphasized, would be given a prominent place in the curriculum, for he saw it as the language of communication among all the Jews of the East, and a symbol of Jewish unity.

As late as 1910, during a visit to the United States, Nathan had voiced priority for Hebrew. When asked by Judge M. Sulzberger about the language at the new institution he had answered briefly: "Hebrew".

What brought about a change in his view? It would appear that there were two factors.

For one thing, the zeal and assertiveness of the Zionist movement was provoking negative reaction from Jews who had a different philosophy of Jewish life. The greater the publicity for Jewish nationalism, the more Nathan was driven on the defensive. He constantly disavowed any association of the Hilfsverein with Zionism. He warned the Jews against committing the "stupid folly" of Zionism. He said that Zionism would lead to catastrophe and Zionists would become their own grave-diggers. He saw Jewish national chauvinism as a crime against civilization but was curiously silent about the dangers of German national chauvinism. The more the controversy flared, the sharper became his criticism of Zionism. And though he had no intrinsic objection to Hebrew as a language, he came to realize that it was the major vehicle for Zionist nationalist expression.

The second factor in the change would appear to be the growing influence of the German Foreign Office. In 1909 James Simon had written to the German Secretary of State assuring him that the teachers at the Technikum would for the most part come from Germany, and the institute would help advance German interests in the area. He made no secret of his hope to procure the backing and support of the German government. In the early years it was always the Hilfsverein which took the initiative, but government response was not warm. And then, some time in 1911 or 1912, the situation seemed to change. The Foreign Ministry emerged as the interested party. It had its own imperial policies in the Middle East, and it expected German organizations to aid in furthering those policies. In a sense, says one observer, the Hilfsverein suddenly became rigidly bound, and now not always free to act as it wished.

The German consulates in Palestine were alerted, and their reports provide much information not only on what happened, but also on their own role in helping things happen.

In Haifa, it should be noted, the German influence made itself felt largely through the Templar colony. Though its members were only about 2½% of the population, they were among the leaders in commerce and industry. The French dominated in the fields of education and culture, and this situation was a challenge to the local German consulate.

Friedrich Keller, for many years German Vice Consul in Haifa, had a most positive attitude toward the Jews. Arthur Ruppin reports that when he visited with Keller in 1907 and noted the Templar development of the city, Keller declared: "We are doing all this for the good of the Jews; has not God vowed to return them to this land?" Before his death in 1913 he told Dr. Auerbach that he had personally planted 300,000 trees on Mount Carmel, which would eventually be for the Children of Israel. The lane on which the good German built his home on the Carmel is still known as Keller Street. Had he still been the active representative of the German government in Haifa during the Technikum controversy the matter might have ended differently, and the course of history might have been changed.

The key figure, serving as Acting Consul, Vice Consul and Consul in Haifa during the crucial years 1911–1914 was Loytved Hardegg. Some believe that it was his agitation and instigation that brought matters to a showdown. He attended the cornerstone ceremony in April, 1912, and reported to the Foreign Office that the principal speaker had been a Zionist. This showed which way the wind was blowing. A copy of his memorandum was sent to James Simon, who shared it with Paul Nathan. The latter minimized the importance of the ceremony, and reassured the President of the Hilfsverein that all the people of importance on the Technikum committee were not only not Zionists, but actually opposed to Zionism. This reply made its way back to Hardegg, and he responded.

He insisted that no matter what the situation in Berlin, the institution in Haifa seemed to be falling into Zionist hands, and he suggested ways of reducing that influence. For one thing, the financial management of the Technikum should be taken away from the Zionist Bank (Anglo-Palestine) and put in the hands of a German bank. He realized that the Technikum was of course a Jewish institution, and was of interest to his government only so long as it served German cultural interests. He therefore urged that insofar as possible only non-Jewish Germans should be employed as teachers.

Hardegg did not let up. In a memorandum directed to the German Chancellor in May, 1913, he warned that steps should be taken to prevent the Zionists from gaining any further influence over the institution. He urged that pressure be applied to the Hilfsverein to obtain a firm decision regarding the use of German as the language of instruction at both the intermediate school and the Technikum proper. This could be done, he wrote, so long as the Committee needed government protection and influence.

Kaiser Wilhelm read this memorandum and approved. The pressure was on.

The Kuratorium had already begun to discuss curriculum and language. James Simon, informed of the Kaiser's interest, felt that a formal decision proclaiming German as the official language was unnecessary, since the teachers would in any event be Germans. Besides, the graduates of the intermediate school coming into the Technikum would all be proficient in German. He could not prevent the growing use of Hebrew as a spoken language in Palestine, he said, but as for the Technikum, it should be obvious that the maintenance of the necessary high standards required the use of German. There could simply be no argument about it.

The differences of opinion, that had hitherto been spoken of only in private correspondence and closed meetings, now broke into the open. The Kuratorium called a crucial meeting to be held in July, 1913, at which decision would be taken. The session was postponed, and finally took place on October 26 of that year. In the months that preceded, the flames of controversy leaped higher.

Schmaryahu Levin pleaded with Nathan, and sought to make the case for Hebrew without basing it on Zionism. He stressed that in the international political struggle then shaping up in the world, the establishment of German at the Technikum would give the impression that Jews as a people were siding with Germany, whereas the Jews should remain neutral. He urged that public opinion should be taken into consideration, especially in Palestine, where feelings ran strong.

Nathan replied, in effect, that he was indifferent to public opinion in Palestine, which had done nothing for the Technikum except give advice. He said that the proper subject for discussion was not the language of instruction, but the goals of the institution. And one purpose was indeed to give the pupils a fluent knowledge of Hebrew. "If that does not satisfy the Jewish chauvinists, I can do without their applause," he said.

In June, 1913, the upper class pupils of the Teachers' College and the School of Commerce in Jerusalem, and the Herzlia Gymnasium in Jaffa, sent a polite letter to the Kuratorium in Berlin. They wrote how eagerly they looked forward to the opening of the Technikum, to be the crown jewel in the Jewish educational system in Palestine. They were disturbed by rumors that the Kuratorium was divided on the subject of the language of instruction. They feared that if a foreign tongue were adopted it would set a tone for the rest of the schools in the country.

"We who are to be the future first students at the Technikum are impatiently awaiting the day of its opening. To us the matter is most important. We therefore respectfully submit these our views, and ask that you take such views into consideration when the time comes for you to make your decision."

Most diplomatically, the letter made no mention whatever of German.

It was signed on behalf of all the pupils by Moshe Shertok, later Sharett, then 19 years of age.

In the meantime, the new director, Alphonse Finkelstein, had come on his first visit. Levin believed that Finkelstein had really been sent to gauge public feeling and to report back to Berlin on the language situation. He decided to handle the visit accordingly, and in a letter to Ahad Ha-Am on May 11, 1913, told what happened.

"I deliberately arranged that he should disembark at Jaffa rather than Haifa. I did not want him to start in a German atmosphere. He was impressed and amazed at all that he saw — the Hebrew spirit, the Zionist spirit, the Hebrew school system. This was all quite the opposite of what he had been led to expect when in Berlin. Hebrew everywhere! But when I spoke to him about the Technikum he gave me to understand that the matter was already concluded: German! That's bad. The verdict may go against us, but I shall never quit the battle. I shall fight on, for the cause is Jewish renaissance. I'll make things difficult for them. So far I have held back public opinion here. I'm very sorry I can't be at the Kuratorium meeting in July, but you can speak on my behalf. Your views are my views."

Three weeks later Levin reported on Finkelstein's visit to the Hilfsverein

Alphonse Finkelstein.

schools in Jerusalem. He had been disappointed by the low standards, and had to admit that even the Hebrew Gymnasium at Jaffa taught a higher level German than the Hilfsverein school in Jerusalem. The Hebrew Language Committee met with him and dinned into Finkelstein's ears their demands for Hebrew at the Technikum. Levin summed up Finkelstein's reaction in two words: They're chauvinists!

There was a momentary distraction of public interest when it was announced that oil had allegedly been discovered in the Tiberias area. Zionists and community leaders began to speak grandiosely of the need to mobilize capital, dig wells, and take all the necessary steps to capitalize on this find. When the reports were shown to be false, the Jewish community reverted to its sense of agitation on the language problem. But not only the Jews were concerned. The French, too, had become disturbed, and the French Vice Consul in Haifa, M. Guy, reported to Paris on developments. As a result, the French Foreign Ministry wrote to the head of the Alliance Israélite Universelle warning that adoption of German by the Technikum, and the general tendencies in this direction "constitute an act very dangerous to French influence."

Others were concerned as well. In August, 1913, Mr. Louis Friedman, an American Jew, returned to New York from a visit to England and brought with him a letter from British Premier Asquith, promising the support of the British government and people for a Jewish National University, to be built in Jerusalem, if the university adopted the English language, and English literature were made a compulsory part of the course of training. Mr. Friedman explained that this was part of a movement to make English the predominant language in the Orient.

Everyone was getting into the act.

Compromise Rejected

The Zionist front was far from united. Differences of opinion were based on practical grounds. Even the most fanatic advocates of Hebrew had no illusions about the feasibility of teaching advanced technology and science in a language which was completely lacking in technical terminology. There were no textbooks whatsoever available, and practically no qualified teachers who knew Hebrew. This was the situation on the university level.

But the proposed secondary school was another matter. Nathan and his group saw it as a preparatory school for the Technikum, and hence the language of instruction there, too, should be German. Ahad Ha-Am and Tschlenow, as members of the Kuratorium, agreed that Hebrew could not be made the principal language in the Technikum. They tried to calm down the Hebrew zealots. Ahad Ha-Am referred to "demagogic agitation". Tschlenow felt his Zionist colleagues were going too far and demonstrating a degree of "chauvinism beyond that permissible to pioneers of a progressive culture."

Levin sought a compromise formula and sent it off to all members of the Kuratorium. In the high school, Hebrew would be used for all subjects, and German would be the major foreign language. On the university level, German would be the tongue of instruction, but one or more of the required subjects would be offered in Hebrew.

This moderate compromise was rejected by Nathan and his associates, whereupon Levin resigned from his membership on the Kuratorium. He could now speak his mind more freely.

He had a platform at the Eleventh Zionist Congress held in Vienna at the beginning of September, and spoke up as a maximalist, demanding Hebrew at every level of teaching in the Technikum. Ahad Ha-Am was willing to fight only for the secondary school. Even Weizmann told the Congress that use of Hebrew on the top level at a time when the language was not yet ready for it might affect the teaching program.

Paul Nathan tried a new tack. He suggested that no decision whatever be taken by the Board. Let the academic staff of the Institute decide. Levin denounced the proposal as a ruse since Nathan was intent on engaging only teachers devoted to "holy German".

What about the American members of the Board? Where did they stand? What about other influential people in the United States? The only one to give an opinion at this stage was Professor Solomon Schechter, and his views did not make the Zionists happy.

"I knew in advance that Schechter would not defend the use of Hebrew," wrote Levin. "One can't change him. I think he equalizes kosher food with everything else in Judaism."

A preliminary meeting was held in Berlin and Levin attended. He reported to Ahad Ha-Am:

"Yesterday we met at Simon's home: Simon, Nathan, Netter, Phillipsohn, Kahn, Franz, Schlesinger, Baerwald, Finkelstein and I. It was decided to open the workshops next Spring. Dr. Nathan wanted to open the first department of the Technikum at the same time, but Simon and the others opposed. They were right. An important question is the choice of teachers. Four names were proposed: Hecker, Wilbushewitz, Dr. Biram and Dr. Lowenherz. Finkelstein asked Kuratorium approval for these appointments at once. I consider it a great victory that Finkelstein was prepared to overlook their Zionism .... I proposed that after the Kuratorium approved the curriculum it should be translated into Hebrew. Nathan objected. Simon suggested that decision be postponed until after the Kuratorium meeting .... Finkelstein admitted they probably would not be able to find suitable teachers willing to go to Haifa, except Zionists .... The atmosphere here is choking, poisonous, malicious gossip."

In the presence of the others the principals were restrained, but the blow-up was yet to come. Levin decided that before leaving Berlin he would call on Nathan, seek to make peace, and perhaps even persuade him to change his stand. The secretary, Dr. Bernhard Kahn, was apparently present. Levin wrote to Ahad Ha-Am what happened:

"Nathan was stubborn and bitter. He accused me of trying to have him kicked out of the picture. When I realized that there was no hope of influencing him, I felt I could throw restraint to the winds and could open up. I reminded him of his answer to Sulzberger regarding the language of instruction. He denied it. I said I had witnesses. He said their testimony was worth nothing .... In his anger he said pointblank that neither I nor any one of us had any share in the institution. He alone was the creator and the director. It was only due to his tender mercies that Tschlenow and I were taken into the Kuratorium. I did not feel obliged to remain silent, and replied as I saw fit under the circumstances.

"He also went further. Since the institution was under the patronage of the German government, we were obliged to let non-Jews pursue their studies there without any obligation about a foreign language. It was obvious that he had already given some promises in this direction. He spoke at length about the German patronage."

Levin gathered that Nathan was trying to influence the Americans his way, and urged that the Zionists, too, should seek to win over Schiff, Marshall, Sulzberger, Cyrus Adler, Samuel Strauss and others. He continued his report to Ahad Ha-Am:

"In the 'program' the name is given as Technisches Institut in Haifa. The word 'Jüdisches' is dropped. I slyly congratulated Dr. Kahn on the change of name and told him that it is true that among our people it was common to change a name in case of illness, but that we usually added to the name, rather than shortened it. Kahn said it was a mistake. I don't believe him."

Consul Hardegg wrote that the decision to be taken by the Kuratorium would be crucial. If the Technikum were to require German as the major language, every school in the country which wanted to afford its graduates opportunity to enter the institution, would have to elevate the status of German in their own curriculum.

There was a report that the German Under Secretary of State, Arthur Zimmermann, had demanded a favorable decision, and if the Kuratorium did not comply, the German government would withdraw its patronage.

Eliezer Ben Yehuda, father of modern Hebrew and author of the Hebrew lexicon, denied that the language was inadequate. He wrote to Nathan that the appropriate terminology could be developed in a year. He warned that the local population was ready for any sacrifice to enforce their demands. "The Technion will not open without bloodshed," he told Nathan. "Don't think this is just the work of a few youths."

The Zionist and general Jewish press in Europe carried articles and letters setting forth the opposing viewpoints. One such polemic is of special interest because of its author. Ing. S. Kaplansky of Cologne (later to serve for twenty years as president of the Technion) wrote that he could speak with authority about the practicability of the Hebrew language since he had been a teacher of mathematics and physics at the Hebrew Gymnasium in Jaffa. Kaplansky made four points:

1. Hebrew is a living, dynamic language, and in time can develop no less than other tongues which a few short years ago had no scientific terminology.
2. No books? Existence of the schools will bring about creation of the texts, not vice versa. Even German is still sadly lacking in basic texts, and besides, we can always translate.
3. No Hebrew-speaking teachers? They can learn, just as the Germans teaching in Russian universities had to learn Russian; and they did.
4. No jobs for the graduates? He pointed out that German and Belgian engineers were getting jobs in Turkey and elsewhere, just as Hebrew engineers would also.

The October 26, 1913, meeting took place in Berlin. The Americans were not present and the Zionists were in a small minority. Paul Nathan laid it on the line. He insisted that the language controversy was a matter of principle, and failure to endorse his stand would be regarded by him as a vote of non-confidence. He and his German associates would resign.

Prof. Schlesinger voiced his bitter opposition to Hebrew and insisted that the school must preserve its German character. He reminded the Board that he had warned them of this when he had ordered the German equipment. The implication was that he had obtained substantial discounts on that basis. The Board must not yield to Zionist fanaticism, he said.

The debate lasted for three hours. The Zionists presented their oft-repeated arguments and again warned that in an era of growing international tension the Technikum should avoid too close identification with one of the major world powers.

Ahad Ha-Am spoke up. On the high school there could be no room for compromise. The language there must be wholly Hebrew. With respect to the Technikum there were indeed practical considerations, though the Board should at least recognize the principle of the priority of Hebrew.

The decision was framed to soften the blow, but there was no doubt that it was a complete victory for the proponents of German:
1. The official language of instruction was not fixed.
2. Hebrew would be given a place in the curriculum suitable to the Jewish character of the Institute.
3. Arabic, Turkish, English and French would also be taught.
4. Natural sciences and technical subjects would be taught in German (This was, of course, the crux of the matter).

The committee which had been working on curriculum presented its plans. In the two upper classes of the secondary school German would be used in 69% of the program, and Hebrew in 17%. Hebrew was preferred for teaching Judaism and singing. On the higher level German took over almost exclusively. Two hours a week were to be offered as an elective in Hebrew, English or French.

Confrontation

The German Board members were so oblivious of the new winds of renaissance blowing over the Jewish people that they had not the slightest understanding of the principles involved. As a consequence they were not at all prepared for the torrent of furious opposition which ensued.

The Zionist members of the Board, after consulting the Small Zionist Actions Committee, accepted the decision of that body that they should resign from the Board. This was significant in that the Zionist Organization, for the first time, undertook a responsibility for the fate and destiny of the Technion — a responsibility which was to waver somewhat in the years to come.

The first appeals of the Hebrew Teachers' Association to the Hilfsverein had been muted. The Association did not wish to provoke a confrontation. But after the Berlin decision all restraint was removed. War was declared. Indeed, the Association and its leaders spoke in military terms: of battle and struggle, of sacrifices, and eventually victory.

They drew up fiery and indignant letters and manifestos addressed to the members of the Kuratorium in Berlin and sent them off — couched in elegant Hebrew which the recipients could not read.

The direct cause of the agitation, of course, was the Technikum in Haifa. "The Technikum must become Hebrew, or it will not exist at all," was one of the slogans. But the battle was not limited to Haifa. The Teachers' Association called for a nationwide boycott by pupils and parents of all the Hilfsverein schools in Palestine. This confronted many teachers with a dilemma. If they did not work, they would not draw their salaries. They were heart and soul in agreement with the principle, but personal problems interfered. The mood and the spirit were such, however, that almost all objections were swept aside. There was an element of near hysteria in the campaign. The Association proclaimed a civil excommunication against the Technikum and everybody and everything connected with it, though the school was still far from opening. Both Levin and Achad Ha-Am thought this was carrying matters too far, but the campaign against German was now almost out of hand.

The Actions Committee of the Zionist Organization came to the conclusion that paralyzing the Hilfsverein educational system, and suspending education for thousands of young people were not helpful to the cause. The Teachers' Association, headed by Dr. Joseph Lurie, gave vehement response: The Zionist opinion, they wrote, "is a national disgrace and serves to undermine our work in Palestine. The Hilfsverein is an anti-Semitic organization, creating controversies among Jews. We hereby proclaim that not only shall we not discontinue the boycott, but declare that all teachers who remain at Hilfsverein schools will be regarded by us as traitors, and for the rest of their lives we shall never consider them teachers."

The boys and girls needed no incitement to enter into the spirit of the battle, and they thought up new and unique ways of expressing their views. The schools in Jerusalem served as the hotbeds of activity. Even before classes were actually suspended the young people did everything they could to give expression to the insistence on priority for Hebrew.

David Avissar recalls that the pupils of the Laemel School and the Teachers' College, both Hilfsverein institutions in Jerusalem, had been prepared for a public celebration of the Kaiser's birthday. The German Consuls were present. The German flag was flying. The music teacher plucked a chord on his violin, nodded his head, and the student body was to raise its voices in singing the German national anthem. Not a sound was heard! Avissar insists that the reaction had not been planned in advance; it came spontaneously, but this did not lessen the disgrace, and the punishment that followed.

At the Teachers' College, Aviezer Yellin refused to post the daily attendance roll in German, when it was his turn to monitor the attendance. There was a confrontation. Students walked out. The compromise: the list was kept in both German and Hebrew.

Young Jehiel Elyachar, a classmate of Yellin's, was another of the hot-blooded young revolutionaries battling the Germans. After graduation from the Teachers' College he went on to further his education at the French College in Beirut. Years later he became president of the American Technion Society and spearheaded a major development — the move from the old campus to Technion City on Mount Carmel.

The community in Haifa called a public mass meeting held on the open plot of what is now the corner of Pevsner and Balfour Streets. Hundreds attended and heard fiery speeches denouncing those who served foreign masters. One foreign master is as bad as another, one commentator noted at the time, and called attention to the fact that on the very day of the meeting six vessels of the French fleet had visited in Haifa. When the Admiral had stepped ashore the Jewish pupils of the local Alliance school, who were dutifully lined up to receive him, had cried out "Vive La France".

The furore caused consternation in Berlin, and Paul Nathan rushed to Palestine at once, hoping to quiet the teachers, the pupils and the parents. In Haifa he met with the workers on the building, seeking to convince them he was not an enemy. Bezalel Shklarsky, the chief bookkeeper, served as their spokesman and replied fervently that the workmen did not feel they were only constructing a physical edifice; they felt they were building toward a national ideal . . . . Hari, the stone mason and draftsman, openly backed him. Within a month both were fired.

Zev Vilnay, then a boy in Haifa, recalls being one of a group of children who were mobilized to throw stones at Nathan, but with discreet instructions to avoid hitting him.

In Jerusalem Nathan received a delegation of the pupils, but there was simply no meeting of minds. The Hilfsverein representative in Jerusalem, Ephraim Cohn, sought to isolate the nationalist Zionists from the rest of the community. Nathan ordered the striking teachers fired and there was a clash. The Turkish police were called in to maintain order. This only poured oil on the fire. Street demonstrations multiplied in Jerusalem. Windows were broken. Nathan was insulted and jeered at as he passed through the streets. The language zealots openly sought to embitter his life. Nathan, still not realizing the depth of the feelings, believed that the storm had been artificially cooked up by a handful of Zionist agitators, no more than 200 at the most. He simply could not understand the opposition — and this, after all he had done for Jewish education in Palestine, he said.

The Pasha of Jerusalem found it necessary to issue a proclamation warning against agitation which sought to prevent parents from sending their children to schools of their choice. He threatened punishment to the full extent of the law against all who were in violation. Ephraim Cohn reported that the German Consul, Eduard Schmidt, sought to maintain the peace, and hence refrained from preferring charges against the perpetrators even in the face of (in Cohn's words) violence, terror, threats, forged telegrams and wild demonstrations. On his return to Germany Nathan published a pamphlet in which he told of his experiences in Jerusalem. He charged that the Zionists were carrying on a campaign of terrorism modelled almost on Russian pogrom lines. The *New York Times* gave headlines to his claims and Louis Marshall and Solomon Schechter found it necessary to reply to the "ridiculous and exaggerated" charges. The controversy had spread to another continent.

From Vilna, Boris Goldberg wrote an encouraging letter expressing support for the teachers in their struggle for the national tongue. He understood the nature of their personal problems and sent a check constituting not only his own contribution but also funds he had collected from others. There would be more, he promised. There were; at a later stage he carried on his shoulders the financial burden of the infant institution. One day his son was to become President of the Technion.

At Beirut the Jewish Student Association held a public protest meeting on the situation and sent a message of solidarity. The Jewish students enrolled at the University in Constantinople also sent a message expressing their indignation against the action of philanthropists who sought to use their largesse as a means of extending their political influence. Signing on behalf of the group were two young law students, David Ben Gurion and Yitzhak Ben Zvi. They were soon to be joined in Constantinople by Shertok.

The orthodox community to an extent stood aside. This was a battle of the Zionists, and they did not, at first, see any implications for them. In 1911 the Ashkenazi rabbi of Haifa, Rabbi Baruch Mayers, had written to Ephraim Cohn warning that contractors employed on the construction should not permit any violation of the Sabbath. He understood that the Hilfsverein operated within the spirit of traditional Judaism.

Rabbi Avraham Yitzhak Hacohen Kook, writing from Safed, expressed his opposition in principle to the use of any foreign language in the schools but at the same time warned that emphasis only on the Hebrew language would not bring about the desired national rejuvenation. There must also be emphasis on the holy principles of the faith as well, he said.

The most original contribution came from Rabbi Moshe Halfon, one of the Sephardi leaders in Haifa, who threw his full support behind the Hebraists on two grounds. First, he thought it very possible that the engineers to be trained at the Technikum might be the ones called upon to rebuild the Temple in Jerusalem. It would be inconceivable for this holy project to be conducted in German.

Secondly, he opposed the use of German in an institution whose name came from such an obviously Hebrew source: *Tanach-Kum* (Rise up, O Bible). In such an institution the speech must be only in the language of the *Tanach*, the Bible.

By now the German consuls in the three major cities were deeply involved. Hardegg in Haifa was the most active. The official records reveal the extent of an involvement which he concealed at the time. In a letter to the Reichschancellor he gave a detailed report on Nathan's visit to Haifa.

"With the permission of Dr. Nathan I influenced the local Mufti and the head of the local court to meet with Nathan in the name of the Moslems and to request him to stand firm against the wishes of the Zionists to banish non-Jews from the Technikum," he wrote. "I also saw to it that several Moslem notables telegraphed a similar request to the Ministry of Education in Constantinople. I influenced the press in similar fashion. I don't have to emphasize that personally I remained behind the scenes, and publicly displayed detachment."

The struggle between the great powers was reflected here too. Hardegg reported that the other consuls were seeking to exploit the controversy for their own ends. He charged that the Russian Consul was secretly helping the Zionist inciters, most of whom were Russians, in the belief that the strengthening of Jewish nationalism would weaken the Turks. And the French Consul, he said, would fight with all the means at his disposal against anything that might advance German influence.

Following one Zionist mass meeting, Hardegg reported back to Berlin that the Zionists were being impractical. They should have realized that what the Germans wanted was to produce efficient and well-educated Jews, and in the long run this would help Zionism. He would have advised the Zionists that it was premature, even from their point of view, to press for the use of Hebrew at the expense of the educational standard. Further, he thought that the Zionist struggles at that time magnified the suspicions of the Arabs and also weakened Jewry by creating internal conflict.

Though he felt that the Jews showed too much arrogance, and lacked tact, he came to the conclusion that Hebrew would eventually become "the living national language of all Jews". He also believed that the Hilfs-verein should pay more attention to Zionist national aspirations — so long as the Technikum followed the German educational pattern. These views did not prevent Hardegg from inciting the Arabs. He encouraged the argument that use of Hebrew would mean exclusion of Arabs from the Technikum. He ignored the fact that as between the two tongues, it would have been easier for the Arabs to learn Hebrew than German.

Hardegg also corresponded with James Simon. He warned that if most of the Technikum students came from Russia, as was expected, they would in any event "capture" the institution, and the Hilfsverein would be advised to seek a way of coming to terms with the nationalist movement.

The reply, coming from Nathan, declared curtly that the Hilfsverein and Zionism were worlds apart.

Eduard Schmidt, the German Consul-General in Jerusalem, was of course in favor of the official line, but felt there was no need to fan the flames. He urged the Hilfsverein not to precipitate a complete rupture, even if this meant reducing the emphasis on the German character of its schools.

Heinrich Brode, Vice Consul in Jaffa, and later in Jerusalem, was the only one of the German Consuls in Palestine who was critical of the Kuratorium decision. In a report to the German Ambassador in Constantinople he opposed the outright confrontation with the Zionists. He warned of the danger of a Jewish boycott against German goods at a time when there was a growing market for agricultural machinery, fertilizer, building materials, and the like. It was the Jews who were the purchasers of these items.

Like Hardegg, he felt that the Zionists were foolish in bringing about a showdown since Hebrew would emerge as the dominant language in any event because of the ever-expanding colonies. He noted that their young generation was learning Hebrew as its mother tongue, and this in the very kindergartens set up by the Hilfsverein.

But Brode also had an official line to follow, and he offered his own tactics. He suggested that if the struggle caused any harm to German political or economic interests, it would be easy to paralyze all Zionist work by calling to the attention of the Turkish authorities the dangers of the separate nation being set up. The Zionists pretended to be loyal Ottoman subjects, he wrote, but they were already beginning to act like masters of the country. The Turks would be quick to react. The Zionist institutions could be closed. Immigration could be halted.

He told the Ambassador that the Jews of Germany should be reassured that the Hilfsverein was not opposed by the majority of the Jews in the country, the old orthodox communities, but by anarchist inciters who had come from Russia.

It should be noted that until 1917 the comprehensive files of the German consulates in Palestine were efficiently maintained. After the War the British Mandatory Government took over all German property and records, and by happy circumstances all these files passed eventually into the hands of the Israel Government Archives.

None of the consuls could understand why the Zionists, in their fervent espousal of Hebrew, directed the struggle only against German. Why not against the Alliance schools, which taught far less Hebrew than the Hilfsverein institutions, and continued to carry on their work without interruption or protest?

As a matter of fact the Zionists were not happy about the Alliance schools. Ahad Ha-Am had at one stage drawn attention to a certain distinction. Before the Technikum crisis broke out he had declared that so long as the Hilfsverein was willing to support schools at which Hebrew was taught, he was willing to support them. Not so the schools of the Alliance and its teachers "who saw no use at all in Hebrew education. We were bound to oppose them tooth and nail and to regard their system of education as worse than nothing at all."

In the case of German, it appears, the confrontation was brought about by the decision of the Kuratorium; all the accumulated Zionist zeal was swept up into the battle against the "enemy" most obvious at the moment. There was also the feeling that changes were being introduced in the German schools, against Hebrew. It was charged that at the Laemel School in Jerusalem, for example, instructions had been issued to replace Hebrew with German in some courses. The same was said to be taking place insidiously at other schools as well.

This Nathan denied vigorously. It may well have been that the atmosphere was charged psychologically. The Hilfsverein's man in Jerusalem, Ephraim Cohn, though he knew Hebrew well, was said to speak German not only in his daily affairs but at home as well. Ahad Ha-Am contemptuously referred to Cohn as being "more royal than the king".

Finkelstein, who came over to take charge of the Technikum, not only knew no Hebrew, but acted with a heavy hand. It was reported that he was dismissing workers who had spoken out for Hebrew. He was said to be vindictively transferring business and orders for materials from Jewish merchants to German companies.

The press reported that he issued an ultimatum to all the Jewish laborers on the site, most of whom had large families, that they must send their children to the Hilfsverein school in Haifa. The implication was that if they did not, they would lose their jobs. There was a howl of protest and he had to back down.

The struggle was bitter because there were sharp issues of principle. To the Zionists it was a matter of national integrity. To the Germans, their own national prestige was at stake. And to Ephraim Cohn and his staff, it was a matter of protecting a system and a bureaucracy (even in the best sense of the word) that they had created.

While the battle royal was going on, the personnel charged with specific tasks continued their work. Target date for opening of the Technikum was April, 1914. In 1912 Schlesinger went to Palestine to familiarize himself with the industrial needs. He drew up a long list of required machinery and equipment, all of German make, naturally. Everything was electric-powered, though there was no central source of electric power anywhere in the country at the time.

It is of some historical interest to note the type of things he ordered. The following is an abridgment of the long lists, but typical: For the

Locomobile steam engine installed in the workshop. 1914?

Carpentry Shop: lathes, precision lathes, drills, workbenches. For the Forge: fans, welding equipment, crane, cooling machine, gas ovens, oil heaters, sledge hammers, air hammers, hearth, anvil and punch. For the Casting Shop: steel forms, hearth, forced air supply. For the General Workshop: radial drills, horizontal and vertical presses, tool grinders, stamp dies, planers, eccentric press, sheet metal machines, shears. For the Electric Shop: batteries, motors, compressors, generators, high pressure pump, fans. For the Laboratory: electrodes, fuses, explosives, glass-blowing equipment.

An artisan was sent from Palestine to Germany and spent two years at various factories familiarizing himself with the equipment so that on return he would be able to take charge of the assembly and installation.

Pressured by the Kuratorium to meet the April, 1914, deadline, Finkelstein proceeded with the engaging of teachers, especially departmental heads. He found that under any circumstances the best people he could get, or those willing to serve in Haifa, were Zionists. Of the seven major posts, five were offered to men who were known supporters of Jewish nationalism: Gedalyahu Wilbushewitz, director of the workshops; Max Hecker, head of the construction department; Dr. Arthur Biram, principal of the secondary school, and two teachers, Arthur Lowenherz and David Tachauer.

This created a new problem. In view of the strike and the boycott, should these young Zionist immigrants, ready to teach in Hebrew, come to Palestine at all, or should they abrogate their contracts? Schmaryahu Levin felt they should come ahead. The striking teachers opposed. The Small Zionist Actions Committee came to the conclusion that the new teachers from abroad should cancel their contracts, as a sign of solidarity with their colleagues in Palestine. Some of them disagreed. Hecker, for example, came to Haifa at once, and while preserving his independent views in the matter of language, worked alongside Finkelstein in getting things ready. The well house and the water tower were completed, so at least there was an assured supply of water on the site. It would now be possible to do landscaping and Avraham Ginsburg, who had done gardening at Mikveh Israel, was offered the post of chief gardener. He came to Haifa seeking his new employer, who he understood was a "Mr. Technikum". Finkelstein engaged him.

Biram went to Haifa in February, 1914, but stood aloof from the Technikum. He subscribed to the decision taken by the Zionist Executive in Berlin in December, 1913, to set up their own Hebrew Reali School, to be managed by the Kaiserman-Auerbach-Pewsner committee. Biram also thought of introducing technical classes into the new school so that if they did not succeed in gaining control of the Technikum they would have the basis for another one of their own. Since Finkelstein was in physical possession of the school building, they sought facilities elsewhere.

The storm had adverse effects on fund-raising, and the flow of money from Berlin slowed down, causing layoffs of workers, and thus further exacerbating the crisis.

The time had come for another distribution of income from the Wissotzky estate, and despite his strong feelings on the language controversy, Ahad Ha-Am persuaded the other trustees to make the second payment of 100,000 rubles to the project. It was not easy. The Russians had returned to their original request that the building as a whole bear the Wissotzky name. Obviously Schiff would never consent. One proposal was that when a chemistry department was to be created it should bear the Wissotzky name. The compromise suggestion was still the erection of a plaque in the lobby. The new money was intended for dormitories, but these were far from realization, and so the funds were "loaned" to the general budget to enable completion of the main building and workshops.

The agitation in Palestine continued at almost frenzied tempo, and the overseas press, especially in Germany, reported the controversy in detail.

The Zionist line was that by its excessive zeal the Hilfsverein had antagonized Jews in many countries and had actually weakened the German position in the Middle East. The Zionists were motivated not so much by opposition to the German language, as by a positive feeling of Jewish nationalism. There need not have been a conflict.

Consul Schmidt, seeking to pour oil on the troubled waters, issued a friendly, diplomatic statement. James Simon wrote an angry note to Dr. Arthur Zimmermann, who handled Palestine affairs in the Foreign Ministry, to the effect that "the Consul has stabbed us in the back." He called for a life and death struggle against the Zionists.

"If they mean war, then let it be war," declared the usually mild-mannered A.D. Gordon, philosopher of the Jewish labor movement. The time had come not to protest but to warn the enemy, he said.

Drastic actions were taken. In Haifa it was reported that one Shmuel Ben-Shabbat, who had taken the public oath at the protest meeting in that city not to set foot in the Hilfsverein schools, had nevertheless gone on teaching for them. A number of leading organizations proclaimed that they regarded him as a traitor and banished him from membership.

During the period of the *Kultur-Kampf* the Technion was projected to the very forefront of public attention as a symbol of the Zionist renaissance. It reached heights of popularity. This was an ironic and fleeting moment of glory, for in the years ahead the institution was to be forgotten and abandoned in moments of great need even by those who had once carried its torch.

Something had to give. The Kuratorium found it necessary to hold a new meeting for reconsideration of the problem, possibly at the urging of the German government, which had been placed in an embarrassing position. For a moment the center of gravity shifted to the United States as both parties to the struggle sought to influence the American members of the Board and the very influential non-Board members, like Schiff.

The American Connection

Since he had been persuaded by Nathan in 1908 to make his first princely gift to the Institute, Jacob Schiff had maintained constant interest. He received copies of various reports issued from Berlin, and kept prodding the committee there to get on with the work.

He was not a Zionist at this stage. He had corresponded with Herzl and had met his emissary, but thought Herzl's plans were "impractical and utopian." In a letter to Solomon Schechter in 1907 he had written that he could not for a moment "concede that one can be at the same time a true

American and an honest adherent of the Zionist movement." It was obvious why he and Nathan found common language.

Yet his interest in the Technion was overriding. In 1911, when it was reported that he had offered to endow a university at Frankfort-am-Main, his native city, he wrote to President Eliot of Harvard denying the story. He said he would not "devote any large sum to educational or altruistic purposes in Europe, when the country of my adoption, to which I owe everything, needs more than any individual can do for these purposes."

Clearly he did not consider the Technikum a foreign institution.

It was Schiff who secured the participation of other Americans in the project. Of the 21 members of the Kuratorium eight were Americans, seven Germans, five Russians and one from London. The eight Americans were Dr. Cyrus Adler, Judge Julian W. Mack, Louis Marshall, Julius Rosenwald, Prof. Solomon Schechter, Mortimer L. Schiff, Samuel Strauss and Judge M. Sulzberger. Their dominance was on paper only, for the members resident in Berlin exercised the day-to-day influence over events.

A fund-raising campaign was launched, with Samuel Strauss as treasurer. Rosenwald led the way with $50,000. "We trust that the Jews of America will grasp the significance of this project which is to inaugurate a new era for the Oriental Jews of today, and for such Jewish students in other lands against whom the doors of education are being closed in increasing number," the appeal stated.

Rosenwald was involved in other ways as well. He had taken a liking to Schmaryahu Levin, and it appears that he was financing Levin's employment in Haifa. In February, 1914, Rosenwald visited Palestine and was escorted by Aaron Aaronsohn, whose agricultural research station at Zichron Yaakov he was supporting. After visiting various schools where Hebrew was in use, Rosenwald wanted to join the struggle. The Kuratorium was soon to meet in Berlin for its reconsideration meeting, and Rosenwald decided to cable his firm opinion that Hebrew should be the language at the Technikum too. He was also against anything that would appear to make this Jewish institution a tool of German imperialism. He asked Baron Edmond de Rothschild, then also in the country, to sign the cable with him. The latter hedged, on the grounds that he had no connection with the Technikum. Rosenwald then suggested that at least he could state in the text of the cable that he had spoken to Rothschild, who agreed with his views.

Attempts had been made to secure the interest of another great American philanthropist, Nathan Strauss, but to no avail. The latter had been in Jerusalem in 1912 and Levin had made desperate efforts to get him to come to Haifa.

Nevertheless others joined, and funds were received which were promptly sent over to Berlin. The Americans were irked that they had not been kept fully in the picture, and that the crucial decision had been taken by the Kuratorium without affording them an opportunity to express an informed opinion. They resented being brushed aside. At the same time, they thought the Zionists were going too far in their revolt.

And so the American members of the Kuratorium held a caucus meeting of their own on January 18, 1914. Jerusalem and Berlin bombarded them with arguments and information. The key personality in influencing the meeting was undoubtedly Dr. Judah L. Magnes. Immediately after the meeting he cabled Jerusalem: "American Kuratorium recognize Hebrew in principle. Within seven years Technikum must be completely Hebraized."

The details he wrote to Dr. Lurie, of the Teachers' Association, pointing out that the resolution was adopted unanimously:

". . . Hebrew has been adopted in principle for the whole institution — Technikum and Mittelschule; Hebrew is to be made the predominant language for the whole institution immediately, whenever in the opinion of the Kuratorium, it is practical; and after a maximum of seven years, the Hebrew language must be the language of instruction in all courses throughout the institution, except where it can be shown that proper teachers and textbooks have not been developed, the burden of proof for this being upon those who hold the opinion that they have not thus been developed. The official languages of the institution within the school, in Palestine and in Turkey are, Hebrew in the first place; Turkish in the second place, that it may be made manifest from the beginning that the institution is entirely loyal to the integrity of the Ottoman Empire; and Arabic in the third place, in order that the good will of the Arabic population be secured.

"As far as foreign countries are concerned, that language is to be used which will be most appropriate for the country in question. In this way, all Teutonic [sic!] ambitions are kept in the background. The Zionist Organization is asked on the basis of this resolution to give its moral and material support; and the Kuratorium is asked to invite Messrs. Ginzberg (Ahad Ha-Am), Levin and Tschlenow to re-enter the Kuratorium."

Dr. Magnes realized that if the Kuratorium were to accept the American resolution, it would be a great victory for the Zionists and the Palestinians. If the Board did not accept the resolution, he was certain that the Americans would at least feel duty-bound to support Hebrew institutions, meaning in effect the network of Hebrew schools being set up by the teachers. The "new and independent Hebrew school system", Dr. Magnes said, should be encouraged, developed and perfected.

Curiously, Nathan's vitriolic attack on Zionism appeared in the *New York Times* on the very day of the American meeting. It was not calculated to win friends, even among the non-Zionists, who were appalled by the tone.

There ensued a not very edifying exchange of charges in the *Times* with respect to who gave more money. The Federation of American Zionists questioned Nathan's right to make accusations when the German role in support of the Technikum was so small. Figures given by the Federation: Wissotzky, 430,000 marks; Schiff, 420,000 marks; Jewish National Fund, 133,000 marks; German Jews, 133,000 marks. In addition, of the 44 scholarships given, 32 were by American Jews and 12 by German Jews.

Ten days later Nathan replied in the *Times*, and offered a different fund schedule. According to him, the Americans gave $160,000; the Russians, $100,000; the Germans $100,000; the J.N.F. loan, $20,000.

He entered into the body of the dispute. He denied that he was activated by German chauvinism, but rather by pedagogical considerations. He cited the lack of teachers in Hebrew for the sciences. He added that there were only 100,000 Jews in Palestine, old men, women, children and beggars, and declared that they were hardly qualified to play the role of pioneers in an independent national language and culture for the whole country.

As for Hebrew at the Technikum, he quoted a section from the text of the American resolution with which he was apparently in agreement: "Hebrew shall be so thoroughly taught that the pupils will be in the position of a student of Hebrew literature, and eventually be enabled to use the language as a medium of conversation."

Louis Marshall believed the Zionists were being indiscreet in referring to the American stand as a victory for them. "It is equally regrettable that our German friends have indicated such extreme rigidity in their ideas as to consider our partial disagreement with them as a reflection."

Ahad Ha-Am had reservations about the American decision, and did not see it as a clear-cut victory. Wissotzky tried to get the two sides to agree to a compromise.

Levin was not happy about what would happen in the seven years, when German and English would still be the languages of instruction. The kind of Hebrew education that Nathan had in mind, Levin feared, would mean the school would turn out etymologists, and not technicians.

The Kuratorium met on February 22, 1914 and adopted new resolutions so ambiguous that both sides could claim victory. As a matter of fact it was victory for neither, but a compromise.

With respect to matters of less importance, it was decided that the Zionist Organization would be officially represented on the Kuratorium by three members, two of them to serve on the Executive Committee. This was a reversal of the previous policy under which there was to be no organizational representation at all. It was agreed that the costs of the Technikum would be shared three ways, one third by the Americans, one third by the Germans, and one third by the Wissotzky family and the Zionist Organization.

The Zionists demanded that Technikum affairs be conducted by a secretariat completely separated from the offices of the Hilfsverein, as had been intended. The Germans realized this would end their de facto control over Technikum affairs and refused. On this, the confrontation remained.

The American and Russian members were in agreement on a separation of the Secondary School from the Technical College, so that the problem of language could be discussed separately for each. It was also decided not to open the Secondary School at this stage; though when it did open Hebrew was to be the unchallengeable language of instruction. The building being constructed for it could in the meantime become a dormitory for the College. At the Technikum proper, mathematics and physics would be taught in Hebrew. Within four years all the teachers had to master Hebrew and at that time further decision would be made with respect to use of the language for other scientific subjects.

Finkelstein, unsure of where he stood and what his directives were, sailed for Berlin in March. The work in Haifa was gradually winding down.

The Hilfsverein held a general meeting in April with Zionist participation. It could have been a peace-making meeting, and it started calmly. Nathan set the tone in an unemotional review of the language controversy. It had really been unnecessary, he said. The Hilfsverein always wanted, and still wanted Hebrew to be the language in all its schools in Palestine. This was necessary because only Hebrew could be a bridge between the Jews coming from various communities. It would also serve to prevent them from leaving Palestine for advanced study. He also favored Hebrew because it was the language of the faith and he believed that orthodox Judaism should be the basis of Jewish education. After all, religion was what bound Jews together. There was room in Palestine for all schools, Alliance, Zionist and others, though this did not mean that he had to subscribe to their philosophies.

Weizmann and Leo Motzkin were chosen as the two Zionist representatives to the Executive Committee. Simon and Wissotzky each pledged to raise their one third share of the money required to finish the buildings. The Americans were not present, and had not been permitted to name proxies. No word had been received with respect to their financial share, without which nothing could be done. Unless positive pledge was received from the Americans within six weeks, the Germans believed the Kuratorium should meet again, dissolve the organization and sell the property.

Arthur Hantke, the third Zionist, spoke quietly and without raising major polemical issues. However, in the closing moments James Simon tactlessly stirred up old wounds. He insisted that despite the February 22 compromise, neither mathematics nor physics could be taught in Hebrew. The meeting ended in a shouting match.

There was much searching of minds in the United States. Schiff learned to his amazement that the Germans looked upon their financial participation as loans, whereas his own gift had been an outright contribution. Loans would be repaid in event of liquidation. The fund-raising drive was continuing. In addition to Rosenwald, there were gifts from the Grand Lodge of Bnai Brith, various organizations in Detroit, Temple Emanuel in New York and a number of other private persons. The total was around 60,000 marks, about $15,000.

Nathan felt that if he were to go to the U.S. he would be able to explain everything to the satisfaction of the American members, and at the same time advance the fund-raising campaign. But things had already gone too far. Schiff made it clear that he did not want Nathan to come.

The news that the Americans were being pressed to the wall, with an ultimatum that either they raise their money or the whole organization would be dissolved, was not received in good grace. It had also been made clear that in addition to the capital funds, the Americans were expected to guarantee a further 80,000 francs annually for maintenance.

Marshall lost patience and suggested to Schiff that all the American members of the Kuratorium resign in protest. Schiff's son had already resigned, independently. Schiff himself agreed with Marshall.

On June 10, 1914, Marshall wrote to James Simon that in view of the ultimatum that the Americans provide 100,000 marks by June 15th, or the project would be liquidated, and preferential treatment in such liquidation would be given to the German contribution, the Americans had come to the conclusion that their "usefulness to the organization is at an end." They resigned en masse.

But Schiff found it difficult to abandon a project which he had to a large extent made possible, and in which he still believed. His feelings with respect to the blame for the situation were mixed. In July he sent 100,000 marks for maintenance of the school and wrote to Nathan: "It is a particular satisfaction to me to receive your assurance that the Technikum will yet be completed so that its opening may be expected in the spring of 1915." In a letter to *The American Hebrew* he spoke of resurrecting the Technikum so that it might fulfill its intended purpose, but added: "The deplorable occurrences that have in the end led to its present breakdown have clearly shown that Palestinian affairs are swayed by what I believe to be a comparatively small group of Jewish nationalists who, while clamoring for the support and consideration of international Jewry for Palestinian work of every character, will not hesitate to stoop to employ the most reprehensible means in order to accomplish by force if needed, their own purposes and designs."

And soon thereafter, to Louis Marshall: "While I am a non-Zionist, not an anti-Zionist, I am perfectly willing and desirous that we should work hand in hand everywhere and in particular in Palestine with the Zionist Organization, provided they will recognize the rights of their fellow Jews who happen to have different views." He fulminated against the nationalist agitators, with Dr. Levin as their spokesman, who when they could not achieve their ends "resorted to these anarchistic means to break up the Technikum and with it the Hilfsverein schools." He had been fed with a running stream of information and commentary from Berlin.

He concluded: "I only wish we had a San Hedrin [sic] which could condemn to death by stoning anyone who tried to sow the seed of dissension within our own ranks."

Strong words indeed, but Schiff was to change his views on Zionism very considerably within a few short years.

The Kuratorium met again, for the last time, on July 17, 1914. Despite the resignation of the American members of the Board, the Berlin administration felt confident because of the moral and financial support from Schiff. At this meeting Weizmann and Motzkin took part for the first time, and repeated the demand for a complete separation of the Technikum directorate from the Hilfsverein. Wissotzky offered to pay for the administrative costs involved. Simon and Nathan refused point blank, and put it to the Board as a matter of confidence in them. The alternative would be to dissolve the organization and sell the property at public auction. The threats no longer frightened the Zionists, and they considered where they might obtain the funds to make the purchase. Baron Rothschild might be interested, and Weizmann arranged to meet with him in Paris.

World War

That meeting did not take place. The war clouds were already gathering. On June 28, 1914, Archduke Ferdinand, heir to the Austrian throne, was assassinated at Sarajevo. On July 28, Austria declared war on Serbia and on August 1 Germany declared war on Russia. The world was plunged into a four-year conflict.

Construction on the site in Haifa came to a halt and most of the workers

On guard in the workshops at the outbreak of the First World War. 1914.

were dismissed. Simon and Nathan at first trod cautiously. America was neutral, and Schiff's support was vital. The threat of liquidation was not implemented at this stage.

In the constellation of foreign affairs, the German government now found it expedient to seek moral support and approval wherever they could find it, including Jews everywhere, among them the Zionists. There were many friendly overtures and an obvious desire not to create antagonism. German Counsellor Richard von Kuhlman told the Zionist leader, Richard Lichtheim, that "Germany would be sufficiently compensated if, besides Hebrew, German would also be cultivated." Speaking in the Reichstag, Paul Rohrbach hoped the Jews of Palestine would take an interest in promoting German language and culture, but he knew that "revival of a Hebrew culture and the creation of a Jewish economic entity in Palestine were no longer in doubt."

Schiff sensed a moral obligation toward the institution, and he expressed his willingness to continue paying 25,000 francs annually toward its maintenance, since that had been an implied pledge. However, he told Marshall that thereafter they should all be very careful not to make any promises which they might later find difficult to fulfill.

An era had come to a close. The Technion, planned to serve a specific educational function in Palestine, had unwittingly served as a catalyst in the determination of major policies and in the creation of movements and agencies which went beyond its own scope. For one thing, the Technion issue had become involved in matters of major international relations. The struggle in Palestine had projected the Zionist Organization into non-Jewish public attention. It had won consideration, even if not always respect, for its ability and willingness to tackle even a mighty opponent like the German government, and fight it to a finish. It had shown that it did indeed have the support of the masses of the Jewish people in the world, if not always of the nominal leadership. These matters were not lost on the German government, and throughout the war overtures were made to the Zionist movement which could very well have led to German sponsorship of Zionist national aspirations had Germany won the war. But the very existence of such activity stimulated a counter-response on the part of the other great powers. There are many who believe that this situation contributed in no small measure to the eventual issuance of the Balfour Declaration by the British government in 1917.

Guardhouse which stood at the entrance to the Technion campus, corner of Herzl and Schmaryahu Levin Streets. 1914.

There were other results of the Technion struggle, certainly not anticipated at the time. In some countries anti-Semites had sought to identify the Jews with German imperialism. Even in the United States, where a number of German Jews had become prominent in finance and industry, there were accusations that their loyalties were with the Kaiser. The Technion/Zionist campaign against the German language served as convincing and striking proof that Zionist interests were, to the contrary, opposed to German interests.

Another, and more positive result was the emergence of the Zionist-sponsored education system in Palestine. Until 1913 the Zionists, concentrating on political activity and colonization, had given little attention to education. This was perhaps due to lack of the necessary funds, rather than lack of interest. The strike of the teachers, and the need to provide educational facilities other than those of the Hilfsverein, changed the whole situation overnight and laid the foundation for the development of the network of Zionist schools. Six new institutions were opened at once in Jerusalem, Haifa and Jaffa, where all subjects were taught in Hebrew. Of the 56 Hilfsverein teachers, 41 unhesitatingly went over to the new schools. Of the 1115 pupils, more than half made the switch at once. By the end of the war there were 27 Zionist schools.

The leadership was taken up by the *Mercaz Hamorim*, the central Teachers' Association. This Association certainly had no governmental or any other legal authority, but it undertook the responsibility for Jewish education, and the teachers accepted its rulings. To be sure, this brought it into clash also with the Alliance and at times with the independent schools of the religious groups. The head of the teachers, Dr. Joseph Lurie, showed initiative and imagination.

Until the Jewish community in Palestine, after the war, organized its own community bodies and education was transferred to these, the Teachers' Association, in addition to its trade union and professional functions, successfully undertook important national and ideological functions. In the interim shaky period, the Zionist Organization provided the funds to keep the new schools going.

After the February, 1914, decision of the Kuratorium, which was a partial victory for the Zionists, the Hilfsverein had asked the teachers to call off their boycott of the German-sponsored schools. It was suggested that the two school systems exist side by side, and there might even have been an implied promise of financial support, this at a time when it was sorely needed. After due deliberation the teachers refused. They had sensed their power, and wanted no compromise. Their faces were set on creating a national Hebrew school system. At one stage the strikers were asked again why they were so fierce in their opposition to the Hilfsverein. Why not battle the Alliance schools, which used almost no Hebrew at all? To this the reply was that parents who sent their children to the Alliance knew exactly what they were getting. But since the Hilfsverein did use a great deal of Hebrew it gave the false impression of satisfying those needs. Hence, this was the enemy.

The early days of the war gave Nathan the feeling that Germany would have the upper hand. In any event, everything connected with the Technikum was now behind German lines. In December he informed the head office of the Jewish National Fund at The Hague that the Technikum Association had gone bankrupt. Though there was no such thing as a mortgage on property in Turkey at that time, he reassured the J.N.F. that it would suffer no loss on its loan.

In the initial bankruptcy proceedings, however, the name of the J.N.F. was stricken off the list of interested creditors because, having been registered in London, it was declared an enemy firm, and hence without rights in a German court. German creditors would of course be given priority. Nathan wrote at length to Schiff explaining what was going on, and Levin, then in New York, voiced his own indignant objections to the procedure and appealed to Schiff to intervene. The latter had had enough, however. "The situation at present is not such that I am willing to get involved once again in a matter which has already caused me great heartache," he wrote to Levin. "However, I am absolutely convinced that at a given time it will

The Technion and Reali School buildings in solitary splendor against the almost desolate Carmel ridge. 1914?

become possible, through renewed cooperation on the part of all factions, to help the Technikum back on its feet, but that without such cooperation nothing of real benefit can be established in Palestine. We should at least learn one thing from the present tragic situation — that to have internal political factions operate within Judaism does injustice to the entire Jewish cause and will have bitter consequences. Only through the cooperation of all true Jews in the cause of international Judasim will something beneficial be achieved . . . ."

Levin saw to it that the American Zionists were brought into the picture, and in February, 1915, Louis D. Brandeis, chairman of the Provisional Executive Committee for General Zionist Affairs, and Dr. Judah L. Magnes, chairman of the J.N.F. Bureau of America, issued the following formal statement dealing with the establishment of the Technikum and the action now proposed in Germany.

"The institution was erected with funds contributed by Jews and Jewish organizations of different countries. The chief contributions were made by Jacob Schiff and by the Wissotzky family of Russia. The Jewish National Fund contributed a site on Mount Carmel. German Jews contributed a relatively small proportion of the whole. The German Hilfsverein is represented as a creditor, having merely made a loan. It is now practically attempting to obtain possession of the institute by first excluding all

The almost-finished, empty building. 1914–1915.

so-called foreign creditors and second by buying in the property in the liquidation proceedings. The Zionists of America, who are among the chief supporters of the National Fund, contributing annually over $40,000 to its income, are in hopes that official steps may be taken by Germany to persuade the German Trustees of the institute to postpone their contemplated action. If the auction takes place, the National Fund, the Zionist contributors and others interested in this institution from a purely philanthropic point of view, will find that their property interest is gone, having been denied a voice in the disposal of the interest.

"Such an official step by the German government as suggested by the American Zionists is the more to be expected since the attitude of the German government toward private property belonging to hostile subjects, has been considerate and correct. If the affairs of the Institute of Technology are to be liquidated, no interest is adversely affected should this action be postponed until the conclusion of the war."

The fine legal hand of Mr. Brandeis can be discerned in this logical and conciliatory proposal. The appeal to the German government was to no avail, and on March 15, 1915, the auction sale took place. The Americans and the Zionists were not represented. Simon utilized an old power of attorney given him by the Wissotzky trustees, which had not been cancelled because of the disruption of communications during the war. There was one bidder, and the Hilfsverein offered 225,000 marks for which it obtained full and complete ownership of all the land and the buildings thereon.

Jewish communities and the Jewish press throughout the world reacted excitedly, each in accordance with its point of view. The Russian Jewish journal, *Razsvet*, charged that what the German Jews had done was nothing short of criminal. *The Jewish Comment*, in Baltimore, headed its editorial on the subject, "Selling Out the Jewish People". *The Jewish Exponent* in Philadelphia, on the other hand, published a long article, translated from the German, justifying the Hilfsverein. As for the future of the Technikum, the paper wrote, that "will depend upon the battles in Belgium, in Poland, in the Dardanelles and in the Suez Canal ...."

Nathan sought to minimize what had happened and wrote that the auction itself was unimportant. Not the sale, but the outcome of the war would really determine the fate of the Technikum. He was correct, but not as he had expected.

Conditions in Palestine during the war did not permit any further work on the buildings. Dr. Finkelstein was called into German uniform and in his place a Haifa resident, Shabbetai Levi, was asked to represent the Hilfsverein in any dealings with the Turkish government. Mr. Y. Rothschild, an artist, who had been engaged as a teacher of drawing, was put in charge of the property, which in effect meant only seeing to it that proper guard was mounted on the abandoned and unfinished structure. Under the circumstances, the formal patronage of the German government was useful in keeping away the Turkish authorities who might otherwise have seized the property outright.

In the fall of 1916 the German army took over the buildings. The workshop was used for sanitary fumigations, and one corner served as an abbatoir. In the following spring the Turkish army took over and transformed it into a hospital. The structure was easily the most prominent and most conspicuous landmark in that part of Haifa, and when Royal Air Force planes bombarded the city they could easily have destroyed the building, had it not been for the Red Crescent flag flying overhead, which protected it. Other buildings in the city were badly hit.

In the early days of the war there occurred an event which, from the point of view of the Hebraists, was a delayed but well-deserved retribution. Consul Hardegg, pursuing his duties as he saw them, had also been hostile to every aspect of French influence in the city. He made life diffi-

Delousing station for German soldiers in a corner of the Technikum workshops. 1916.

Abbatoir of the German army in the workshop building. 1916.

cult for the Carmelite Order and most of the monks were deported on the charge that they used their monastery to signal to the enemy. He was blamed for the action of a group of Turkish soldiers who in May, 1915, dug up the memorial to Napoleon's soldiers set up outside the monastery under a pyramid of cannon balls and a large iron cross. A few days later a French warship, the *Ernst Renan*, sailed into Haifa port and calmly shelled Hardegg's home on the hillside, in what was clearly an act of retaliation.

Levy and Rothschild had a difficult task. No funds came from abroad and they had no resources from which to pay even the skeleton watchman staff. There was no choice but to convert assets, namely, to sell some of the brand new machinery that had been shipped from Germany and had been stored pending opening of the school. The Turkish government bought some for its workshops in Damascus. The Hedjaz Railroad firm also acquired equipment. The goods were taken, but full payment was deferred, and became a matter for litigation after the war. Kibbutz Merhavia bought some of the equipment. The meager income was used to purchase food supplies, which were running low and severely rationed. At one stage the Technikum staff was padded with additional "employees" who thus qualified for rations and the survivors later testified that in this way their lives had been saved.

The local committee was seriously disturbed when, in 1917, a carrier pigeon fell into the hands of the Turks and the activities of the NILI spy ring were disclosed. A frantic cable was sent to James Simon in Berlin stating that if the accusations against the Jewish community were well founded, the results for the Yishuv could be most serious. Simon was urged to do everything possible to make clear to the Turks that the Technikum was a German institution. Instructions were given to the German Ambassador in Constantinople as a result of which German protection was assured, and deportation of many Jews prevented.

The British conquest of Palestine moved into high gear in 1917. On October 31 Beersheba fell. Jaffa was captured in mid-November, and on December 9 Allenby entered Jerusalem. The north of the country was still in Turkish hands. The battle of Megiddo on September 19, 1918, brought about the collapse of the Turkish resistance and the following month the British army moved into Haifa. It required base facilities, and at once occupied the grounds of the Technikum, which had been abandoned in a hurry by the Turkish forces. However, the main building and the workshops were in such a filthy and infested condition as to be beyond immediate cleaning. Shacks and tents were set up in the courtyard, and the secondary school building was fumigated to permit its use as a hospital to care for the numerous British casualties.

Post-War

Back in Berlin the leaders of the Hilfsverein found themselves in an untenable position. As the legal owners of the Haifa property they were responsible for it. The defeat of Germany and the occupation of Palestine by the British meant the end of German influence in that country. There was no point holding on to property which could serve only as a financial drain. Shabbetai Levi sent urgent requests for money with which to pay the guards, and small sums were doled out. Typical of the correspondence which ensued at the time was Bernhard Kahn's letter of December 5, 1919, from Berlin to Levi:

"We are sending you 100 pounds which can be utilized for necessary payments. We hope within a short time the whole question of the Technikum will be settled. Of course we have to accept your judgment that salaries of the personnel were raised by 100%. I must emphasize that it will not be possible for us to send any further money from here ...."

Levy's problems were compounded by the fact that the property was being increasingly used by others though he was responsible for its maintenance. The Zionist Organization's Palestine office had stepped up its colonization program in the aftermath of the war, and the Technikum

Shabbetai Levi (1876–1956).

buildings made excellent storage space for lumber and other supplies coming in at the port of Haifa. It was not only convenient; it cost nothing. Dozens of trivialities were transformed almost into major issues. The following is but one typical event.

The watchman and the gardener engaged by Levy could fulfill their tasks only if they were given a place to live on the premises, and he reluctantly let them move in. The hospital authorities were quick to complain that in the absence of a latrine and adequate cooking facilities, the families were fouling up the grounds, and this fairly close to the hospital kitchen.

Levy pointed out that the central latrine was in the hands of the British, and not available to his staff. Everything would be much better if only the British would vacate. They did not, and instead the "Director of the Jewish Technikum" was served with a formal warning, signed by the Military Governors in Jerusalem:

"Notice to Abate a Nuisance. You are required to abate the nuisance arising on your premises from the lack of latrine arrangements and receptacle for rubbish . . . . Failing compliance with the above, legal proceedings will be taken."

Levy countered with formal request to the British to get out, addressing his application, as required, to "The District Governate of Phoenicia". The authorities advised that the matter was being looked into.

In the meantime, the Zionist Organization, with headquarters in London, decided to ask the British government to cancel the forced sale of 1915 on the grounds that it was illegal and because it violated the rights of the J.N.F. and the Wissotzky family. But as early as July, word had reached Weizmann that the Hilfsverein was ready to pull out against a payment of 240,000 marks. Matters began to reach a head at the September 11, 1919, meeting of the Zionist General Council. Menahem Ussishkin urged speedy action, and asked for a power of attorney to acquire the Technikum if the matter reached that stage. With the buildings in Zionist hands, it was pointed out, they could be used by the University. After discussion, however, it was agreed that the Technikum should be developed as a Technical College on its own.

Jacob Schiff had heard from Berlin and on November 5, 1919, wrote to the President of the Zionist Organization of America, Judge Julian W. Mack, referring to the willingness of the Hilfsverein to sell, and said: "If this be correct, I am inclined to acquire the Technikum and turn it over to the Zionist Organization, provided it wants it and can put it to good and advantageous use." This was a responsibility which the American Zionists were not sure they could shoulder, and they reached no immediate decision. Schiff wrote to Nathan and received confirmation that the property was indeed up for sale. He again approached Mack on January 5, 1920, and asked to know if the Zionist Organization was ready to accept the buildings as a gift from him, in which case he would make an offer to the Hilfsverein.

There ensued a rather odd flurry of negotiations which can be explained only in terms of the political confrontation then already shaping up between the World Zionist Organization led by Weizmann and the European leaders on the one hand, and the American organization led by Mack and Brandeis on the other.

The Americans had not neglected Schiff's offer. Just a few days after Schiff's November 5th letter Mack had a cable from the Zionists in London. It appears they had learned from Berlin of Schiff's interest, and asked Mack to head off the purchase attempt. They regarded the 1915 seizure of the Technikum property illegal and were planning action to have it returned without payment. They were looking into the legal aspects.

Mack replied that he had in effect already accepted the Schiff proposal, and it was subject only to legal obstacles. The Zionist officers in London apparently were informed thereafter that there had been nothing illegal about the 1915 sale and there were no grounds for the British authorities in Palestine to set the sale aside. Hence they cabled Mack that the property would be purchased — but they preferred to have it transferred directly to the World Organization. They did not want to lose the Schiff money, however, and urged Mack to use his influence with the donor to provide the necessary equipment instead. The reason offered was rather weak: since the Zionist Organization had contributed the first money to buy the land, they therefore felt it would look better if they again purchased it "themselves", than if they permitted someone else to buy it and then make a gift of it to the Organization.

The Hilfsverein had also cabled Schiff directly about the offer from London. Schiff told Mack that if the Zionists in London wanted to deal directly with Berlin, he was perfectly willing to stand back. ". . . It must be understood that the offer I have made, to take over the Haifa Technikum with the view of giving it to the Zionist Organization of America is withdrawn in consequence . . . ." He was obviously hurt by the attempt to bypass his generous offer.

Mack reacted ingenuously. "In your letter of yesterday you mention for the first time the Zionist Organization of America. Heretofore you have said simply 'Zionist Organization' and I have assumed, I trust properly, that it is immaterial to you whether the American or the World Organization takes the title." Mack pressed his argument. "If it is bought direct, the Jewish National Fund, which holds all of the land in Palestine, would be a perfectly proper purchaser. But while all of us on this side of the ocean believe that the Technikum, like the University, should be controlled by the World Organization, and not by any single branch, even the American, we on this side were particularly desirous and are still particularly desirous that you should make the purchase and the gift. We too are actuated by sentimental reasons although I am frank to say that there are practical considerations that have also influenced me. Every one of us here have felt that your offer not only evidenced your generosity, but it evidenced what we regarded as of even greater importance, both directly and indirectly, in its influence upon others: viz., your desire to cooperate in ever-growing measure with every practical work in the upbuilding of Palestine."

Schiff was appeased, and after word from London indicated that the Zionists there would be perfectly willing to have the Technikum presented to them by Schiff, so long as the actual purchase would be made by the Zionist Organization, he wrote to Mack on February 5: "I should have preferred to make the purchase of the Technikum building personally, as I originally put $100,000 into that enterprise, and a personal purchase would have rounded this out more appropriately and satisfactorily to myself. However, I want to help the matter along and shall have no objection to having the Zionist Organization buy it and I to furnish the funds, 530,000 marks, for the purpose of effectuating this." He cabled to Dr. Simon that he had no objection to having the sale made directly to the Zionist Organization. The actual payment was effected on May 5, and since there had been a fluctuation in exchange rates since the Organization had made the purchase, Schiff paid 550,000 marks.

The Zionist General Council in London, apprised of Schiff's decision (in the minutes he is referred to as "Otto" Schiff), formally approved the purchase with its own money on January 12. The London minutes also referred to the Technion in Jaffa!

More on Jacob Schiff

Considering his role in the establishment of the Technion it is appropriate to devote a few more lines to this remarkable man, and his attachment to the Jewish people. As he so often reiterated, he was not a political Zionist but did not consider himself an anti-Zionist. His attitude to Jewish life was positive. As early as 1914, reacting to Israel Zangwill's play, *The Melting Pot*, he wrote that he did not believe in the desirability of assimilation, and therefore did not subscribe to the melting pot theory. His generosity got the better of him and he gave funds to help the play nevertheless.

His attitude toward Zionism underwent a gradual change with the progress of the war. By 1917 he began to realize what the war meant to whole Jewish communities in Europe and he said publicly: "It has come to me while thinking over events of recent weeks — and the statement may surprise many — that the Jewish people should at last have a homeland of their own." He had not yet travelled the whole path, and he hastened to explain that he meant a cultural homeland where "there would be a great reservoir of Jewish learning." To Zangwill he wrote at the time of "Palestine as a Jewish homeland — not a Jewish nation, which in my opinion is a Utopia." To another friend, Elisha Friedman, he wrote that he would be willing to embrace Zionism were it not that Zionism and Jewish nationalism had become synonymous.

His theoretical objections did not prevent him from extending aid whenever it was required. His support of the Technikum was the major project, but certainly not the only one. When Pewsner came to the U.S. in the early days of the war to help find a market for the oranges which Jewish farmers had picked but could not send to their usual European markets, Schiff threw himself into a twin effort to help. First he cooperated in the floating of a Palestine Orange Growers' Loan to tide the farmers over, and then he launched a sales campaign in the United States to sell the stored Jaffa oranges.

In 1918 he contributed $25,000 to the Palestine Restoration Fund of the Zionist Organization for the purpose of restoring colonies or creating new ones, and he repeated this gift annually for several years. He gave $25,000 to found and erect a Jewish University in Palestine.

By the end of that year he completed his ideological journey, and firmly rejected a call by a leading American rabbi to rally American Jews to combat "the menace of Zionism". Though he had originally not favored Zionism, he confessed, "I feel now that the creation of the Jewish Homeland in Palestine is most desirable . . . ."

It was not long before the Zionist Organization of America could publicize a statement in which he not only espoused Zionism, but announced himself an enrolled member of the Organization. He went further. He sought to get his friends interested. Asked by Judge Mack to draw Julius Rosenwald into the movement, in view of Rosenwald's interest in both the Technion and the Aaronsohn Agricultural Station, Schiff revealed: "I have only very recently talked to Mr. Rosenwald in an effort to convince him of the necessity and advisability that American Jewry unite in the restoration of Palestine, but I am afraid that I have not made very much of an impression and I fear if I write him now again it will have just the contrary effect that you wish."

He was willing to lend himself to almost anything that would help advance Jewish settlement in Palestine. In mid-1920 he received a proposal from Sir Herbert Samuel, new High Commissioner of Palestine, to throw his weight behind a plan to sell to the Jews of America £ 3,000,000 of

Bonds issued by the Government of Palestine. This was, in effect, a premature forerunner of what later became Israel Bonds. He was ready to help, but submitted the proposal for the careful scrutiny of his friend, Louis Marshall. The latter had to dampen his ardor. He pointed out that the "so-called Government of Palestine" was still vague, indefinite and intangible in nature, and its continuance still shrouded in doubt and uncertainty. It would be impossible to sell the Bonds without a guarantee from the British government. "If an attempt were made to bring out an issue of unguaranteed bonds of this character," he wrote Schiff, "there is every likelihood that it would end in a dismal failure. The impression to which such a result would give rise would be most deplorable, and would go far to crush the hopes of those who would like to see Palestine restored to prosperity, with flourishing Jewish colonies and a happy people ...."

Yet there were some matters in which Schiff drew the line, even for Palestine. He explained to Dr. Magnes that he refused to contribute to the Jaffa Gymnasium (Herzlia) because he understood the Jewish religion was foreign to the institution, and the school was conducted along non-religious lines.

Jacob Schiff died on September 25, 1920, three days after Yom Kippur, on which he had fasted faithfully, as he had all his life. His Will contained no bequest to the Technion.

It is idle to conjecture what further role Schiff would have played in the rebuilding of Palestine, had he lived longer. His place in the history of the Technion was already firmly assured during his lifetime.

Exit Hilfsverein

The bill of sale for the Haifa property was signed in Berlin on February 9, 1920. It contained three paragraphs.

1. The Hilfsverein conveyed to the Zionists all items and rights which it had acquired in the auction on March 15, 1915. Everything was handed over on an "as is" basis, and the buyer undertook to take possession at his own risk.
2. The price was 538,042.01 marks, which included the Hilfsverein's expenses for operations up to the date the British took over. Payment was by check.
3. Provision was made for mutual payments for expenses and for the value of securities held by the Hilfsverein on behalf of previous donors. However, the rate of exchange had shrunk to such an extent that the paper was worth very little. The Hilfsverein also had a claim against the Turkish government for 100,000 marks, plus interest, for machinery which it had taken, presumably for the Hedjaz Railway. Professor War-

burg, of the Zionist Organization, undertook to make the claim and seek to obtain payment for the Hilfsverein. He also assumed all obligations toward employees.

The document was signed by Otto Warburg, James Simon and Paul Nathan. Felix Makower was Notary.

Technically, title was at this stage transferred to the British Controller of Enemy Property. It was not until 1927 that the land was at last formally registered in the name of the Jewish National Fund and the buildings in the name of the Keren Hayesod, both Zionist agencies.

Word was at once flashed to Shabbetai Levi in Haifa. First Simon notified him that the Hilfsverein was hereafter no longer responsible, and he should get his instructions from Prof. Warburg. The latter cabled separately asking Levy to continue to manage the affairs locally for the new owners. "Don't sell anything without permission," he was admonished.

With this the Hilfsverein der deutschen Juden stepped out of the picture. The conflict with respect to language had been bitterly fought. Feelings ran deep on both sides. There is little doubt that in the heat of the polemics both parties had made intemperate statements. Each had been ideologically motivated, and certainly no one can question the integrity of either side. In world affairs history is always written by the victor, and in this instance the Zionist account has usually emphasized principles and issues as seen from that point of view. One Zionist historian, writing some years later, gave his version of the Hilfsverein stand: "If they did not love the Hebrew language less, they favored their own German language more, and their teutomaniac proclivities made it possible for them to lend themselves as tools in the hands of their government."

Still, full credit must be given to the role which the Hilfsverein played in the founding of the Technion. A publication issued by the Technion in 1953 acknowledged without qualification the credit due to Paul Nathan. The idea was born in his heart and reached fruition only because of his boundless devotion. Only his vision and energy brought about the creation of the institution. "Paul Nathan is assured of an honored place in the annals of the Technion," the publication concluded.

There has also been recognition of the fact that besides the German political influences that were at work in the period preceding the First World War, the leaders of the Hilfsverein also honestly believed that they were contributing to the elevation of the standards of the new school by insisting on the use of the German language. And there are some who believe that if the first decision had been clearly in favor of Hebrew, the Zionists would have been unable to open the Technion, for lack of teachers and texts.

Further, even if the battle had not been pressed with such unrelenting

Left to right: Ephraim Cohn,
Mr. Dickenman, Dr. Paul Nathan,
Alphonse Finkelstein, Dr. Ritter. 1914.

fury, the fate of the German language in Palestine would have been settled in any event by the outcome of the World War, and everyone would have been spared the unhappy experiences which left their shadow upon the Technion for years to come. Cohn felt that the War of Languages had brought a demoralization to the Jewish community; it had introduced a spirit of poisonous hatred, bitterness, and civil war, turning one Jew against another. Children rebelled against their parents, and families were split.

The Zionists had established their new schools, but the Hilfsverein educational network quickly recovered. Despite the withdrawal of teachers, new instructors were found. The pupils who had left were replaced, to a large extent from religious families which had hitherto avoided the German-sponsored schools. At the time of the outbreak of the war enrollment began to exceed its previous total, but war had its effects on the whole program. Jews were evacuated from Jaffa, and the schools there were moved to Jerusalem. When the British established their control they took over the German schools and transferred them to the Zionist Commission.

Paul Nathan's world crumbled about him. His attachment to the Technikum was paternal in the best sense of the word, and he felt that he had lost a beloved son. He found it difficult to forgive those whom he held responsible for stealing his project, and also robbing him of the dream to build a small house on the slopes of Mount Carmel from which he would be able to watch the growth of the institution that was so dear to him.

The Zionists sought to maintain his connection with the institution. When the ceremony marking formal beginning of classes was held in 1925 he was invited. He did not respond.

When he died, in 1927, the Zionist Executive wrote to the Hilfsverein: "It will never be forgotten that the deceased was the first to found Hebrew kindergartens in Palestine and that he worked most successfully for the revival of our language. We shall never forget how much he contributed to the cultivation of Judaic learning and the Hebrew language in the schools founded by the Hilfsverein. If there was a period when differences of opinion existed between him and us, then it was a fight for noble ideas, and there was no doubt that his intentions and thoughts were pure, for during his whole life he aspired only for the good of his people."

A plaque in memory of Paul Nathan with an inscription paying tribute to him as the "Founding Father" of the Technion was erected at the entrance to the historic building in Hadar Hacarmel early in 1955. Plaques to Schiff and Wissotzky had been erected many years before.

Germany's defeat in the war also meant the collapse of the Hilfsverein complex of schools in Palestine. The organization continued in existence in Germany and renewed its program of philanthropy and civil defense for Jews. In the late 1920's it explored the possibility of resuming support of Jewish education in Palestine, perhaps to help combat the missionary movement. Some of the Hilfsverein people were helpful in a fund-raising campaign for the Technion launched in Germany a few years later. In the early 1930's it made small grants to the Reali School in Haifa, and a few other institutions. One was a grant to the Hebrew University in Jerusalem — for the teaching of German. The Hilfsverein went out of existence in 1941.

Ephraim Cohn, a native of Jerusalem, who had served as the Hilfsverein's representative in the country from the beginning, endured the harshest treatment of all. He was the major target of Zionist criticism during the language controversy, and he never forgave his opponents. During the war he continued to write Paul Nathan, reporting on events in the country and voicing his complaints about the Zionists. Some of these letters were sent on to the German Foreign Ministry and from there found their way also to the Turkish authorities. There is no evidence that Cohn ever "informed" against any individuals, but his comments, uttered in time of war, were certainly provocative. The record shows that he justified the deportation of Jews from Palestine and especially from Tel Aviv which, he said, was run by the Russian Zionists, "Ussishkin's Cossacks". He told Nathan that someone ought to explain to the Turks that only 5% of the Jews of Palestine, and only 2% of those in Jerusalem, were associated with Zionism. Most of the Jews would rather be saved from the terror of the Russian Zionist chauvinists, he added. If the government were to deport all these *"agents provocateurs"* from Palestine and from Constantinople, nobody would object, he said. He advised that it would be wise of the Turks to close up the Zionist press and the Zionist schools.

These views were given wide circulation in Berlin and in Constantinople and of course they reached Zionist ears too.

After the conquest of Palestine the British drew up a black list of seven Germans who were held responsible for espionage, anti-British propaganda and the like. Of the seven, Cohn was the only Jew. His wife and daughter were arrested, taken to Egypt, and held in detention for a year and a half. Cohn himself was in Europe, but he was barred from ever returning to Palestine. On his 70th birthday in 1933, he did go back, and was feted at parties and receptions by former colleagues, pupils and even erstwhile adversaries who preferred to remember his contributions as a pioneer of modern Hebrew education in the country. His visit to the Technion in that year was perfunctory.

The Reali School building, under construction. 1913.

*   *   *

CHAPTER THREE

Under Zionist Administration

Clean-up and Reconstruction

The formal agreement in Berlin for the sale of the institution was followed by a simple ceremony in Haifa, at 8:30 in the morning of April 26, 1920, at which Shabbetai Levi, representing the Hilfsverein, handed over the property to Max Hecker, who had been designated to represent the Zionist Commission in Palestine. The document specified plot, building, contents without list, and a notation to the effect that the Reali School, a large part of the grounds, the gardens, the well, the pump and the locomobile were still in the hands of the British army. There followed an accounting of the meager funds on hand.

Immediately thereafter the Provisional Committee for Management of the Technion held its first meeting. The Committee was composed of the Chairman, Dr. Hillel Jaffe, Shabbetai Levi, Dr. Arthur Biram, and Baruch Bina, who represented the Zionist Commission in Haifa. Their first action was to address a request to the British authorities to vacate the premises. They then dismissed all employees, and proceeded to re-engage some of them. As a result the Committee was plunged into a series of disputes with the employees on such matters as accumulated back wages that had not been paid, severance compensation, occupancy of quarters on the site, etc. There were court suits and public appeals for "justice" and "mercy". Wrote one dismissed worker, who had financial claims: "I need the money, my wife is about to give birth." Another, a former watchman, wrote that his family was hungry and naked, and he had sold all his furniture to keep alive. "If there is a God in Heaven, he will succor me in this hour of need . . . ."

The harried Committee was filled with good will but had no funds, and none had yet been received from the new owners, the Zionist office in London. Some of the employees sued the Technion, and the Committee counter-sued to evict them from the rooms their families were occupying in the building. It was a most inauspicious beginning.

A second problem was the condition of the building. The facilities not occupied by the British were in a state of near-wreck. Everything movable had been stolen. One report had it that the leather straps in the German machinery left in the workshops had been removed and were used to help set up a shoe industry in Haifa! What the vandalism of the Turkish army and the pilfering of local inhabitants had not done, the exposure to weather took care of. Plaster had fallen off the walls; doors, windows and even window frames were gone. The wooden staircase leading to the upper floor of the main building had been carted away. The upper stories had in any event never been completed. Nothing could be done without funds.

There had been temptation to sell educational equipment in order to meet at least the most pressing demands of the hard-hit workers. Shabbetai Levi had resisted such temptation even before he received the warning cable from Berlin, though at one stage there had been a tempting offer. In January he had received a letter from the Town Council of Tel Abib [sic], signed by Meir Dizengoff. At the time Tel Aviv had a few hundred buildings and a population of less than 5,000. There was no electricity, and illumination was from kerosene lamps or Lux lights. Here and there a factory or cinema had its own small generator. Dizengoff had heard that the Technion had some Diesel engines and dynamos, intended for the Institute's power plant, but not yet in use. If Tel Aviv had such equipment it could produce its own electricity. To be sure, new equipment could have been ordered from Europe, but delivery could be made only in about 12 to 15 months, and Tel Aviv did not want to wait that long. Since the Technion had no immediate use for these items, Dizengoff suggested two alternate proposals: one, Tel Aviv would buy the equipment outright at a price to be mutually determined; two, the Technion would deliver the equipment at once to Tel Aviv, and when the same or similar equipment arrived from Europe, as ordered, the Technion could take possession of the new items. The result in either case would be to provide Tel Aviv with electricity almost immediately.

Central staircase of the old building.

The deal did not go through, and it was not until June, 1923, that Pinhas Rutenberg put up the first oil-powered electric plant in the all-Jewish city. Thus the Technion lost an early opportunity to "electrify" Tel Aviv.

The Committee in Haifa had other things to worry about. Squatters had to be evicted from the Technion buildings. In addition to the former employees, most of whom had some right to reside on the premises, there were strangers as well, who had simply moved in. Considering the dire shortage of housing in Haifa in the post-war period, the Committee had to temper its demands with some degree of mercy. Thus a typical notice to one Mr. Medina read as follows: "We hereby respectfully inform you that you must vacate your apartment on the Technion site *immediately and forthwith*, without protest of any kind. Nevertheless, we are prepared to wait a short period of time until you find another apartment."

Furthermore, communal institutions, lacking premises of their own, looked upon the big buildings as public property. The Maccabi sports organization, two kindergartens, a synagogue, the Neighborhood Council and others had made the place their headquarters. The Zionist Commission utilized the workshop for transient housing for new immigrants arriving at the Port of Haifa. During Hanukah in 1920 the founding convention of the Histadrut, the General Federation of Jewish Labor in Palestine, took place in the main machine hall of the workshop building. The historic event is duly marked by a plaque.

British and Australian army tents on the Technion premises. 1919.

The biggest occupant of all was the British army, which was in no hurry to move. It was suggested that the Technion demand that all occupants, including the army, pay rent for their quarters. Levy pointed out that if they accepted rent they would be subject to property tax, and the tax would be much higher than any possible income.

In December, 1919, the Brigadier General commanding the North Force informed Shabbetai Levi that the military would shortly withdraw from some of the facilities "at the Jewish University, Haifa," where the army was operating a dental clinic and laboratories. Two weeks later there was formal notice of evacuation, and Ussishkin asked Selig Weizmann, in Haifa, to take possession on behalf of the Zionist Commission. As it turned out, the withdrawal was only partial.

There was a naive indignation in the efforts to get the British out. A strongly worded letter was addressed to the High Commissioner in Jerusalem, reviewing the situation and the many promises made and broken. "Must it really come so far that we will have to communicate to the whole world that this school, after eight years of the most troublesome existence, even during times of war and of hard political persecution, must definitely be broken up because it has no home of its own, because its home is withheld against right and justice?" Three months after this letter, in November, 1921, the British forces finally handed their "occupied territories" back to the Committee.

Attempts were repeatedly made to collect from the Turkish government and from the Hedjaz Railroad for the equipment which had been "borrowed" or "bought" during the war. Receipts signed by the requisitioning authorities were found, together with detailed lists of the items taken, but somehow the parallel records in Damascus seemed to have disappeared. A lawyer was engaged. For claims against the German army there were no receipts at all. Prof. Otto Warburg had promised to recover as much as he could, as property of the Hilfsverein, and indeed desperate efforts were made. The case was hopeless. Technion's archives still contain the impressive bundles of lists and receipts on which payment could never be collected.

The Zionist authorities in London, familiar only in a general way with the problems in Haifa, wanted the school opened immediately. A circular letter was sent in 1920 to various technical universities inviting applications for qualified academic staff. "Qualified" meant anyone familiar with a technical profession, and able to teach in Hebrew. Replies were received, especially from Eastern Europe. There were many engineers and technicians ready to come to Haifa and take up posts as "professors".

The Provisional Committee in Haifa was strengthened with the addition of Shmuel Pewsner, now head of the Hadar Hacarmel Neighborhood Council, and the distinguished Russian Zionist leader, Boris Goldberg, who had taken up residency in Haifa. Dr. Biram, principal of the Reali School, wanted to get on with his school, and also to establish an organic relationship between the Technion and the Reali School, which had in the meantime been functioning on its own. Biram, it will be recalled, had been engaged before the war, while still in Berlin. He had been a teacher at a government high school and an avid Zionist. When he heard that the Hebraists in Haifa had decided to set up a separate, Hebrew-speaking Reali School, instead of the "German Reali" School, he went out on a private fund-raising campaign to support the new school. It met in temporary quarters, and in December, 1913, was officially recognized by the Zionist Small Actions Committee, which voted it a budget. Biram came to Palestine on February 13, 1914, and took up his post as Principal. He found the school located in a dilapidated Arab house with broken windows and no equipment beyond some battered tables and chairs. He faced immediate difficulties with the old-line, Russian-trained teachers. His concept of education was German in spirit, emphasizing discipline, thoroughness, depth, physical education and what would today be called a puritanical code. He was shocked at the relatively free and easy methods introduced in the schools by teachers who drew their educational philosophy from the revolutionary currents of Eastern Europe.

During the war Biram returned to Germany to serve in the armed forces.

Dr. Arthur Biram (1878–1967).

He was on the Russian front for a while, then had himself sent to the Middle East and was eventually stationed in Jerusalem. Before the fall of that city he was sent to Affuleh and became station master of the railroad depot there. He could not be responsible for the time of arrival of trains in Affuleh, but he insisted on punctuality with respect to their departure from there, a characteristic familiar to all who knew Biram personally. He remained with the German forces until the end of the war and late in 1919 returned to Haifa to resume his administration of the Reali School. The institution had continued to operate throughout the war, despite all the obvious problems, including a severe shortage of food.

His return to what was called "Prussian" methods of education did not earn him the affection of the labor movement, but with a strong hand he fashioned a high school which became a byword in the country's educational system. When funds were required, Biram did not hesitate to travel abroad to raise them, and his name and face became known in all the capitals of Europe. During the mid-1920's he obtained more than half of his operating budget from contributed funds.

The Reali School eventually established a contractual relationship with the Technion whereby the two agreed on certain cooperative efforts, and on division between them of the plot of land in Hadar Hacarmel.

All the plans to repair, complete construction, acquire new equipment, engage staff and at last open the Technion to students, could get nowhere in the absence of funds. Much hope had been attached to the Zionist Commission which was sent to Palestine in 1918. Headed by Weizmann, it included technical experts and representatives of British, French and Italian Jewry. The Commission remained in Palestine until 1921 when it was succeeded by the Palestine Executive named by the Zionist Congress of that year. However, the Commission never got to visit the Technion, even when joined by American members who recommended such a visit.

In December, 1920, Professor Warburg, Chairman of the Zionist Executive, visited Haifa and met with the local committee to discuss the future. They urged that the Technion be opened as quickly as possible, not as a limited vocational school, but on university level. They asked early appointment of a managing director to take things in hand. They requested that the Keren Hayesod allocate funds for the physical rehabilitation and provide for annual maintenance. If the latter was not feasible, then permission should be given to the Technion to raise its own money. Warburg listened, apparently with sympathy, but made no commitment. He had to discuss matters with the Zionist Commission in Jerusalem, and with the Executive in London.

The days and weeks went by. Expenses continued and debts were mounting, but no money was received. A small check sent by Professor

Warburg in January to pay severance compensation and pressing miscellaneous expenses, remained uncashed for some time. No bank or individual could be found who was willing to honor it!

A few weeks after Warburg's visit, Dr. Chaim Weizmann came to the Technion, accompanied by Sir Alfred Mond (later Lord Melchett), who had shown some interest in the institution. Weizmann made several specific promises:

1. The first funds received by the Keren Hayesod for the Hebrew University would be given to the Technion because it was to be part of the University and because it already existed, whereas the University itself had not yet been established. 2. He would endeavor to procure from various sources the sum of £10,000* required to finish the building. 3. He authorized Dr. Biram to request of the South African Zionists that the money they had raised for the University be given to the Technion. 4. The Committee should also feel free to turn to Dr. Magnes and other friends for the purpose of getting them interested in the Technion. Weizmann further promised that when he reached the United States soon thereafter, he would procure further help.

The pace of activity quickened. Goldberg, now an active member of the Committee, made contact with Wissotzky and sought to induce him to visit the Technion. Hecker made a study of the physical plant and estimated that £24,000 would enable reconstruction and thorough rehabilitation. Application was made to the Zionist Executive in London.

Despite the financial crunch, things were beginning to look good. Studying the records, Goldberg discovered that the Zionist Organization already held some £4,400 pounds in its Hebrew University fund. The Technion Committee took immediate steps to claim the money, in line with Weizmann's promise.

Visitors to Haifa were brought to the site, and their interest sought. One was Emilio Stock, head of the Spalato Cement Co., with head offices in Trieste and what is now Split. Himself an engineer and devoted Zionist, he had been much impressed with the Technion's possibilities. Noting the broken down condition of the building, he pledged supplies of cement. He kept his word, and the cement began coming in by the hundreds of barrels. The customs authorities charged import duty.

Goldberg continued to needle the Committee. He felt they were not doing enough in the way of public relations. He suggested issuing periodic reports to the press. Using Stock as example he urged that a systematic effort be made to receive overseas visitors properly, promising that it

---

* The currency referred to in Palestine was stated in terms of pounds Sterling or Egyptian pounds, or later Palestine pounds. For all of these the symbol £ is used here.

would pay off. For this they would need a suitable, responsible person. The name of Dr. Ze'ev Beigel, of Austria, was suggested. There had been good reports about him. Goldberg took the bull by the horns and proposed that the man be brought from Austria and engaged for a six-month trial period on condition that within three months he master the Hebrew language and show that he could do a proper public relations job. The other members objected. They pointed out that they were not free to incur expenses. The Zionist Commission in Jerusalem might not approve. Goldberg was impatient. He urged that action be taken first, and approval sought afterwards. He offered to undertake personal responsibility for payment of Beigel's salary and loaned the Technion £1,000. Within two months Beigel arrived.

New American Interest

From the United States Ussishkin reported that an association of Zionist engineers there had *promised* (his emphasis) that when the buildings were finished they would undertake to take care of the interior — machinery, laboratories, etc. Recalling the great generosity of Schiff and his colleagues, the Haifa Committee was encouraged. Things did not develop as expected.

A Zionist Society of Engineers was established in 1917 by a group of engineers attending the Baltimore convention of the Zionist Organization of America. The initiative was taken by Benjamin Halpern, a Zionist who had gone to the U.S. in 1915 after studies at the Polytechnic Institute of Warsaw and the Technical School at Grenoble. First officers were Max A. Greenberg, chairman; Peretz W. Etkes (later to be city engineer of Jaffa), treasurer and secretary. Boris Kazmann later came to New York from Michigan and became Coordinating Secretary of the Society, which changed its name to the Zionist Society of Engineers and Agriculturists. The Z.O.A. voted them a generous budget to develop practical plans for the reconstruction of Palestine.

On December 11, 1918, they submitted a formal proposal to their first national conference "to accumulate all the necessary information and data for the establishment in Palestine of a national settlement of a self-sufficient economic unit which may be localized in one section of Palestine or may be distributed throughout Palestine. This settlement is to be provided with the full equipment essential for its natural development upon a self-sufficient economic basis." The statement was signed by Isaac J. Stander, James Haines, Henry J. Nurick, J. Cooperstock, M. Rosen, A. Rosenzweig, Boris Kazmann, Coordinating Secretary, and E.H. Mohl, Executive Secretary.

By their second convention in December, 1919, they were deeply involved in a discussion of a project for setting up a colonization unit with a population of 100,000. They also proposed sending an engineering commission to Palestine to survey the country, with the object of constructing harbors, railroads and highways and reclaiming swamp lands.

Aware of the need for locally trained personnel, they drew up an organization chart which called for two major departments, one in engineering, and the other in agriculture. Engineering included sub-departments in civil, mechanical, electrical, chemical, marine and mining engineering as well as architecture. The Agriculture Department called for agronomy, agricultural engineering, animal husbandry, poultry husbandry, horticulture, veterinary and bacteriological units. Each of these was in turn broken down into sub-units and minor divisions. The organization was detailed and thorough.

The Society built itself up to a membership of about 300 at the most, and continued to meet for several years, but was unable to find any responsible body that would commission implementation of its program. "It finally became difficult to maintain an organization engaged purely in theory and debates, and fed by enthusiasm only," Halpern reported.

An effort was made on a different level. Some fifty Jewish manufacturers and businessmen were invited to meet with Weizmann and Ussishkin in the interests of the Technion. Only three showed up.

A few years thereafter Louis Cantor reported that he had addressed meetings in various American cities and found some interest, especially on the part of Mrs. Mary Fels, J. Pincus and Dr. Lowenstein, a G.E. Consulting Engineer.

Pincus, then chairman of the Zionist Engineers Society, asked for pictures and information to help publicize the Technion. Isaac Kalugai, serving as an industrial chemist with the Starr Chemical Laboratories, also asked for information for members of the Society. When will Technion open? What will be its standing? There were more and similar questions that could not yet be answered. "What do you expect of the Society here?" could be easily answered, but again nothing happened. The few who were interested remained in touch and eventually formed the nucleus of the American Technion Society which came into being some 20 years later. Kalugai served for many years as a professor of chemistry at the Technion.

The *Jewish Technical Bulletin*, published in New York, and said to go to more than a thousand "Jewish agriculturists, architects, engineers, journalists and manufacturers", published an editorial in November, 1923, hailing the imminent opening of the Haifa Technical Institute, and conceding that part of the fault for the delay was theirs because they had not provided the funds required.

Fund-raising efforts in Great Britain met with no better luck. The Jewish Colonization Association, which had been expected to play an active role, saw no value in the Technion and refused even to serve on the new Board of Governors which was soon to be set up.

In June, 1921, Ussishkin and Isaac Naiditch, then director of the Keren Hayesod, cabled that a construction budget of £10,000 had been approved, payable at the rate of £2,000 per month beginning in July. There was jubilation in Haifa. On the basis of the cable, Goldberg loaned the Committee another £2,000 so that work could be commenced at once. Thought was given to establishment of a permanent endowment fund, to assure the annual maintenance of the institution. There was also need to put the administration on a more stable basis, since the Committee was still provisional in nature. This would require setting up a Board of Governors, a Kuratorium, as in the days of the Hilfsverein. The need for a managing director was again felt.

Hecker, head of the Technical Department of the Zionist Organization in Jerusalem, was apprised of the budget and authorized the Committee to proceed at once with the drawing up of plans, within the budget, of course. Even in small things there was every indication that the Technion was at last in business. The Commission in Jerusalem sent a shipment of office supplies, including such things as ink, paper, pen-holders, erasers, thumb tacks, and water colors. There was talk of applying to the government for damages caused to the building during the occupation.

Contracts were signed for the construction work to be done, many of them with a newly organized Office for Public Works, a collective, which later developed into Solel Boneh. The workers were high in ambition and idealism, but their technical skill did not always reach the same heights. Complaints that the laborers were wasting material and doing a poor job were countered with demands for higher pay. When work was scarce on the kibbutzim, the *halutzim* were brought into Haifa to earn some money at the Technion. In 1923, Max Sharp recalls, he and ten other pioneers from Kibbutz Beitania helped get the buildings into shape. His function was to do fresh caulking on the stone walls. If his work then was not fully satisfactory, he more than made up for it fifty-five years later when, as a successful Toronto builder, he provided spacious facilities for students in the Canada Village area of Technion City.

The Haifa Labor Council warned that workers could be employed only through the Histadrut, and that the workers must set up a works committee to represent them. No dismissals could take place without consultation with the Council. At times it appeared that the "organized labor" aspect of the project was more important than the work to be done. The Committee protested that the labor costs were mounting too high. The

Histadrut, for its part, complained that payments were falling behind and were being made only after long delay, and often in deferred notes.

For many months the Committee met almost weekly. There was a great deal to do. Dr. Biram was indefatigable in his efforts, but there was no doubt that the dominating, driving force was Boris Goldberg. He recalled that during the days of the War of the Languages he had sent financial help from Vilna to the striking teachers. With his wife and two sons, Alexander and Elias, he had left Russia in 1914, bound for Palestine via England, where he had business interests. The family was trapped there by the war. After the Kerensky Revolution he went back to Russia to reorganize the Zionist Organization and to raise funds for Palestine. With the collapse of Kerensky, he and Eliezer Kaplan escaped across the border, carrying with them the contributions they had managed to accumulate, in the form of gold and jewels. In 1919 he took his family to Palestine, settling at Haifa where he had initiated the first steps leading to establishment of the Nesher Cement Works. Alexander remained in London, at school, but from time to time visited his family on vacations. Goldberg's stay in Haifa had a very great influence on the institution in those early critical days, and he retained that interest even after he moved to Tel Aviv. He was hurt in the Jaffa riots of May, 1921, but continued to be active both in business and in Technion and Zionist affairs. He passed away a year later as a direct result of his injuries.

The Committee was hard at work. One meeting lasted over seven hours, and was devoted to such prosaic matters as labor problems, building contracts, and materials for flooring and walls. Hecker guided them in the technical aspects. They had to become experts also on doors, windows, sand, gravel, lime and other raw materials.

The pump had worked adequately throughout the war but had sustained damage. It had been repaired and improved by the British, since the well served as the source of water for their hospital and encampment. More and more the Technion well was providing water for private homes built in the neighborhood, and this constituted a source of income. The water level had dropped, however, and some work was necessary on the well.

The entire power structure of the Technion had to be rebuilt from scratch, since practically all electrical fixtures and installations had disappeared. In 1924 a talented young technician, Eliyahu Sochazewer, was asked to undertake the job. He was working at a center for agricultural machinery in Merhavia that was secretly producing arms for the Jews. He looked over the Technion, saw that it was in pretty bad shape, and turned down the contract. He went back to Merhavia and to his infant son, Amos, later to be known as Amos Horev.

Boris Goldberg (1865–1922).

The wellhouse in the courtyard. 1925.

Period of Illusion

The members of the Committee soon discovered that they were living in an unreal world of illusion. The lack of cash should have been a warning, but Goldberg's help had postponed the moment of realization. He had in the meantime loaned another £1,000. On July 21, 1921, the Provisional Committee sent an urgent cable to the Zionist Executive asking when the promised monthly payments of £2,000 would begin, and where the money would be coming from. The reply from London was a bombshell: "Neither our Executive nor the Finance and Budget Committee ever decided to grant to your committee the amount claimed by you for the Technion; nor has our treasurer any knowledge of the existence of such a grant."

Word had come from London also that there was no money in the Hebrew University Fund; to the contrary, the Fund owed money to the Zionist Executive. The supreme authority, the Zionist Congress, was to hold its next meeting, the twelfth, at Carlsbad in the first two weeks of September and preparations were made to put the case before the Congress. Goldberg attended. Hecker drew up a pamphlet telling the Technion story and the needs. The Congress was warned that all the work would have to be stopped and all the workmen dismissed unless resources were provided. It would make a terrible impression.

The Congress had other matters to worry about. There were important debates on colonization policies, on the Mandate which the League of Nations was to confer, on organization and broad financial problems. Furthermore, there was a strong opposition to the Zionist leadership, headed by Julius Simon and Nehemiah de Lieme. Weizmann was elected President of the Zionist Organization. And the Technion? A budget of £10,000 was approved. The Congress also decided the time had come to establish a new Kuratorium, with broad membership, to run the affairs of the Institute. The first meeting actually took place only about a year later (November 7, 1922) but in the meantime what would happen in Haifa? What about the decision to open the school in October, 1922? Schmaryahu Levin had given three good reasons for getting on with the matter quickly:

1. The need for a technical institution; 2. The impetus which the school would give to the development of Haifa; 3. The need to save face with the British, who had been forced out of the premises on the grounds that the school was about to open. Ussishkin added his weight, and expressed the view that opening of the Technion was more important even than creation of another agricultural settlement.

But still, without money nothing could be done. A month after the Congress an ingenuous answer came from Ussishkin. He informed the Committee that all its frantic cables had been received, but no answer had been sent because the authorities did not know what to say. He repeated the news of the approved budget and added "You've already got £3,000, and the rest will come. I know it's hard to work under such conditions, but at least it's better than not working at all. Don't be upset with me . . . ."

The £3,000 so glibly referred to by Ussishkin were none other than the £3,000 loaned to the Committee by their own Boris Goldberg!

Allocations of a few hundred pounds were made from time to time, but they were not sufficient to keep the work going. Goldberg's money had long since been used up. The Zionist Organization also warned that it was beginning with a clean slate, and assumed no responsibility for any debts that had been accumulated.

Under the circumstances, the Haifa Committee considered the possibility of inaugurating a fund-raising campaign of its own. Hecker offered to go to Europe on such a mission, should it be approved.

During all this period and for several years thereafter, it should be noted, the cumbersome channels of communication made it difficult to procure decisions. The Haifa Committee was provisional, without any authority, and was responsible to the Zionist Commission in Jerusalem. The latter had paper authority, but had to take its directives from Zionist headquarters in London, which provided the funds. The situation was not unlike that which had existed ten years earlier, when the pivots had been Haifa, Jerusalem and Berlin.

The problems of remote control were felt in other ways as well. The office in Jerusalem was in charge of auditing the Haifa books. This was done simply by collecting all the accounting and financial records and shipping them off to Jerusalem for study. There they would be held sometimes for over a month, and only after imploring pleas from Haifa would they be returned with the explanation that the accountants in Jerusalem had not had time to do so earlier. The effect this had on the administration in Haifa can be imagined.

The Zionist Organization continued to exploit the building whenever necessary. The Provisional Committee did not mind; after all, the Organization was providing some funds, but there was objection to the way it was done.

A letter to Ussishkin in Jerusalem early in January, 1922, told the story: "This morning we received a phone call from your office in Haifa that this afternoon we would have to provide housing for 300 *halutzim* (newly arrived). No advance notice? Not even asking us if it would be possible? We shall not be able to underwrite the expenses involved and want to make sure these expenses are not charged up against Technion's budget. Also please make sure the immigrants do not stay more than a week, because their presence will hinder the construction work."

Toward the end of April, 1922, the Finance Committee of the Keren Hayesod in London finally cabled to the Commission in Jerusalem: "Cannot accept responsibility inauguration Technion next year; further oppose appeal different institutions stop agree Hecker's mission provided confined collection of machinery donations excluding money appeals and Hecker willing work close contact with Berlin office."

The reply was completely unsatisfactory, but at least it was a decision. The Committee now knew where it stood.

The Provisional Committee had been pressing ahead with work on the site, and had been making progress despite the obstacles. On June 1 the Committee informed the Jerusalem office that the building was approaching completion and the time had come to plan the opening of the institution. It would be a shame to let the big structure stand empty. There were political implications to the opening as well. Some Arab families had expressed their desire to register their sons at the Technion, and it would be better to have them at Haifa since the alternative was that they would go to schools operated by the extremists. There followed a review of the need to produce skilled manpower and the effect this would have on the industrial development of the country.

The next thunderbolt was already in the mail from Jerusalem, and crossed the optimistic letter making plans for the future. The new instructions were all too clear: "As a result of the present financial situation, and the last news from London about the failure of the Keren Hayesod campaign in New York thus far, we are compelled to limit and reduce all budgets even further, to the minimum. Please discontinue immediately all work at the Technion, and do not undertake any new obligations until fresh instructions from us . . . ."

In all fairness it should be noted that the paucity of funds in those days affected not only the Technion, but also the general settlement and colonization program of the Zionist Organization.

Reorganization

The pendulum swung back to London. An informal meeting had been held on May 7, 1922, pursuant to the decision of the 1921 Congress to look into the development of the Institute. Those attending were O. D'Avigdor Goldsmid, Arthur Blok, Leonard Cohen, Dr. Montague D. Eder, M. Hecker, H. Hirst, Dr. Weizmann and L. Kohn, Secretary. Weizmann revealed that a contribution of $10,000 had been received from Dr. Emile Berliner, of Washington (inventor of the microphone), and the need now was for a responsible Board of Trustees. The importance of the Technion, Weizmann told the gathering, went beyond its contribution to the upbuilding of the country. It would also assist in large measure in establishing friendly relations with the non-Jewish population of the country, who would send their sons there for training. The meeting heard suggestions that a further £25,000 was needed to put the buildings into operating condition. The first class would begin with 50 students. The new Board would be responsible for the general management and financial administration of the Institute, including the search for new sources of income. In the meantime, the Zionist Organization would provide the buildings and part of the budget.

The Zionist Executive met in August and blithely decided to proceed with the opening of the Technion and the setting up of the administrative body. A budget of £3,000 was allocated. Weizmann undertook to handle the organization side, and since the Mandate over Palestine had a month earlier been given to Great Britain, London was now undisputably the center of the Zionist world. The Technion's headquarters would therefore continue to be located there.

A further meeting of Weizmann's informal group took place in November with the added participation of individuals from the Jewish Colonization Association, the Anglo-Jewish Association, and representatives of the Zionist Executive. It was agreed that the W.Z.O. and the J.N.F. should transfer their rights to a Board of Trustees, to be appointed by a new association to be called the Board of Governors, half of whose members would represent the Zionist Organization. Among other things, the Board would be expected to assure the budget of the Technion. In Palestine there would be a smaller body to be known as the Palestine Committee of Management to deal with day to day affairs in the operations of the school. It was also felt advisable to set up an additional International Council to be composed of the major overseas contributors and other persons of influence and prestige. The Technion would certainly not lack for organization and administrative framework. Legal counsel drew up an involved and detailed constitution. It was quickly discovered that the proposed framework was much too cumbersome, and in practice it boiled down to the Board of Governors and the Palestine Committee of Management, with final authority remaining in London. Ussishkin and most of the Palestinians present objected, but to no avail.

Another six months went by, and it was not until May 1, 1923, that the new Board took shape. At Weizmann's request, Sir Alfred Mond became first chairman. Other members were, representing the Zionist Organization, Dr. Weizmann, Dr. M. Eder. Dr. B. Feiwel, Dr. G. Halpern and Joseph Cowen; representing the J.N.F., Dr. M. Soloweitchik. The other members, in addition to Mond, were Goldsmid, Blok and Lt. Col. H. Solomon. Three places were held open for Americans, when a society of friends would be organized in the U.S.

The full Board seldom met, for the members were busy people. Most of the work, it appears, was done by Eder and Blok. The Committee was remarkably unaware of all that had transpired in the previous decade, for one of its decisions was to the effect that while the language of instruction should be Hebrew, other languages could be used to the extent that it was considered necessary or desirable. The Committee also decided to call the school the Haifa Technical Institute. Both these decisions upset the Palestinians, the latter especially because it in no way reflected the Hebrew or Jewish character of the school. The Institute continued to be known by a variety of names. A 1918 letterhead, for example, identified the institution as the Jüdisches Technisches Institut zu Haifa, and then, in Hebrew, *Bet Haroshet L'Limud* (Factory for Study)–Technikum.

The Technion problem was brought before the 13th Zionist Congress held in August, 1923, again at Carlsbad. There Ussishkin presented the

Technion case. The Zionist representatives on the Board were instructed to see to it that Hebrew should be specified as the language of instruction, without qualifications. The Board, at its subsequent meeting, straddled. It eliminated the paragraph about language from its constitution, and among the general resolutions declared that Hebrew should be the language of instruction in all regular classes, but use of a foreign language would be permitted upon special decision of the Board. The purpose was to enable distinguished guest lecturers to speak, no matter what their language.

Nahum Sokolow proposed a new name for the Institute, the Hebrew Techniah, and this name was used in the Congress deliberations. The Congress went out of its way to emphasize that the Techniah was an integral part of the Zionist structure. Again an allocation of £3,000 per year for each of the next two years was approved, on condition that other funds be found to balance the operating budget. This was not easy since windfall gifts for the Technion that fell into the hands of the Zionist Organization were used to cover the Zionist allocation, as happened with the Emile Berliner contribution. It was clear that the Technikum or Techniah or Technion would not open in 1923 either.

Back to Germany

Despite the restrictions placed on Hecker's proposed fund-raising mission to Europe, the Haifa Committee felt it had no choice but to send him. His expenses were to be paid out of the normal budget of the Institute. On this risky basis Hecker went to Berlin in the summer of 1922, and within three months was frantically cabling for remittances to help pay his expenses. The reply was they had no money to send. As required he maintained close contact with the Zionist office in Berlin, which made sure that he was not soliciting and diverting contributions which might otherwise have gone into the usual Zionist funds. It was at any rate against the law to send money out of Germany. Funds could be transferred only in the form of local goods, and the Technion drew up a list of its physical needs. Whatever was shipped by the Zionist office was deducted from the promised budget.

Hecker remained in Germany until April, 1924. His fund-raising activity was handicapped by the dreadful post-war inflation, and the figures on his collection lists must be read and understood in inflationary terms. For example, he reported contributions of ten million marks each from Herrn L. Zimels, Nathan Braude, Feltenstein and Weissbrem, John Frankel and Nathan Rawraway. On the other hand Herrn Leo Abramowitz, Walter Rablinsky, John Rabinowitz, Herman Lewin and Arthur Propp gave only five million marks each. It should be added that 10 million marks were worth $50 in those days.

Nonetheless, Hecker continued his prodigious, almost herculean, efforts during his two years in Germany. He constantly called on prospects, wrote letters, pursued every possibility for help. He received support from more than a dozen communities in the country. He had the competent help of a devoted assistant, Lotte Herrmann.

The Keren Hayesod was disturbed by his activity and complained that he was competing with the normal Zionist fund-raising, especially in Königsberg. Siegfried Kanowitz, of the Keren Hayesod office in that city, replied to Berlin that the contributions had been obtained from "radical anti-Zionists" who would never have given to the Zionist funds anyhow. Others made their contributions to the Technion after they had fulfilled their obligations to the Keren Hayesod. Individual Zionist leaders, who understood the importance of the Technion, gave their full cooperation. In the Free State of Danzig a physician, Dr. Yitzhak Landau, who headed the Zionist office there, was a pillar of support. His son, Moshe Landau, was to become not only President of Israel's Supreme Court, but for many years also Chairman of the Technion's Board of Governors.

Hecker received help also from Poland and Czechoslovakia. However, most contributions were of material or equipment rather than money. Some were practical and useful, like tool-making machines given by industrialists in Berlin and Düsseldorf. The manager of a copper mine supplied all copper needs and electrical wiring. Some donors, many requesting anonymity, provided motors, pumps and various machines. There was lumber for construction of furniture. No less than eight truckloads of wood reached Haifa from various parts of Europe.

There were also gifts, presented with a maximum of goodwill, but of questionable value to an embryonic school of technology. An industrialist in Freiburg, Conrad Goldman, presented a painting of Moses, executed by the artist, Prochownik. One Herr Hirssenberg promised to loan another painting. There were several lovely porcelain vases, gifts of the proprietor of a building materials factory. The vases, incidentally, still adorn some of the administrative offices of the Technion today. A physician in Leipzig donated a number of lithographs. Hecker admitted it was difficult to set up an institute of technology with these items, but he accepted with good grace everything offered. Friends from Chemnitz contributed equipment for a physics laboratory in memory of one Apisdorf who had recently passed away. Other friends in Hamburg and Berlin gave a chemistry laboratory and the means for a lecture hall. The Wolffsohn concern and others in Berlin undertook to equip a carpentry workshop. From Leipzig there were gifts for the metallurgy department; from Schlesvians, tools in memory of a Herr Dobrzynski. The value of all these items was estimated at many tens of thousands of gold marks, as against the inflated paper marks.

In some places Hecker found that the energetic Dr. Biram had already preceded him, and had procured generous help for the Reali School, which by then was already in full operation.

The machinery, equipment and supplies began to reach Haifa. The German Foreign Ministry gave the required export permits, but at Haifa Port customs duties were imposed. For lack of money to pay, the material often remained in the port for long periods.

Hecker rendered other services as well. He wrote articles and pamphlets on the Technion, studied technical education methods in Germany and England and even drew up detailed plans for the layout of laboratories and workshops. Hechalutz sought his advice on opportunities in Israel for mining engineers.

He consulted experts on Technion's needs. He turned to one Prof. Rudolf Samuel, of Göttingen, for advice on equipment for the physics laboratory. Samuel wrote details of everything needed, including a reminder to provide for a source of gas for the Bunsen burners. He offered to be of further help. He was, some years later, not only as a professor at the Technion, but as an overseas organizer responsible for the establishment of the American Technion Society.

There were also applications for teaching positions at the Technion. In December, 1923, one Samuel Sambursky, then 23 years old, wrote Hecker from Königsberg, asking when the Technion would open. He wanted to know what the prospects were for a teacher of mathematics or physics at the Institute. Hecker replied that there were no immediate prospects at the Technion, and while he could not give advice, he told the young man to go to Palestine and see for himself. The following year Sambursky did go. Eventually he joined the faculty of the Hebrew University and became one of Israel's most distinguished scientists.

Some of the active members of the Hilfsverein were of help to Hecker, but he decided to go after the leadership. Paul Nathan was willing to receive him and listened sympathetically as Hecker told of the fortunes of his old Technikum. However, Nathan pleaded that he was unable to participate in any way. Hecker then turned to James Simon and invited him to join a new committee of friends of the Technion he was setting up in Germany. When Simon refused to join, Hecker invited him to help set up a section of the new architecture department to provide students with an aesthetic appreciation of beauty. This was catering to Simon's known interests. The latter replied that he was over-extended already. Architect Erich Mendelsohn expressed an interest and offered to help get together

Dr. Albert Einstein, on his visit to the Technion. 1923.

Jewish engineers and architects who could render their assistance.

Hecker's time in Germany was running out, and he finished his activity there in 1924 by setting up a *Deutsches Komitee für das Technische Institut* in Haifa, a German Committee for the Hebrew Techniah in Haifa. A number of Jewish industrialists joined. A noted German scientist, who had visited Haifa with his wife in February, 1923, and had planted two palm trees on the grounds of the Technion, agreed to head the committee. Thus Dr. Albert Einstein became the first chairman of the first Technion Society in the world.

Hecker wanted to go back to Haifa to become the managing director of the Institute. Through his connections with the London Board he had in August, 1923, been appointed acting Technical Director, and his precise title thereafter was left open. In April, 1924, he returned to Palestine.

Vocational School or University?

During all this time of fund-raising and physical preparations, the academic side of the Technion had not been neglected. In November, 1921, the Palestine Executive of the Zionist Organization in Jerusalem held a full-dress meeting with the participation of representatives of all interested bodies. The purpose was to make recommendations with respect to the major problems affecting the educational program, and to send these on to the Greater Actions Committee in London. It quickly became apparent that questions which had been left unanswered under German administration before the war were no closer to solution after the war under Zionist administration.

Ussishkin presided over the meeting with a strong hand, and the various problems were discussed seriously and to the point. The first issue went straight to the crux of the problem. What level of technical manpower should the Technion train? For the sake of clarity three levels were put up for discussion:

1. Engineers and architects, on a university level; 2. Technicians, assistants to engineers and work foremen; 3. Artisans, craftsmen and industrial laborers.

It was obvious that the overwhelming majority of those present favored a low level. The reasons advanced were varied, among them: The country did not need engineers; enough were coming in as immigrants. There was no academic personnel available to make the Technion a university, and the students would therefore be inadequately trained. The Engineers' Associations of Jerusalem, Jaffa and Haifa had each, separately and individually, adopted resolutions which their representatives presented to the meeting, to the effect that they firmly opposed the training of engineers. One of the spokesmen perhaps revealed the inner feelings behind the rationally worded resolutions when he blurted out: "We don't need more engineers. We'd just be creating competition for ourselves." There were some who felt Palestine would never have to train engineers. And there were others who stated that when the time did come, the level of instruction could be raised, but that was still in the future.

Only two favored opening the Technion as an institution of higher learning. Prof. Aharon Tcherniavsky, who was already teaching physics at the Reali School, spoke fervently and at length. He stated that there would always be young people who would want a higher education, and if it were not available in Palestine, they would leave the country. A university would be established sooner or later, and if the Zionists did not do so, the British or the Arabs would. He scoffed at the view that enough engineers were coming in as immigrants. With that logic, he said, there was no need to prepare anybody for anything — just rely on immigration. But among the new arrivals would be many students who, barred from educational opportunities elsewhere, would come to Palestine for their studies. There should be no fear of over-producing engineers. The course would be difficult, and those that did not make it to the very end would be the technicians and skilled artisans of whom there was also need. Prof. Tcherniavsky predicted that the school would train technical manpower not only for Palestine but for the entire Middle East, thus giving a boost to local industry. In this he was echoing an argument advanced by Paul Nathan more than ten years earlier.

"It takes five years to turn out a first-class engineer," he concluded, "and even if we start now, who knows, we may already be too late. I'm for the highest level of university that our budget will permit."

The only other advocate of the top level was Pewsner, who was not present, but sent a letter. He maintained that no reliance could be placed on immigrant engineers who knew neither the language nor the country's needs. He pointed out that most of the government's construction projects of roads, bridges and the port, were being assigned to foreigners because of the lack of locally trained manpower. He criticized the low level of many of the engineers practicing in the country. If for any reason the opening of a full polytechnicum was not possible, then he was in favor of a middle-level school, like that at *Mitweide*.

Ing. J. Breuer had a novel objection to the middle category. Its graduates frequently sought to do engineering work for which they were not qualified. And when they were asked to do artisan work, they would say they were not laborers. He was therefore for the lower level.

In summary it appeared that there was a general consensus at least on the second level, for the training of technicians and professional craftsmen. Ussishkin refused to close the door altogether, and added that as the need and possibilities permitted, they could shift over to a full institution of higher learning.

The second item on the agenda was: What subjects should the Technion introduce in its curriculum? As a point of departure they listed: Building construction; Roads and highways; Mechanical and electromechanical trades; Surveying; Chemical technology; Irrigation.

There was almost unanimous agreement on the importance of the first two. Most of the speakers gave next preference to subjects related to agriculture such as irrigation, land amelioration and farm machinery. Almost everybody rejected chemical technology. Since the Technion would be limited in budget, there was no point trying to teach everything, one of the participants declared. As for chemical technology, he saw no prospects for it in the foreseeable future. Surveying was opposed, at least as a separate department. Suggestions were made to add to the list: architecture, sanitary engineering, telephone and telegraph, and railroads.

The remaining subjects on the agenda reflected the differences of opinion already expressed. How long should the course of studies be? Two or three years, with a mixture of theoretical and practical, was felt to be enough for the lower level. If indeed a university were to be decided on, then three or four or five years were suggested.

What should be the age of entering students? For the secondary school, 16; for university, 18. It was agreed that previous practical experience should not be a required qualification.

The final questions had to do with selection of teachers, appointment of a managing director and setting up of a Management Committee. Again the suggestions ran a broad gamut. The Jaffa Engineers' Association was of the opinion that the teachers be chosen competitively, and they would in turn name the Director. Another participant proposed that the Director should be named by the Zionist Executive and he should be given the power to engage the teachers. Ing. Wilbushewitz had words of sage advice: "Let's not save money when it comes to hiring teachers. Good teachers should be engaged even if it costs money. If we hire 'cheap' teachers, we'll simply destroy the institution."

The matter of budget was not neglected, but a committee was named to draw up the figures and present them to the Zionist Executive.

Ussishkin closed the meeting, the first serious and comprehensive study of the Institute's academic and administrative problems, with the words: "May we witness the opening of the Technion soon."

As we have already seen, the physical condition of the buildings and the lack of funds made it impossible to do anything about the proposals emanating from this November, 1921, meeting. A year and a half later the whole procedure was repeated, but this time a questionnaire was sent to ten leading engineers. The result was the opening of a new controversy which at times seemed to approach in its intensity the famous war of the languages.

Many of the old arguments were repeated, but a number of new ideas were also introduced. Ing. Samuel Tolkowsky, replying to the questionnaire on behalf of the township of Tel Aviv, stressed that the problem was not just lack of training, but the absence of any desire for training. "Workmen who have lived in the country for only a few months, and have not even satisfactorily worked with others, consider themselves 'skilled laborers'," he said, "so that they no longer wish to be really educated." In his opinion, the answer was to open as many evening courses as possible. Dr. A. Ciffron, Municipal Engineer of Haifa/Acre, stressed that the need was for a "man of function", between the European research worker and the Palestinian laborer.

A Jerusalem architect voted in favor of immediate opening of a department for building tradesmen, such as house painters, tinsmiths, stucco workers, etc. The Director of the Government Public Works Department in Jerusalem was in favor of turning the Technion into a factory at which the practical training could be conducted on a commercial basis.

The noted architect, Richard Kauffmann, was also among those who favored training workers of the lower ranks, with only secondary consideration given to the middle ranks.

On the other hand, there were some who had more vision and faith in the future. Ing. Simon Reich, of Haifa, spoke of the export opportunities which would open up, including to the Far East. If local industry was not developing properly it was because the flood of immigrant engineers who had brought over the "technical dogmas of Europe" were not acquainted with local conditions. The result was that costs were unnecessarily high. The solution would be a polytechnic institute, and attached to it a school for tradesmen. If there were a choice between the two, he would start with the college. Ing. Joseph Loewy, also of Haifa, likewise called for a Technical College to serve the rising generation of technically gifted young people.

Ing. Markus Reiner, of the Government Public Works Department, who was later to become a distinguished professor at the Technion, called not only for high-level instruction in construction, civil and mechanical engineering, but also for the opening of laboratories for scientific engineering research. Both he and Loewy spoke of the opportunities for economic expansion in the neighboring countries. One is tempted to speculate on what political consequences might have ensued had this line of thought been seized and vigorously prosecuted by the Zionist authorities.

The one who came closest to spelling out the program that was eventually adopted was Ing. Hugh Elkes, who also served with the Government P.W.D. He favored a full four-year course of studies, with the first two years to be common to all professions, and the last two years specialized. Civil Engineering and Mechanical Engineering were favored for the first two departments, but at the same time the Technion should not neglect the need for supervisors, foremen, skilled artisans and craftsmen.

Finally Architect Barsky, whose association with the Technion went back to the days of the Hilfsverein, echoed a suggestion that had first been made by Wilbushewitz in 1921. He called for establishment of a Building Materials Experimental Institute. He noted that architects and engineers were compelled to grope about with regard to local building materials and judge them only according to individual tests, Arab traditions or theories from abroad. There was no well-grounded research and no reliable statistics. Such an Institute, he said, could be set up only as part of a technical college.

From the historical perspective it may seem surprising that at this stage the major assault on the Technion was launched by the Palestine Association of Engineers and Architects. It repeated the views expressed by its representatives in 1921, and summoned all its members to do battle against the opening of a technical university. In July, 1923, immediately after the questionnaire to the ten engineers, the Association dispatched a formal memorandum to the 13th Zionist Congress held that year. It repeated that there was no need for an institution, no matter what it was called, to train top-level technicians. The need was only for workmen and artisans. Any money spent on more than that would be wasted. The Association had some practical advice. In the whole country there was need for only 12 courses, nine for builders and three for carpenters and plumbers. If the Zionist Congress would budget £2,000 a year the problem of technical education would be satisfactorily solved for the coming years. There would be no need to set up special laboratories or workshops. The memorandum concluded solemnly: "We recognize our responsibility before the people of Israel and before the world, and issue this warning in due time!"

Gedalyahu Wilbushewitz added his voice to the anti-Technion campaign. In a three-column article in *Haaretz* on July 27, 1923, under the heading, "Is There Need for the Technikum?" he ticked off the usual negative arguments one by one and added a few more. For one thing, it would be impossible to get the kind of teachers required. Next, there was not enough local industry to assist. Further, Jewish youth, unlike those of other nations, were not the sons of technical people with technical traditions, and hence were remote from the field. He reviewed the situation of local construction companies. Two had been set up in the last three years, he said; one had already failed, and the other was about to go under. The Histadrut company was still in existence, but it had already lost more money than both of the others together. Their failure had not been due to lack of engineers. The fault was not that of the laborers, the devoted, diligent *halutzim*. It was not the fault of management, competent and conscientious. The blame must be placed on the lack of artisans and craftsmen. The Technion could not solve this problem. What was needed were evening classes to train labor. For specialized training, the best of the workmen could be sent abroad. To set up a Technion as an institute of higher education would only continue the Diaspora mentality which put university training on a pedestal, and drew youth away from physical labor. He conceded that there was need for research and testing laboratories for building materials and chemicals, a technical library, technical publications and exhibitions, but these could be set up with the funds available, rather than waste the money on a costly and unnecessary Technion.

Support for a high-level Technion came from the Association of Technicians. The latter questioned the right of the engineers to speak in the name of all who were interested in technical education. They made no bones about their feelings that the engineers were interested only in eliminating future competition. The technicians and the craftsmen, too, they said, wanted the opportunity to study to be engineers.

Between Jerusalem and London there was vacillation with respect to the proposed level. The views of the Provisional Committee in Haifa were given no consideration.

It was clear that the Technion was floundering for lack of leadership, both in terms of a strong individual at the head, and in terms of a properly constituted and responsible board with authority. The Kuratorium in London proved to be ineffective, for most of its members had no interest in the Haifa institution. The search for a managing director, or principal, was intensified.

Hecker saw himself as the outstanding and logical candidate. In July, 1923, the Kuratorium in London had named him "Acting Technical Manager" for a six-month period. His duties were to arrange for construction work on the building and render it fit for use, and to order and install equipment. The appointment implied no pledge with respect to the future, but it was also "understood that Mr. Hecker is not debarred from applying for the position of Director, should he so desire."

It will be recalled that he had come to Palestine in 1913 to take up a position with the Technikum. He returned after the war as a technical expert on the staff of the Zionist Organization. He had been responsible for the drainage of the Jordan swamp near Kinneret and had supervised sewage works in Jerusalem's northwest quarter, as well as construction of more than 50 houses in Nachlat Yehuda, Ein Ganim, Kfar Saba, Machane Yehuda and Kalandia. He had directed the installation of the water supply for the colonies Gan Shmuel, Kfar Saba and Kiryat Anavim, and had been responsible for repairs and new construction work at the Bezalel School. He is said to have been the one who encouraged the construction workers to organize their own contracting firm, thus laying the foundation for the building activities of Solel Boneh. He campaigned vigorously for the appointment to the Technion.

The Kuratorium published its first advertisement for a Principal in 1923, and November 21 was set as deadline for submitting applications. The only requirement specified was a good knowledge of Hebrew. The functions of the Institute were stated to be "to train technical workers of the lower and middle grades, in particular artisans, craftsmen, supervisors and foremen." Opening date of the Technion was set for April, 1924.

Since no progress whatever had been made, the school did not open as scheduled, nor was a Principal named. A new advertisement was published, this one listing new criteria. The successful candidate would have to have a university education, experience in technical education and organizational ability. Familiarity with sports activities was listed as an added desirable qualification. Nineteen applications were received, and the list was narrowed down to five.

Pending consultation with the Palestine Executive, no decision was made in London.

Ussishkin was unhappy with the modest plans for a simple technical school. He continued to press for an institution of higher learning and recommended that the head should be a personage of scientific distinction, whose name would add to the prestige and reputation of the school.

The London Board favored the candidacy of Arthur Blok, whose views on the function of the Technion more closely coincided with theirs. He held an engineering degree from University College, London, had been head of a small trade school and was principal examiner in the Patent Office of the London Board of Trade. He knew no Hebrew, but had been close to Technion affairs, having served, at Weizmann's request, as a member of the first Technion Committee formed in London in 1920. London's decision was overriding, and the choice fell on him. Already deeply involved in Technion affairs, Blok intensified his efforts and sought to raise funds. Writing to Hecker in April of the problems he had run into he noted: "It is a matter of extraordinary difficulty to get money for Palestine in England. It is shameful but true that the whole burden of every appeal falls on a few people — the same few each time — and these men are squeezed dry . . . ."

Blok made application to the Board of Trade for a one-year leave of absence. Consent was delayed, and at one stage, in July, it seemed that the request would be turned down. Thereupon L. Kohn, Secretary of the Board of Governors, formally notified the Palestine Committee of Management that the appointment of Blok as Principal had been dropped. Further, the arrangement made with Hecker whereby he was to have acted for one year as Deputy Principal and Head of the Building Department, and to have been appointed Principal in the subsequent year, "also falls through, it having been made expressly dependent on Mr. Blok obtaining leave of absence." It was suggested that Hecker be asked if he would consider staying on at the Technion in any other capacity.

No more than two days later the Board of Trade approved Blok's leave, and the original situation was restored. The lack of confidence in Hecker was not lost upon the latter. The agreement with him provided that he would be Principal for two years after Blok, executing the programs and policies already determined.

In this atmosphere the Provisional Committee in Haifa, which had been maximalist in its views, resigned, and Dr. Biram filled the administrative vacuum until Blok's arrival. In the meantime a new national committee had been formed, the Palestine Council for the Technion, with headquarters in Jerusalem, and drawing its authority, like its predecessors, from the Board in London. The members were Col. Frederick Kisch, representing the Zionist Executive; Dr. Ussishkin, for the *Vaad Leumi*, the National Council of Palestine Jewry; David Remez, for the Histadrut; Biram, Pewsner and Ing. Y. Rosenfeld, from Haifa. Dr. Magnes also promised to help. It was decided not to have a president, but Ussishkin was to be presiding chairman at meetings.

As was to be expected, the Haifa members, with the addition of Hecker, constituted themselves a small executive committee *(Vaad Menahel)* to direct activities at the building and to prepare things for the opening of the school.

The new leadership in Jerusalem made itself felt, and at two key meetings the decisions were taken which at last made it possible for the Technion to open. On June 22–23, 1924, the Committee got down to fundamentals and adopted a series of basic resolutions, the most important of which were:

1. "The Palestine Executive Committee sees the institution as a middle-level technical school combining science with practical work. This should develop in the future into a high-level technical school, and a center for technology and industry in the country. The immediate activities shall begin with the development of a department of construction and road building in a three-year course, and with establishment of theoretical and practical courses for the training of workers and artisans as required."

2. A budget of £8,100 was adopted for the year 1924–25. Dr. Magnes was asked to press for this budget and further to assure the financial stability of the Technion by arranging for joint Technion–Hebrew University fund-raising as previously promised by Weizmann and Sokolow, and also by arranging with the Anglo-Jewish Association for a contribution, after the Association received its share of the Kadoorie estate.

3. The Administrative Council was asked to take all necessary steps to open the school in the coming year, and to invite applications for staff positions, academic and administrative.

The second meeting took place in Haifa on October 18, and the final decisions were made on standards, budget, appointment of teachers, acceptance of students and tuition fees. There was much discussion regarding the name of the institution. The German version, Technikum, was still being loosely used. The Zionist Congress had spoken of it as the Techniah. Ahad Ha-Am suggested Technicon. The name Technion had originally been proposed by the poet, Bialik. Another year was to go by before the name was finally fixed.

In the meantime it was reported from Berlin by Dr. Max Berlowitz that the balance of shipments of materials purchased or acquired as donations by Mr. Hecker had been completed, and since there had been no further requests from Haifa, the committee engaged in these activities had suspended its operations.

The Technion gardens restored. 1924.

One sure sign of progress was the restoration of the garden. Ever since Avraham Ginsburg had been engaged as Technikum gardener in 1913 the appearance of the grounds had reflected the general state of the institution. Ginsburg had lovingly planted the original trees and laid out the landscaping. Much of it had been neglected during the war years. Now, on the eve of the school's opening, he renewed the gardens. Signs were put up asking the public not to walk on the grass and not to pluck the flowers. Ginsburg remained on the job until 1935 and founded the famous Haifa florist firm which still exists at the corner of the Hadar campus.

There were other and even more significant signs that the period of anarchy was approaching its end. Hecker drew up a basic curriculum of studies.

The Lights Go On

The actual beginning had yet to be made. Inertia had to be overcome. Labor groups in Haifa were getting impatient, and asked for the opening at least of evening courses for construction workers, who were in short supply. Dr. Biram grasped at the opportunity to broaden the scope of his Reali School, and launched the first course. For lack of both funds and pupils it lasted less than a year. But the ice had been broken. On December 2, 1923, two new courses were begun, one for the building trades, and the students were all Solel Boneh workers; the other was for carpenters. At the request of employees in the railroad workshops another course was started on January 7, 1924, in the mechanical trades. The pupils were on varying levels of skill, and hardly suited to be all in one class. The teachers were inexperienced. The construction course called for eleven hours a week; mechanics, 10 hours; carpenters, four hours. The important thing was that there was movement.

At the end of 1924 the evening program for laborers and technicians was expanded. There were two programs in metal-working, one a continuation from the first year; another in railroad boilers, and still another for builders and stone masons. Woodworking was added, and then two more courses, one for electricians and the other for telephone and telegraph workers. Supervision of the workshops was put in the hands of Yaakov Erlik, who before the war had headed a trade school in Pinsk. He reassembled some of his pupils in the interior of Russia and continued teaching them, eventually reaching Palestine in 1924. He was appointed a member of the Technion staff and also undertook to conduct courses for the railroad students. A scholar, he was already familiar with Hebrew. For some years thereafter he offered various day and evening courses in technical subjects and became headmaster of the Technical High School. Much later his son, Prof. David Erlik, was to become the founder and first dean of the Technion's Faculty of Medicine.

On its higher, university level the Technion's first program was begun on December 14, 1924. Instruction was provided in the evening. Day classes in construction and road building, on a high technical level, started on January 7, 1925. On that very day a press dispatch from Cairo reported, from French sources, that Moscow was preparing a program of anti-British propaganda, with the aim of bringing about revolutions in all countries of the Middle East, including Palestine.

A new and exciting era had begun for both the Technion and the Middle East.

*   *   *

The Institute Opens CHAPTER FOUR

The Arthur Blok Administration

If there had been initial doubts about the location of the Technion in Haifa, and in that remote area which had been selected, halfway up the hill, such doubts were effectively set at rest with the passage of time. By the mid-1920's Haifa was already beginning to emerge as a busy industrial center. The Mandatory Government set in motion large public works projects, including a network of roads and the expansion of the port. And the once isolated buildings were now surrounded by a growing business and residential area. The city was more nearly beginning to approach the description of it given by Theodor Herzl in his 1902 utopian novel, *Altneuland*, when he envisioned the Haifa of the future as a great park, with thousands of white villas gleaming out of the luxuriant green gardens . . . a center of international commerce, with an overhead electrical train (Haifa's went underground instead) . . . a city of magnificent homes and public institutions, all made possible by applied science, engineering and technology.

The Technion undoubtedly served as the magnet around which the middle city was built, and Herzl, Balfour, Nordau, Jerusalem and Herzlia Streets marked the growth of Jewish population in the Hadar Hacarmel.

Interviewed in London on the eve of his departure for Palestine, Blok had made his views on the scope of the institution clear: "We shall begin slowly. At first no attempt will be made to produce skilled engineers, architects or advanced specialists of any kind, as they are already available either in the country itself or in the Diaspora. The immediate task is to train skilled operatives up to foremen and supervisors' rank: Carpenters, plumbers, masons, fitters, turners, tin- and locksmiths and so on. Later, no doubt, an electrical engineering department will be developed, but the first faculties will be those of the building and mechanical trades."

Blok made his maiden appearance at the meeting of the national Management Committee on October 18, 1924. His title was Principal.* In his remarks he again expressed his understanding of the scope of the Technion. These views, reflecting the position of the London Board, were not wholly in consonance with what the Palestine Committee desired, but everyone agreed that the important thing was to begin. Writing to Weizmann, Blok reported that "if the means and opportunities were in proportion to the enthusiasm, we should have Charlottenburg on Carmel within

---

* The Hebrew title was *menahel*. This was usually rendered in English as Principal or Director. In the late 1940's it became the practice to refer to the chief executive officer in English as "President", but it was not until 1962 that formal decision was taken to identify him as *Nassi*, President.

Arthur Blok (1882–1974).

a year . . . (there are) new problems every day and plenty of extemporization to get around them. But these are teething problems, no doubt, and I am satisfied that the Institute has come to stay . . . ."

Blok knew what the buildings had gone through, and he was aware of Hecker's heroic efforts at restoration, but coming from neat and orderly Britain, he was unprepared for what he found. Not only doors were missing, but in some cases even staircases. The roofs leaked. There was little furniture. He found only one plumber in the whole country who was qualified to install the needed facilities in the chemistry laboratory. The motor in the well worked, but not always. One night he found the whole steel frame of the well house electrically alive. It might have been that way for a long time, although there was no record of anyone ever having been electrocuted.

He was still faced by the various communal squatters, all claiming "rights". Anyone could walk into the workshops and make free use of the machinery. It was reported that for years Hadar Hacarmel housewives were accustomed to putting up their pickles in the large glass jars which they had "salvaged" from the supplies intended for the chemistry department. The very idea of putting a gate on the yard to keep out donkeys and thieves raised an outcry. One can picture the quiet, gentle, soft-spoken Blok patiently seeking to create order out of Middle East chaos.

The building had also become a popular overnight hostel. Excursionists and hikers always knew they could find a dry corner in the Technion where they could bed down with their blankets. After all, was it not empty, and public property?

These were not his only problems. Lethargy and inertia were even more difficult to overcome. Who was to give the responsible order to begin? As late as October 30 the Palestine Management Committee cabled London that it had approved the proposed budget and had set opening dates in November and December but required ratification of the budget, authority to complete the teaching staff, and a remittance to enable the doors to open. The reply was not altogether encouraging. Dr. Montague D. Eder cabled back that there was only £250 on hand, but in any event they had to await Weizmann's return for a meeting of the Board. Haifa fretted. If there were to be any appreciable delay it could mean the loss of another year. Some say it was Shmuel Pewsner who pounded the table and insisted that they go ahead and *begin*, and thus compel a favorable post-facto decision. The first classes met on the evening of December 14, 1924. And on the morning of January 7 the future civil engineers gathered for their first lecture.

When the belated approval finally arrived it contained a warning. The rock-bottom operating budget had been cut further in London, and a number of items had been eliminated. "The Board was forced to make these alterations in order to avoid a deficit this year, and I am directed to point out that the Board must absolutely insist on these alterations being strictly adhered to," the secretary wrote.

The engaging of teachers was of course another major problem. Fortunately some of the staff members of the Reali School were qualified, and arrangements were made for them to teach in both places, although there were the inevitable problems with respect to hours and scheduling. Prof. Jeremias Grossman, formerly of the University of Ekaterinoslav, was fluent in Hebrew and was named professor of mathematics. Dr. Aharon Tcherniavsky, former lecturer at the University of Geneva, who had been on the faculty of the Herzlia Gymnasium, also knew Hebrew; he became the first professor of physics. From the Polytechnicum of Kharkov came Prof. A. Ilioff, in chemistry and geology. Hecker, too, joined the staff as teacher of engineering and technical mechanics. The fifth senior member was Barsky, another veteran of the old days, who taught construction and building materials. Soon thereafter Prof. Baerwald was invited to join the staff.

Prof. Grossman had studied at Odessa, Heidelberg and Göttingen, receiving his Ph.D. in 1912. He became a Zionist, went to Palestine in 1924, and decided to turn his back on the academic world. He wanted to become a *halutz*. However, the rugged, physical work was too much for him and he became a shoemaker, happy in the thought that he would be serving the *halutzim*. It was only with difficulty that Dr. Biram convinced him that by teaching mathematics at the Reali School he would also be making a contribution to the building of the Jewish State. Grossman was an idealist, a perfectionist, and also an eccentric. His unfailing distinguishing identification mark was his black bow tie. His shy smile and frail frame belied a great strength of character and a stubbornness on behalf of principles as he saw them. These characteristics were yet to involve him in major difficulties with his colleagues.

Prof. Tcherniavsky had been one of those, some years earlier, who had come out strongly for a high-level Technion. He had first been to the old building in 1920 and had always felt close to the institution. When Einstein had visited in 1923, the famous scientist recorded in his diary how impressed he had been with Tcherniavsky's talk at the reception. Of stable temperament, he was usually the peace-maker when academic or administrative arguments broke out.

Prof. Ilioff was over 60 when he came to the Technion. He knew no Hebrew, but was engaged with the understanding that he would work hard at acquiring a knowledge of the language. In the meantime, he had all his lectures translated into Hebrew, which he then laboriously tran-

Prof. Yitzhak Kalugai (1889–   ).

scribed into Russian script, and he read the material phonetically to the class. He never believed that his students understood a word of what he was saying, and so he would go over the material thereafter, illegally, in Russian. Since most of the students were from Eastern Europe, they managed to understand. One day it was explained to Ilioff that a new student was from German-speaking Danzig, and did not know Russian. Ilioff reassured them. "Don't worry, he will!" And he did! Prof. Ilioff died in 1943.

The precedent set by Ilioff was followed later by others. Dr. Peretz (Fritz) Naphtali, who lectured in economics, wrote out his Hebrew text in Latin characters.

Of all the academic staff, Baerwald had had the longest association with the Technion. He was born in Berlin in 1877 in an assimilationist Jewish background. He loved the sea, was an officer in the German navy and served during the First World War. He studied architecture at Charlottenburg and at Munich. His Jewish feelings were first aroused when he went to Palestine for the planning of the Technion's buildings and as a result he became a Zionist. His home in Germany served as a cultural center, and he was often host to a musical quartet in which he played the cello and Einstein the violin. In 1925 he was happy to accept the Technion's appointment to the chair of architecture and took up his home in Haifa. He never learned Hebrew, but then he never lectured. He taught by drawings and by plans. He loved his teaching and wanted the students to get as much out of him as possible. "Children, exploit me," he used to say to them. By the third year the coterie of seven students which formed around him was the nucleus of what later became the Department of Architecture.

A friend characterized him as a naive idealist with faith in everybody, impractical with money, a lover of beauty, and most un-German in his warmth and friendship for his students. He died in 1930.

Other teachers included Y. Erlik, who taught mechanical engineering and directed the workshops, E. Strich, free-hand drawing, and two Hebrew teachers, Saadia Goldberg and Dr. Y. Cohen. Dr. Yitzhak Kalugai, whose first contact with the Technion had been from the U.S. some fifteen years earlier, began teaching chemistry in the evening classes and served as assistant to Prof. Ilioff. He retired in 1956 with the rank of professor. Prof. Bernard Shenburg joined the faculty in 1927, established the Department of Geodesy, and headed it until his retirement in 1957.

Arthur Blok sought to broaden the curriculum. He announced a course in English, given twice a week by M. Kanovitz. Attendance was sparse and the course was discontinued. Meir Dingott took over and continued as teacher of English until 1937.

Dr. Ernst Simon, a medical man who taught gymnastics at the Reali

School, was engaged to do the same at the Technion. The program lasted for only about a year because the students stopped appearing. The nucleus of a tennis club remained for a few years, using a court on the grounds.

A course in hygiene and first aid was given by Dr. Hillel Jaffe, one lecture a week, but the students showed little interest and it was discontinued after one year.

Strich, the art teacher, was characterized by Blok: "A real academician from Russia, teaching Biram's boys and girls how to draw — verily a royal academician hired to paint the garden fence."

Administrative staff was also named. Ben-Zion Cohen, a graduate of the London School of Economics, was the first Secretary, an omnibus title which covered multiple duties, since he also lectured in economics. He was followed in quick succession by A.L. Felman and Yosef Lipshitz. Early in 1927 Hayim A. Krupnick took over and during his period the functions of the post crystallized to include academic procedures, manpower, fiscal matters, etc. He served in that position for 16 years until his death in 1942.

Yehoshua Nessyahu (1900–1978).

Yehoshua Suchman (later to be known as Nessyahu) had been doing odd jobs in the Technion yard as gardener and builder, but for lack of work decided to resume the study of medicine which he had begun before coming to Palestine. He was accepted at the University of Rome. Just before his departure in 1924 he was offered the job of assistant secretary at the Technion and took it. He became Secretary, Administrative Secretary and eventually Director of the Division of Finance and Control, remaining with the Technion for 50 years until his full retirement in 1974. He died in 1978. Through many difficult years and under various administrations it was Nessyahu who was responsible not only for preparation of budgets, but often also for procuring the funds to meet those budgets. More than anyone else, too, he saw to the systematic and orderly preservation of the Technion's historical archives which later were to bear his name.

Patiently, persistently, Blok got things started. His lack of familiarity with Hebrew did not make things easy, but Hecker was of competent assistance. It was common knowledge that Blok was to stay for only one year and so even those who disagreed with some of his policies felt it was not worth forcing any issue.

He was very British, in the best sense of the word, a gentleman and a stickler for order and form, as well as for petty detail. A case in point was the set of regulations which he drew up governing conditions of appointment for instructors.

"1. Each teacher shall devote, in addition to the time actually required for his work in the class-room, such time as may be necessary for preparation of the apparatus or appliances required for use in the classes so that the classes may begin punctually at the appointed hours. After the class he will see that the apparatus, etc., is safely put away in the proper place.

"2. Each teacher shall keep proper records of students' attendances and marks and shall, if required by the Principal, furnish a report or other particulars concerning the working of any classes for which he is responsible.

"3. Each teacher shall devote such time as may be necessary for the marking of students' homework and to the setting and marking of their examination papers.

"4. A teacher if prevented by any cause from attending to his duty shall give sufficient notice to enable the Principal to find a substitute if possible in time to conduct the class or classes.

"5. Evening part-time teachers will be engaged on a monthly basis and the Principal may terminate an appointment by one month's notice if the number of students attending the class is considered by him too small to justify its continuation."

The regulations sounded just as grade-schoolish in the Hebrew into which they were translated.

There were also nationalist echoes reminiscent of the German days in Blok's letter to British firms asking for samples of various construction materials. Perhaps he only sought an approach calculated to bring the best results, when he wrote that the Board of Governors wanted the collections "to show British practice as well as that of other countries."

The actual beginning of formal instruction in the day classes was marked with only the simplest of ceremonies. The student body gathered at 8 A.M. in the presence of the teachers and Messrs. Pewsner and Rosenfeld of the Board. The speakers were Blok, Hecker, and Pewsner. A student acknowledged, and the classes began.

However, the need was felt for a more impressive, public ceremony at which the Technion could put its best foot forward, and it was decided to hold such a function on February 9, 1925.

Letters of invitation were sent far and wide. To Paul Nathan, Blok wrote that the opening should give him great satisfaction, in view of all that Nathan and his associates had done to get the project under way. James Simon was also invited.

Another invitation went to Wissotzky, who responded graciously expressing his pleasure that at long last the institution founded in memory of his father was being opened. "I have a feeling of satisfaction that ten years ago, together with my friends, we fought the ideological and national battle to preserve the sanctity of the Hebrew tongue," he wrote.

Invitation to the Opening Ceremony, February 9, 1925.

Weizmann and Schmaryahu Levin responded with polite messages of greeting. Ahad Ha-Am recalled that they had once thought the Technion had been lost forever, but fate had determined that it should return to Zionist hands. In this he found a moral for the general situation: "Be the circumstances what they may, we must not despair. What looks today as though it were completely lost, may change for the better — just as the Institute was returned to us."

The President of the American University in Beirut regretted that he could not attend. "We wish you the utmost success, and if there is any opportunity for cooperation with you, it would be most welcome to us," he wrote.

Expecting a large turnout, as indeed there was, Blok thoughtfully arranged for the preservation of order. He asked the Haifa police to assign special constables and since the majority of the guests would be Jews, he asked that the task be given to Jewish police officers.

Dignitaries seated on the platform during the ceremony included Bialik, Dizengoff, Ruppin and British and Arab notables. There were twelve speakers and 44 messages of greeting. The celebration lasted all day, and included also scientific lectures by Hecker, Tcherniavsky and Grossman. For the weary audience Prof. Grossman's lecture must have seemed the most appropriate. He spoke on "Infinity".

Ussishkin, who presided, thanked all who had helped to create the Technion and he did not neglect the Hilfsverein. "Though we opposed that body in a struggle of contending ideals," he said, "we are not ungrateful to it." Blok recalled later that as Ussishkin went on and on it became clear that the timetable which had been so carefully determined in advance was being completely upset. Blok leaned forward from his seat on the platform and unobtrusively pinched Ussishkin's leg. The latter stopped almost in mid-sentence and looked at Blok in astonishment, whereupon the proper Englishman motioned to him to call on the next speaker — which he did.

Sir Alfred Mond, chairman of the Board of Governors, used the occasion to criticize the Mandatory Government for not providing financial help to the school. Col. Symes, British Governor of the Northern District, could not resist referring to the major addresses as being either "oratory" or "appeals to the purse strings". To another observer this showed how well the Yishuv combined idealism and practicality.

"Yet one more link in the chain of Jewish achievements in Palestine" was the way the London *Jewish Chronicle* characterized the opening of the Technion at "a ceremony which certainly was the most impressive and dignified ever known in Haifa."

The program concluded with an evening banquet. The festive menu included golden clear soup, fish in mayonnaise, tongue with vegetables, asparagus in gravy, stuffed turkey, chocolate pudding with vanilla cream, fruits, liqueurs, coffee and various wines. Proudly attending this dinner were also the students, but under unusual circumstances which will be described later. Blok's thrifty budget for the day's ceremonies was £10. An accounting later showed that it cost ten times as much.

The Student Body

Applications for admission as students were received as soon as it was certain the Technion would open. There were 22 candidates already in Palestine and applications from abroad numbered 35. In Blok's opinion, at least eight of the youths would first have to pass qualifying examinations in mathematics, physics or Hebrew. As for those overseas, not only were their qualifications unknown, but it was not even certain they would be able to obtain the entry visas to get to Palestine.

The physical presence of a reasonable number of students was necessary to make a beginning, but Blok did not want to accept everybody who applied. "The Jewish thirst for knowledge made the task of selecting and rejecting among the applicants for admission an unenviable one," he wrote. "Watchmakers wanted to study electricity, tinsmiths wished to learn stonecutting, and some people with no particular feeling about subject matter clamored to attend classes of any kind so long as they could hear lectures of some sort."

The record shows that the first applicant officially registered was Shlomo Zakai. It was decided that twelve students would be the minimum necessary to start a first-year class. Applicants appeared in Haifa, made inquiries, and then disappeared. The problem was to get at least twelve students to file formal registration. Then young Nahum (Nyoma) Levin, who had arrived in Palestine only five months earlier from Russia, took the initiative. At a local restaurant in Haifa he met Zvi Frenkel, who was about to leave for France to study engineering there because of the uncertainty about the Technion. That made two; only ten more were needed.

Frenkel added the name of his brother, Avraham. The restaurant owner, Ephraim Shiloni, gave them the names of others whom he had observed coming and going. Within a week Levin was able to assemble 15 candidates most of whom signed a letter empowering him to speak for them, and undertaking to enroll as students. Back to the Technion he went, and gave Suchman the fifteen names; his own made sixteen.

Prof. Tcherniavsky set Tuesday, December 16, as the date for the entrance examinations. The students organized as a bloc and presented an ultimatum. Unless all of them were *guaranteed* admission in advance,

The British High Commissioner visits the Technion. Left to right: Sir Herbert Samuel, Mrs. Samuel, Baruch Bina, Arthur Blok. December 17, 1924.

no matter what the results of the entrance examinations, nobody would sit for the exams. Blok ruled that no entrance examinations would be given at all, unless all candidates agreed to abide by the rules and regulations.

The Principal was patient. "In judging this strike," he wrote later, "it must be remembered that these young people had grown up during the World War in the cauldrons of Russia and Central Europe where life was often a contest of wits with authority. The Technion may well have been their first contact with any discipline other than military."

The first student strike in the history of the Technion (not the last by any means) passed without any further crisis. Levin presented documents showing he had been a third-year student of engineering in Moscow and had already passed the requirements there in advanced mathematics and physics, with high grades. He therefore requested exemption from the entrance tests. Prof. Tcherniavsky decided to give him a special examination. He wrote out an integral equation on a piece of paper and passed it to the boy. Without reply Levin wrote the solution.

"You have passed the examination," said Tcherniavsky. Others did not fare as well, and some were not accepted.

There was one girl in the group, Zipporah Neufeld, who wanted to study architecture. Her problems were unique since in the planning of the building no provision had been made for the necessary elementary facilities to accommodate girls. A temporary arrangement was made to meet the needs. Her only housing in Haifa was a photographer's studio. He let her sleep there at night after he had finished his professional duties, but for some 16 hours daily she was homeless. Blok appointed her Technion's first librarian. As custodian of the key to the library room she at last had a place in which she could work and study all day long.

The next crisis had to do with the formal opening ceremony on February 9. The students noted that the evening dinner, in honor of the chairman of the Board, Sir Alfred Mond, was to be attended by administrative officers, professors and Haifa dignitaries. What, no students? Levin summoned an emergency meeting and it was unanimously agreed that if the students were not invited to the dinner, none of them would show up the next morning for the physics class which Sir Alfred was to visit officially with his entourage.

Blok and Hecker were furious. They called in Tcherniavsky and Pewsner, and the four tried to persuade the students to withdraw their ultimatum. The young people stood firm, whereupon the authorities relented and agreed to invite the leaders of the student committee, three in number. This was unacceptable. The students wanted all of them to be invited to the dinner — or no one would show up the following day.

At this table, in the restaurant operated on Herzl Street, Haifa, by Ephraim Shiloni and his wife, the first group of students organized for admission to the Technion. 1924.

Class in drafting. 1925.

At 5 o'clock in the afternoon a note came from the office: all were invited. Now it was the students' turn for consternation. None of them had any clothing suitable for the formal occasion. There was much scurrying around and they all managed to appear in dark blue suits. Zipporah Neufeld came in a blue dress. Each of them managed to exchange a few words with Sir Alfred, and at the end of the evening they formed a hora circle with the guest of honor in the middle. "Everyone had to admit we were the hit of the evening," Levin reported.

And the next morning all the students were at Prof. Tcherniavsky's physics laboratory when Sir Alfred came to visit.

Applications from overseas presented a problem. The British Mandatory Government agreed to provide certificates of admission to Palestine for students from abroad if the institution would post a bond. In consideration of the Government's permission to the student to enter the country and take up his studies, the institution undertook to be responsible for his maintenance, failing which, if he were declared destitute, the institution would forfeit its bond in the amount of £300.

Blok appealed and managed to have the bond reduced to £100, or a total of £1,000 collectively for a maximum of 20 students. Attempts were

Eliezer Strich (standing at left) directs a class in art. 1925.

made to have the Zionist Organization give its guarantees instead of actually posting bond.

A few applications were received from local Arabs, but none of them knew enough Hebrew, and they were not accepted.

Tuition fee in the first year was £20. This went up to £24 the following year, but was reduced to £18 in the third year. There were constant negotiations with the students about tuition payments, delays in payments, advances, and in many cases eventual compromise. Alexander Hassin, first chairman of the Student Association, recalled that whereas most of the students paid advances of only £3 he had innocently and naively advanced £10.

The course of instruction was set at 38 hours per week, with an additional 6 hours optional. The full curriculum for the first year comprised the following courses:

Mathematics, Physics, Chemistry, Mineralogy and Geology, Technical Mechanics, Descriptive Geometry, Drawing and Modelling, Building Construction, Building Materials, Geodesy, Building Plans, Architecture, Highways and Railways, Hydraulic and Sanitary Installations, Job Estimating and Management, Bookkeeping and Fundamentals of Business, Languages (Hebrew and English), Hygiene and First Aid, Physical Exercise.

Tentatively it was decided that the full program would take three years, but when the controversy about the Technion's level was reopened soon thereafter, the program was lengthened to four years. Pewsner complained to the Committee that the lay members were hardly qualified to pass judgment on academic matters and he suggested that a committee of senior teachers be designated to handle such affairs. That required distinction between senior teachers and assistant teachers. Academic forms were beginning to crystallize.

The students continued to assert themselves, and let no opportunity pass to challenge an authority which they felt was circumscribing their status. Blok called them pupils; they referred to themselves as students. Blok ruled that pupils were forbidden to smoke in the building. They agreed that *pupils* should not smoke, but as students they had the right to smoke if they wished, though not in class, of course.

Blok sought to have attendance taken in every class. The students insisted on their right to attend class or not as they chose, but agreed to a roll call only at exercises and laboratory sessions.

Some of Blok's reactions to student aggressiveness were naive, based on his lack of familiarity with the life of the country. On April 30, 1925, the student committee sent him a note, in Hebrew, informing him that since the majority of the students were members of the workers' union, the Histadrut, they planned to mark the workers' holiday on May 1 and stay away from classes. To this Blok appended a note in English to his staff: "Hold classes as usual and inform the signatories." Needless to say, no classes were held.

Looking back on the struggle years later, Levin admitted that the battle was cruel, at least from the students' side, but they would not yield an inch. They also had strong opinions about the quality of their teachers. They respected Barsky as a good architect and engineer. He had capped a distinguished career by supervising the building of the Bat Galim suburb of Haifa and the colony of Nahalal, but he had had no teaching experience, and the students demanded his replacement. Blok put his foot down. Students could have no voice in the appointment of teachers, he said. The students then instituted a form of sanctions which expressed itself by limiting attendance at Barsky's classes to no more than three students at a time, each time different ones. This kept up all year. After Blok's departure changes were made.

Finances were a perpetual problem for the students both individually and collectively. Many of them gave private lessons. They set up a mutual aid fund and determined to run a party to raise money for their empty treasury. The first function, held on March 7, 1925, was the first of a long tradition of Student Balls that were for many years the major event on the entertainment calendar of the country. The first party was modest. Furniture, rugs, pictures and various decorations were borrowed from neighbors and completely transformed the appearance of the Technion's lobbies and rooms. Mrs. Tcherniavsky undertook to conduct the first raffle, then an innovation, and also solicited contributions of prizes. Tickets of admission were sold throughout the country, and guests came from Tel Aviv and Jerusalem as well. The publicity was on the whole favorable, though there were some who capitalized on the occasion to observe that Palestine did not need a Technion nor engineers — only farmers and laborers.

The party was a huge success and lasted till dawn. Then everything that had been borrowed was returned to its owners. The students at last had a modest treasury.

The next big event of the year was the reception for Lord Balfour. He had come to Palestine for the formal opening of the Hebrew University on Mount Scopus on April 1, and on his tour about the country included a visit to the Technion. The audience which he addressed from the steps of the Technion was estimated at several thousand. He was mindful of the feelings of local patriotism and in his address coupled the Hebrew University and the Technion, each in its own field. For the Technion he foresaw a future of service to mankind everywhere in fulfillment of its great aspirations. Ironically, the visiting British statesman appeared to have a much higher regard for the Technion and its possibilities than many of the leaders of the Zionist movement.

Lord Balfour speaks from the steps of the Technion. 1925.

The self-defense movement, the Haganah, made one of its first appearances on the occasion as its members, dressed in blue and white, formed an honor guard for the distinguished visitor. They were prepared for any possible disturbance of the peace, and came to the conclusion that a public show of their strength would be the best deterrent. The visit passed off peacefully, in contrast to the unruly demonstrations that greeted Balfour on his subsequent visit to Damascus where the guard was ineffective and he had to flee through the back door of his hotel to avoid Arab demonstrators.

The news that the Technion was already functioning brought new students, and several more who could produce the necessary qualifications were admitted. Before the year was up there were 26 students. There would have been more but many who sent in their registration from abroad were unable to obtain entry visas. The evening classes had an enrollment of about 100.

Unexpectedly, the language of instruction again became a subject of controversy. In view of the continuing demand from Arab railroad workers

Honor guard of students with bicycles for visit by Lord Balfour. 1925.

that courses be provided for them, it was suggested that such classes, to be given in the evenings, should be in Arabic. The Committee held a lively discussion. Some members maintained that the principle in favor of Hebrew meant that no European tongue, like German, should be used, but Arabic, after all, was a local language. The vote went against it.

The local Committee was in effect the administrative mechanism of the Technion. It dealt not only with major matters like budgets and engaging of teachers, but also with relative trivia such as the cases of individual students who did not pay their tuition fees, or determination of the rates to be charged the neighbors for the water they drew from Technion's well. Deliberations on such subjects consumed precious hours.

The Hebrew University

Technion's problems were discussed by the Palestine Management Committee on April 12, 1925, in the presence of Dr. Weizmann. In the light of the apparently insoluble financial crisis Dr. Weizmann suggested that perhaps it might be best for all concerned if the Technion were simply absorbed within the new Hebrew University.

The Board of Governors in London, floundering in its own inability to do anything, grasped at the idea. The resolution they adopted seemed to solve all problems:

Whereas the Board was of the opinion they could not undertake the responsibility for the further financial maintenance of the Institute, and whereas the Board considered it was not practicable to launch a public appeal for funds for the Institute, but that it would in their opinion appear appropriate that the administration of the Institute be taken over by the Governors of the Hebrew University, the Institute being regarded as the teaching nucleus of the Faculty of Engineering, it was resolved:

a. That the administration of the Institute be handed back to the Zionist Organization.

b. That the Zionist Organization place the administration of the Institute in the hands of the Board of Governors of the Hebrew University.

c. That the proposed budget of £8,000 seemed proper, and the Board recommended its approval.

The Board further passed a vote of appreciation to Blok for his services. It denied Hecker's request that as the new incoming Director he be permitted, like the teachers, to do private consultation work.

And with that the Board of Governors resigned. It had thrown up its hands and returned the Technion to the Zionist Executive, saying in effect: "It's your baby!"

On August 6 a letter was written to the University Chancellor, Dr. Magnes, exploring the possibility of the union of the two institutions, with the University to assume responsibility for the Technion's budget.

At this point Arthur Blok solved his own problem. His eleven-month mission having ended, he returned to London, though this was not to be the end of his connection with the Technion. When a British Technion Society was organized he became one of its leaders, and when the international Board of Governors was set up some 27 years later he became a member and participated actively in almost all its meetings in Haifa practically until his death in 1974 at the age of 92.

Max Hecker assumed direction of the Technion in accordance with the agreement reached a year earlier. Back in London, Blok continued to pursue the Hebrew University angle, and was joined by Dr. Eder of the Zionist Executive in addressing another letter to Dr. Magnes in which they spelled out a specific proposal:

"1. That the Haifa Technical Institute now be made an affiliated school of the Hebrew University with a view to its ultimate development into the Engineering Faculty of the University.

"2. That the Board of Governors of the University appoint a committee to direct the educational policy and control the finances of the Institute.

"3. That the Board of Governors of the University in all appeals for financial support include the Institute as an institute which is under its direction.

"4. That the Board of Governors of the University allocate £5,000 for the budget of the Institute for the year commencing October 1st next."

Since the proposal now seemed serious, the Palestine Management Committee, at its meeting on September 22, formally supported the recommendation and called for the conduct of negotiations for the absorption of the Technion into the University.

At almost the very time that this resolution was being adopted (Sept. 23–24) the Hebrew University Administrative Committee met in Munich. The proposal was presented and discussed.

Dr. Magnes gave the decision in his letter of October 11 to the Technion. The Committee had "dealt with the future administration of the Technion, and to our regret did not find it possible to take the Technion under our administration and to provide its budget."

The same proposal was taken up again in a few short years. Hecker had to concentrate on his immediate problems, and they were many.

Hecker's Problems

The students renewed their demand for changes in the teaching staff. They respected the pioneers who had made it possible to open the school, but they wanted to make sure that the standards were high. The basic factor, they felt, was the teaching level. When Baerwald replaced Barsky as teacher of architecture they considered it a result of their activity, and resumed the battle in other directions.

It was more difficult to change the teacher of Descriptive Geometry for this was none other than Hecker, now head of the institution. The students recognized his talents as an engineer, but found him completely unsatisfactory as an instructor. A committee of the students appeared before the Director of the Institute, Mr. Hecker, and asked that the teacher of Descriptive Geometry, Mr. Hecker, be replaced. The request was refused, and again the students began their partial boycott, with attendance of no more than three students at a time. Hecker's failure as a teacher undoubtedly contributed to undermining and destroying his authority as the Director. He had to yield, and a year later Avner Badian, who came from Galicia, undertook the course.

Other new faculty members who joined, then and later, included Dr. H. Neumann, from Dresden, S. Ettingen, who had been educated in Moscow and London, Eugene (Yohanan) Ratner, from Russia, with education in architecture at Karlsruhe, M. Ladyjensky, A. Avigdor, D. Kutzinsky and others. Most of them knew little or no Hebrew and had to learn as they went along. Neumann used to make frequent use of Russian terms, thinking they were Hebrew.

Learning to use plastic materials. 1927?

But Hecker had even more serious problems than those with the students. The budget recommended by the Board at its dying meeting had not yet been approved by the Zionist Executive. The opening of the second academic year was delayed until November 1 by which time it was hoped to ascertain if the Technion would continue or not. A begrudging approval came through.

During all this period the Institute was called by various names, and the local committee decided to recommend to the Executive that it be known as the Hebrew Technion.

As the second year commenced Hecker could report that enrollment had reached 55, and 20 more were expected from overseas. Classes were under way, but the treasury was empty. The skimpy £200 received from the Keren Hayesod was gone, and despite frantic appeals to the Palestine Committee of Management in Jerusalem and the Board of Governors in London, no more seemed to be forthcoming. The Jerusalem committee responded that if there were no funds, why was the school engaging more teachers? One member proposed taking a loan from a bank — if it would be given. The Mandatory Government offered annual support in the amount of £43 on condition that it could control the Institute. And Hecker wailed that in all its publicity and propaganda the Zionist Organization ignored the Technion as if it did not exist.

The student representatives asked for a meeting with Hecker and told him that the majority of students were unable to pay. They therefore asked for full exemption from tuition fees, or very sizeable reductions. This would have resulted in a serious reduction in anticipated income. Hecker explained that remission of fees would be given only in cases of obvious need. He then instituted a new form which students were asked to sign at the time of registration, obligating themselves to pay the fees. Over the years many scholarship and loan funds were established to help needy students, but at the beginning such funds were few. The first was the contribution of some hundreds of pounds Sterling in 1926 by friends of the late Frederick Spiers in England. That fund provided income for an annual prize for many, many years until the value of the fund shrank to such an extent that it was folded into Technion's pool of student aid. From 1927 and for a period of more than twenty years a loan fund set up in the name of Judge Julian Mack, and administered at the Hebrew University, provided help to scores of Technion students.

Through the final months of 1925 the Zionist Executive was too busy to pay any attention to the Technion, and there was no individual sufficiently interested to put the Institute on the agenda. Not until the end of December, long after classes were already under way, did the Executive inform the Palestine Committee that it would assume responsibility for the school. Despite the uncertainty, the Technion staff and students continued with their work. The flooring in the long neglected building was at last completed, and the wooden furniture for the classrooms was made in the Technion workshops.

More important, the academic framework of the institution was being forged in the fire of need and experience. In July the Teachers' Council adopted the first two sets of regulations and procedures to assure orderly and systematic operations. Both documents underwent many changes in the months and years that followed.

First was a set of Regulations for Students. Chief among the ten listed items were:

1. Students must attend classes and participate in laboratory exercises regularly and punctually.
2. Students must abide by the discipline of the institution and assure that their behavior, both within and without the Technion, shall not reflect on the honor of the institution.
3. The Technion is not responsible for any personal property losses sustained by the students, nor for any injuries.
4. Promotion from one class to another will be dependent on the result of the final examinations.
5. Students may be suspended or expelled for flagrant violations of the Institute's regulations.

Regulations for Teachers consisted of seven short paragraphs, chief among them:

1. The Council of Teachers was to be an independent and autonomous body operating within the framework approved by the *Vaad Menahel.*
2. The Council would deal with internal matters of the Technion such as scientific affairs, proposals for acquisition of teaching equipment, curriculum and scheduling, assignment of the subjects among the various teachers, acceptance of new students, examinations, etc.
3. A representative of the Council would participate in meetings of the *Vaad Menahel.*

At subsequent meetings the Teachers' Council continued to add to the several provisions. In December it was reiterated that teachers had to make sure that students in the Technion used only the Hebrew language. In January it was decided to raise the entrance standards and require that candidates must be graduates of a secondary school.

In the course of time these various regulations and procedures developed into the full-scale Academic By-laws which today run to 345 paragraphs over some 51 pages.

Hecker's proposed operating budget for the academic year 1925–26 was £8,875. The Palestine Committee of Management was now composed of seven members: Ussishkin, Magnes and Col. Frederick Kisch, of Jerusalem; David Remez, Tel Aviv; Biram, Pewsner and Y. Rosenfeld, of Haifa. For a while Shlomo Kaplansky replaced Kisch as head of the "Technion Desk" for the Zionist Executive. This committee received an order from the Keren Hayesod, which was the source of the operating funds, to cut Hecker's budget down to £6,000. There was no alternative but to oblige, and plans to open a department of mechanical engineering as well as a laboratory for testing of materials, were shelved.

Even the promised monthly payments were slow in coming, and by May, 1926, a cable was dispatched to Weizmann to the effect that if the Zionist Organization were not interested it should close down the Technion entirely and give it a decent burial.

Academic Level Again

In the meantime the question of Technion's educational standards had again been projected to the fore. Again it was the students who took the initiative. They were determined that upon completion of their studies they should receive proper diplomas certifying them as engineers and architects. If this meant adding a fourth year to the program, then they demanded the fourth year. They were actuated by a sincere interest in the future of the country, and were prepared to endanger even their own personal careers for the sake of the principle. "A common vision united us, as well as absolute faith that Palestine would not forever remain static," Levin wrote later. They were convinced that the population would increase, industry would develop, and there would be need for high-level skills.

On January 20, 1926, they presented a detailed memorandum setting out the case for assuring the Technion's level as an institution of higher learning. They put their case before the Council of Teachers (*Moetzet Hamorim*) which had now been organized to deal with academic matters, as Pewsner had suggested. Members were Professors Ilioff, Badian, Baerwald, Grossman, Hecker, Neumann and Tcherniavsky. The Council seized the initiative and affirmed that the academic program should be extended to four years and the curriculum should be on university level. The matter was then brought before the Management Committee in Jerusalem on May 24, 1926. That body practically echoed the recommendations of the Council that since the Technion would have to meet the professional needs not only of Palestine but of the whole Middle East as well, as of next year it should be an institute of higher learning and the course of study should be extended to four years. The students were jubilant.

A month later an expanded Committee, serving as a Board of Governors, met in Jerusalem under the chairmanship of Ussishkin. The Board was immediately asked by what authority the Management Committee had decided to transform the Technion into an institute of higher learning without consulting the Zionist Executive. Ussishkin, who had approved of the decision, pointed out that a representative of the Executive was officially a member of the Committee, but he did not come to all meetings. He agreed to submit the matter for approval in London.

Dr. Magnes was not satisfied and charged that the Committee had knuckled under to student pressure. In his opinion the Technion should remain modest in scope. Those who spoke in terms of providing engineers for neighboring countries were premature, he said. The immediate purpose should be to meet the requirements of Palestine, and those were still small. There was plenty of time to think about helping other countries. Indeed, he added, everyone hoped that the time would come when all the institutions of Palestine would be able to help the Diaspora.

The Board dealt with the financial situation as well. Hecker presented a budget for the coming year totalling £12,900 and complained that payments due on account of the current year's budget had not been received; the till was empty. He defended his higher budget on the grounds that faced with the choice between "doing" and "not doing," he preferred that things be done. Furthermore, he reminded the members that higher technological education was expensive.

Magnes led the attack on the budget. When it was proposed to level it off at £10,000, he urged that it be reduced to £6,000. Other members pointed out that the Technion was growing from year to year; each year a new freshman class entered as other classes moved up, and the budget had to keep pace accordingly. There were still only two Departments, Building Engineering and Architecture.

Many hours were spent discussing teachers' salaries. Badian wanted more money? Let him go back to Lwow, was the unsympathetic reply. Magnes favored reducing the teaching staff if there were no funds.

Ussishkin presented the case for the Technion. Since Magnes headed an Institute for Jewish Studies he was using the wrong criteria, Ussishkin said. The Jewish studies had no fixed program. They could be expanded to the extent that money was available, or reduced without harm. But in the training of engineers every effort had to be made to assure that the product was superior. If the hours of instruction were cut, the institution would be ruined and the engineering graduates would be second rate.

Domestic income was small. They anticipated £750 from tuition fees; £100 from sale of water; £10 from the Government; £100, miscellaneous.

Recalling the relative success he had had in Germany, Hecker continued his fund-raising efforts even from Haifa. Through the years 1924–1927 he kept up a constant campaign, following up every visitor to the campus, and writing to every name suggested as a prospective contributor. For a while the work in Germany was carried on under the auspices of a Deutsches Landes-Komitee des Technischen Instituts Haifa, with Dr. Max Berlowitz providing the active leadership. Its activity was eclipsed and soon swallowed up by a parallel organization operating on behalf of the Hebrew University. Even Einstein, who maintained his nominal interest, was too busy to help. During this period Hecker's letters went out also to Canada, South Africa, Austria, Czechoslovakia, Australia, Egypt and of course the United States. The results were meager. There was hope that the Max Pamm estate of $50,000 might be obtained for the Technion, with the help of Col. Kisch, but it never materialized. News that L. Moisseiff had undertaken to serve as Chairman of a Committee for Technical and Scientific Studies at the Hebrew University and the Technion elicited a prompt reaction from Hecker. The contact was later followed up, but without success. In April, 1927, A. J. Freiman reported from Ottawa that Weizmann, during his visit there, had obtained contributions of $2,500 for the Technion.

An indication of what could be expected was provided by Blok, who reported to Hecker from London in mid-1926 regarding the Society of the Friends of the Hebrew University, which had been asked to include the Technion also as a beneficiary. Wrote Blok: "The definite intention of the

Society is more to stimulate intellectual cooperation with the University and the Technion rather than to raise money . . . ."

The question of budget had to be put before the Zionist Executive which was to meet in London. The memorandum Hecker had drawn up was said to be much too long; nobody would read it. In condensed form it was approved and was to be printed for presentation to the Executive. Ussishkin promised to speak up there, as he had in Jerusalem.

What happened in London? On August 3 Ussishkin reported briefly to Hecker that the meeting was over. Nobody had diplayed any interest in the Technion except for Eder, and even before Ussishkin had arrived, the Executive had already decided to approve the £6,000 budget.

The bad news from London arrived just as Hecker was faced with a new student revolt. They had heard of the discussions in Jerusalem and feared that the four-year policy would be reversed. They were now approaching the end of their third year, and wanted to know about their future. Thirty new students had registered as freshmen, all in anticipation that they would have university status. The students brought up every argument. And to those who maintained that the Technion should not try to do more than meet the immediate modest needs of the country they replied that these people lacked vision. The students preferred the Zionist point of view, that the country would develop and grow and would be a center to which Jews would come from all over the world. Parenthetically it was noted that most of the Technion students were immigrants; Palestinian Jews for the most part preferred to go abroad to study.

One of the factors contributing to the student effervescence was the fact that the student body was divided between those who came from Poland and those who came from Russia. Some of the latter were older, had been in the country longer and were being supported by their wives while they studied. There was a competitive clannishness in the two groups, carried over from the Diaspora, which sometimes led to overt antagonism.

"Nobody in authority displayed an interest" was in effect the key to the general situation as Hecker realized. His memorandum was published. After reviewing the program, budget and operations he stated the case in almost plaintive terms:

"What the Institute requires now is not only financial support and a more effective organization — it wants moral encouragement as well. While the (Hebrew) University has been for years considered the center of all cultural problems of Palestine and has enjoyed all kinds of official and non-official help, the Hebrew Technical Institute has remained in the shadow. There is no person of authority to support it; the official organs of the Zionist Organization hardly ever mention it; and it is small wonder that the activities of the Institute are nearly unknown even in Zionist circles. Such indifference precludes the creation of a warm atmosphere of active sympathy without which no growing institution can keep on progressing."

Matters finally reached a head at the meeting of the Zionist Actions Committee in London on May 12, 1927. It was a crucial meeting, and Ussishkin was again relied upon to present the case for a Technion of university level. The fate of the institution hung in the balance. Technion was put at the tail end of a long agenda and finally, just before the onset of the Sabbath, Ussishkin was given an opportunity to state the Technion case. As he later reported to Pewsner, only fifteen minutes remained before the meeting had to be adjourned. When he finished Weizmann spoke up and minced no words. He lauded the Technion with faint praise, but opposed the proposal that it be conducted on a university level. Its purpose, he said, was to prepare technically trained personnel, but not university graduates. To seek to have a technical university would be nothing less than a "bluff," he maintained. The necessary facilities did not exist. There were no funds for the purpose, and an exaggerated claim at that time would merely serve to ruin prospects for what the Technion could best do. He scoffed at the attempt to make comparisons with the Hebrew University. And since the Actions Committee was not equipped to deal with scientific matters, he asked that the question be removed from the agenda altogether.

By way of compromise it was decided to appoint a committee which would study the matter and consult with Weizmann on his next visit to Palestine. Thereafter a report would be prepared for the next meeting of the Actions Committee. Weizmann agreed but only on condition that the administrative authorities of the Technion make no changes in the status of the school in the meantime.

Nevertheless, before the month was up the office in Haifa sent inquiries to some 40 of the leading technological institutions in the world, requesting their catalogues, constitutions and other material to help the Technion develop its own regulations and curriculum. American institutions listed included the University of California, Stanford, Yale, Alabama Polytech and Throop Institute in Pasadena. The lists in France and England were more comprehensive.

The controversy became heated and erupted into the press, both in Palestine and overseas. Pinhas Rutenberg, approached for advice, minced no words. He gave three reasons why he considered conversion of the Technion to a school of university rank not only inadvisable, but even harmful to the interest of Palestine:

1. Technion would in no circumstances be in a position in the near

Max Hecker (1879–1964).

future to graduate competent engineers to meet the demands of growing industry.

2. Palestine industry would not be able to absorb the large number of experienced and really competent engineers and architects then arriving from abroad.

3. Such an institution would become a heavy financial burden on both the Zionist Organization and the local community, and this would prejudice the possibility of ultimately transforming it into a university when the time would be ripe.

The Russian Zionist paper, *Razsvet*, featured a striking article entitled "Stubbornness", charging that Weizmann and others opposed the Technion only to protect the Hebrew University, and out of envy. They therefore sought to throttle the Technion. To the writer, the Technion was closer to a full-scale university than the "parody" in Jerusalem.

Naturally the students in Haifa did not remain silent. They adopted indignant resolutions in which they bitterly attacked the Zionist Actions Committee. Fearful lest such unbridled attacks further antagonize the Zionist bodies, from which the funds came, the Technion administration asked the local press not to publish the student statements, since they could only lead to misunderstandings.

Hecker Departs

Hecker was in the middle. He did not see the need for immediate university status, and so had no support from the teachers or students. Neither did he go as far as Weizmann and the Zionist leadership, who therefore distrusted him. In Haifa he was the embodiment of the short-sighted Zionists who had appointed him Director. When there was no money for salary increases, or even to pay current salaries, he was held responsible, and his relations with his academic colleagues deteriorated. The two-year term which he had been promised was almost up and at its meeting on July 24, 1927, the Palestine Management Committee decided to dismiss him.

He noted that the action was not a surprise to him, but asked that the reasons be given. He was told, briefly, due to lack of normal relations between him and the teachers and students. He was asked to continue as a member of the teaching staff, but he refused.

He drew up an accounting of the Technion under his stewardship, and then lashed out at all those whom he held responsible for the situation. He directed his first fury against the Zionist leadership in London which had failed to clarify the status of the Technion, keeping the whole project in mid-air. Weizmann kept sniping at the Technion, he charged. If there was chaos in Haifa, it was because of this attitude on the part of those who should have been responsible. Hecker said that he, as Director, was the only one in Haifa who sought to some extent to defend the actions of the Zionist Executive, yet that very body turned around and made him the scapegoat for the controversy, blaming him for not getting along with the staff and students. When he had asked for a chance to present his case personally before the Executive, it had been denied him. The withholding of funds was another major factor contributing to the disturbed situation.

He put his finger also on the old problem of absentee management. The Zionist Executive in London was completely detached from the realities of Palestine, but still sought to exercise authority. The Jerusalem Committee was not active; many of its members seldom came to meetings and they could not even agree on a chairman. The students became demoralized, and even those who could afford to do so got out of the habit of paying tuition fees.

In his bitterness Hecker castigated his associates. He accused Pewsner of misappropriating funds. He maintained that Biram was seeking to exploit the Technion for the good of his Reali School. He charged that Ussishkin was interested only in the grand external appearances. Magnes, he said, was constantly negative because he saw in the Technion a potential competitor of the University. The Council of Teachers constantly interfered in the day-to-day administration. The students were basically good material, but were incited against him, Hecker felt, because of his stand on the higher learning issue. They adopted impertinent resolutions. Discipline was absolutely lacking.

His last words to the Technion were written in a private letter in which he asked if payment could be accelerated on his salary, now already three months in arrears.

Following submission of his 12-page lamentation entitled "The Fate of the Technion", Hecker withdrew completely from Technion affairs. In 1944, under a new administration, he accepted membership on the Board of Governors, and remained a member almost until his death in 1964.

Controversies and confrontations marked the Hecker administration but they served to obscure progress and positive developments. Despite the many problems, constructive things were done.

The evening courses continued to flourish. They catered to some 150 pupils in an age range between 16 and 51 years. There were courses in carpentry, mechanical and electrical trades, fitters' work, telegraphy and telephony, and for building foremen. From the point of view of numbers served, the evening program at times seemed more important than the day courses. Most of the evening pupils were older, working people, *halutzim* among them, some of whom had been in the country for a relatively long time. Many of them knew Hebrew better than the day students, and better than most of the teachers. There were also some ten Arabs enrolled, mostly in a carpentry class.

As the Hadar Hacarmel neighborhood grew, the Technion became its natural cultural center. In a more orderly manner than in the past, it made its facilities available for art exhibits, public lectures, receptions for distinguished visitors, and meetings of scientific and cultural bodies. Its large hall had seating space for 300, the largest in Haifa. A proposal to make this hall available as a commercial cinema was turned down.

As the dominant institution of the Jewish community it was also subjected to bizarre pressures. The local dairy interests asked permission to use its recessed cellars for storage of their cheeses, which required cool surroundings. When musicians who had rented the hall infringed on musical copyrights without paying composers' fees, it was the Technion, as host, which was sued. A group of teachers visiting from Egypt was lodged for some time in Technion's spacious quarters. A synagogue was permitted to become a permanent tenant.

There may have been a degree of pretentiousness in the claim in the Technion's 1926 promotional brochure that its "museum, library and reading room are open to the public," but the library did claim a stock of 2,500 books and 50 periodicals.

Surprisingly, contributions trickled in, despite the unfavorable publicity which the Institute was constantly receiving. In 1927 word came that the Zionist Organization of Lithuania had allocated a scholarship of £20 for a worthy student, and the teachers selected Alexander Hassin, who had earned the highest grades in his class. When the money finally came through, via Keren Hayesod channels, it turned out to be only £10.

Early in the same year Mrs. Felix Warburg, of New York, had visited the kitchen which the students had set up in a basement room. Wives of the married students were employed to do the cooking, and the students took turns on a voluntary basis to wait on tables, so that hot, kosher meals could be available. The facilities were very modest. Hecker wrote to Mrs. Warburg, and she responded with a check for $500, the first of several contributions she was to give. An importuning letter to Nathan Strauss in New York brought only a printed rejection, obviously prepared for the many such requests which the philanthropist received.

A committee on Hebrew technical terminology, which Hecker had first set up in November, 1924, had completed lists of all the major words and terms used in construction and highway building and the list was submitted to the *Vaad Halashon*, the National Language Committee. The committee on technical terminology was later taken over and administered with diligence and ability by Saadia Goldberg.

Electric power was provided by the Palestine Electric Corporation, but the Technion had its own emergency generator. It had a small gas plant to supply gas fuel for the laboratories.

The number of girl students had increased. Rahel Znamirowska, a student at the Warsaw Polytechnic Institute, had come to Palestine as a tourist in 1925 with a large group of students to be present at the opening of the Hebrew University on April 1 of that year. What she saw of the country fascinated her, and she decided to remain, despite the objections of her parents. She studied Hebrew and was accepted at the Technion. Years later, as Rahel Shalon, she became the first Technion graduate to reach the rank of full professor and the only woman to serve as a Technion Vice President.

In the following year two other girls enrolled, and Mrs. Myra Jacobovitz was engaged to give gymnasium lessons to the four girls. The program was soon discontinued for lack of funds.

Yaakov Pat, later to become head of the Haganah in Haifa, was put in charge of guard duties. His tasks included ensuring that the campus gates were closed at 10 in the evening and opened again at 6 A.M. He also had to

Plaques in tribute to Jacob Schiff and Kalonymous Zeev Wissotzky, unveiled in the lobby of the old building in 1926.

make sure all lights were out, and to engage only trustworthy guards. His own duty hours were from 6 to 6.

The second year ended with an enrollment of 67 students. Of the 45 who were in the freshman class only two were graduates of a Palestine high school. Allowing for drop-outs, in its third year the Technion had 86 students. This broke down into 19 juniors, 32 sophomores and 35 freshmen.

The teaching staff was now composed of ten senior members and 14 others in junior ranks. They were a motley group, and from diverse backgrounds. Some of them were not the best of pedagogues, but almost all had a great redeeming feature. They were devoted to the students and worked hard. Their working hours embraced morning, noon and night. There was little or no research; the teachers concentrated on teaching.

Experiments with the curriculum continued. In addition to English, Arabic was offered as an elective subject, but it was discontinued for lack of student interest.

The pledge to honor Wissotzky's memory was finally kept. At an impressive ceremony on February 3, 1926, two memorial plaques to Wissotzky and Schiff were dedicated in the main lobby of the building. Ussishkin, as chairman of the Palestine Committee, spoke on Wissotzky, and Reuben Brainin spoke on Schiff. The two metal tablets, both in Hebrew, were designed by Strich.

The text of the Wissotzky marker was written by Ahad Ha-Am: "In memory of the philanthropist, Kalonymous Zeev Wissotzky, of Moscow, whose heirs drew upon his 1905 bequest left for charitable purposes and gave 100,000 rubles for the founding of this technical school, thereby enabling the beginning of construction; and in 1913 the heirs again gave 100,000 rubles from the estate for the continuance of construction." To this day it is the only plaque anywhere in the Technion which specifies the amount of the donor's contribution.

The other plaque was written by Schmaryahu Levin: "In memory of the noted philanthropist, Jacob Schiff, of New York, whose broad and generous participation in the construction of the Technion made it possible to establish the Institute on a firm and enduring basis. May his memory be blessed from generation unto generation."

Later that year Hecker invited Levin to lecture to the students and the teachers on the eventful years in the early history of the Technion.

In the Spring of 1927 the land on which the Technion stood was finally and formally recorded in the Land Registry in the name of the Jewish National Fund. Despite all efforts to throttle it, the Technion lived.

*   *   *

Stormy Years

CHAPTER FIVE

"Temporary" Dismissals

A replacement for Max Hecker was not easy to find. One suggestion was Prof. Baerwald, on the grounds that he was above all the politics and the intrigues. Another suggestion was to unite the Technion with the Reali School under the direction of Dr. Biram. It was finally decided to name joint directors, one representing the teachers, and the other the *Vaad Menahel*, the administrative council. The two were Prof. Tcherniavsky and Shmuel Pewsner, Chairman of the Teachers' Council, and the Treasurer of the Technion, respectively. The irony of the situation was that both were advocates of maximal standards, a fact which did not ensure harmonious relations with the minimalists who dominated the Zionist Executive. The choice of Tcherniavsky was especially pleasing to the students, and on many occasions he served as a calming influence on the more tempestuous spirits.

The new administration inherited a series of complex problems including relations with the Reali School, teachers' status and salaries, clashes with the Engineers' Association and student discipline. The immediate problem, however, was budget. The 15th Zionist Congress was to take place in Basle at the beginning of September, 1927, and that was the body which would authorize the Technion's budget for the academic year 1927–28. It was clear that little could be expected from Basle, both because the Technion had no influential friends at court, and because the Zionist Organization itself was going through a critical period due to inadequate funds to carry on its normal activities.

Tcherniavsky and Pewsner brought economy suggestions to the *Vaad Menahel* but it was difficult to decide whether to dismiss some teachers and ask the others to carry extra hours, or to have everyone work overtime without pay. Would the teachers agree to either alternative, and how much money would the Zionist Executive recommend? Both Tcherniavsky and Pewsner felt that they were only stopgap directors, and that a permanent administrator would be named shortly. As a matter of fact they both served through two stormy years.

The replies to their questions were not slow in coming. The office of the Zionist Executive in Jerusalem wrote a formal note on September 26, 1927, that instructions had been received from the Zionist headquarters in London requiring that all teachers in the Institute be informed immediately of their dismissals. Such notices would be "temporary"; after the Technion budget for the coming year had been determined it would be possible to re-employ some of the teachers within the funds available.

The Technion Committee replied at once. Technically, they said, the letter had been received only on September 30, one day before the beginning of the new fiscal year, and after agreement had already been reached with most of the teachers for the ensuing year. Furthermore, the Palestine Management Committee had at its March 9 meeting informed most of the teachers that they had tenure, and the mass dismissal now proposed was therefore impossible. Many of the teachers were eminent men of science who could not be hired or fired so lightly. They would not remain at the Technion under such conditions.

With respect to the budget, the committee went on, any further reduction would be disastrous. The Zionist budgeters should understand that because of the unique nature of the institution the budget could not be cut automatically down the line.

Classes continued, but salaries were not paid until a budget had been approved.

In January, 1928, a joint meeting was held with the participation of representatives of the Zionist Executive, the Palestine Management Committee, and the Technion administration. The Technion was accused of having run up a deficit, but Pewsner denied there was a deficit. The real problem was that the Keren Hayesod had not made the payments which had been promised. The atmosphere was cool, and Tcherniavsky again put his finger on the fundamental reason when he complained that the Technion was a baby without parents. As the time neared for the first graduation, he wanted to know who would sign the diplomas. The Zionist Executive?

Reference was made to an article by Hecker in the press, alleging student unrest, but Tcherniavsky insisted the contrary was true. There was now discipline and good relations between the students and the administration.

The same, unfortunately, could not be said about relations between the academic staff and the administration. The delays in salary payments and the constant threat of closure hanging over the institution had heightened the state of tension in which personality clashes between talented and temperamental men took place. Inequalities in employment conditions further contributed to the ferment.

When someone proposed reducing the hours of teaching in order to save money, one of the professors suggested that the funds so saved be added to the salaries of the senior teachers. Professor Badian raged when his salary was delayed for months, and did not hesitate to air his complaints even in front of the students. In December, 1927, his attorney wrote to the Zionist Executive threatening court suit unless back salary was paid by the 20th of that month. His colleagues, who felt the pinch no less than he, urged him to consider the welfare of the institution as well. In the meantime Prof. Grossman was embittered when he discovered that he was receiving only £27 a month whereas Badian was getting £40.

A corner of the main lobby in the old building.

The teachers, feeling that their interests were not being properly looked after by the *Vaad Menahel*, requested that the Teachers' Council be given a voice in the determination of financial matters. The administration ruled that the Council could deal only with academic and pedagogic affairs, but could not intervene in matters purely administrative. The Council could receive information, but that was all. The teachers complained that even information had been withheld; when the negotiations were taking place with the Hebrew University they had not been informed officially, but of course there were no secrets.

If the Teachers' Council had no authority to discuss financial matters officially, there was no reason why an unofficial social gathering of the teachers could not be free to talk about anything they wanted, and such a private meeting was accordingly held at Baerwald's home. The financial plight of the Technion was analyzed and the teachers reached the obvious conclusion that the only difficulty was lack of money. It was proposed to send Grossman to the United States to seek funds. Perhaps he could extract contributions from rich people there — Julius Rosenwald in Chicago, and Henry Ford in Detroit were naively mentioned. It was also decided that letters should be written to the Zionist Executive in Jerusalem demanding that Technion's financial affairs be put on a more secure basis.

Professor Tcherniavsky, who was present in his capacity also as a teacher, protested that this was anarchy, and that only the directorate had the authority to manage Technion's affairs. He was overruled. What had begun at the private party soon spilled over into the official meetings, and more and more the sessions of the Teachers' Council were devoted to wrangles over budget, disparity of teachers' salaries, and administrative matters, all at the expense of discussions on academic affairs.

Revolt of the Teachers

Affairs reached a head in June, 1928, and the Teachers' Council voted lack of confidence in the administration. They demanded that the managing director be chosen by the teachers themselves, and they nominated Baerwald and Neumann.

Tcherniavsky and Pewsner were prepared to leave their posts since life had become most unpleasant for them, but the Zionist Executive did not consent. On October 1 the Council sent a delegation of teachers to Jerusalem to repeat their insistence that the two directors be relieved of their positions. They realized it was not their official duty to name the administration, they said, but they were concerned for the welfare of the institution. The Zionist Executive refused to heed, and formally reappointed

Tcherniavsky and Pewsner for a second year. They sought to decline but the Organization imposed Zionist discipline on them to accept.

Obviously this did not lead to any improvement of relations within the Technion. The spearhead of opposition was composed of Badian, Grossman, Neumann and Baerwald. They decided on a policy of non-cooperation with the directorate in an attempt to bring it down. Not all teachers agreed with this policy but had no forum in which to voice their views since Baerwald, now serving as chairman of the Council, refused to call meetings. There was danger that the academic year 1928–29 would not open for lack of a properly prepared academic program. Critics of the rebels referred to the campaign as "terror and sabotage".

Jerusalem sent Henrietta Szold and Dr. Joseph Lurie to Haifa to reason with the teachers, but they could not get the Council to function properly. Thereupon the Zionist Executive, in its capacity as the "owner" of the Technion, revoked the authority of the Teachers' Council and placed control of academic affairs in the hands of the Technion administrative authorities. In October the *Vaad Menahel* was also broadened to give it more influence and prestige. The new membership was composed of: representatives of the Zionist Executive, Dr. Lurie and Miss Szold; Technion management, Tcherniavsky and Pewsner; a representative of the teachers, to which post Prof. Breuer was eventually chosen; M. Aleinikoff, director of a bank in Tel Aviv, and a former Zionist leader in Russia; and Shmuel Itzkovich, chairman of the Haifa branch of the Engineers' and Architects' Association. Dr. Lurie served as chairman. For some time this committee functioned well, and the two Jerusalem representatives came frequently to attend the meetings. As a result of the leadership and authority which the group was able to exert, the Palestine Management Committee in Jerusalem, which had in any event not played a role in the daily functioning of the institution, receded even more into the background.

It appeared that only three of the teachers, Baerwald, Badian and Neumann, had written contracts, and these were due to expire on September 30, 1928. They were informed by the Zionist Executive that the contracts would not be renewed, but they could continue working on an open basis thereafter, like all the other teachers. Of the three only Badian refused to agree. He continued to press for the special arrangements which had originally been made with him and carried his fight into the public press. After lengthy deliberations the Zionist Executive decided to terminate his employment completely as of December 1. This immediately brought matters to a crisis. The final examinations were due to be held at this time and he was relieved of any role in these examinations. Professors Grossman, Neumann and Baerwald protested that their colleague had been dismissed in mid-year without notice and without provocation. They served notice that unless Badian was permitted to take part in the examinations they would refuse to participate themselves. As of December 1 they walked out on strike. Badian's lawyer sent a formal letter threatening that unless his client's salary was paid in full he would bring suit against the Technion. There was indignant reaction in the *Vaad Menahel*. The country was at the time in the grip of an economic depression, and members pointed out that during that very month some 500 teachers and civil servants had not received their salaries, yet no one had threatened court action. The unofficial reaction to Badian was that if he could not stand the conditions in the country he could leave, and no one would throw stones at him. The official response was a request to him to withdraw the threat.

A truce was signed with appointment of a neutral investigating committee composed of the Attorney-General, Norman Bentwich, Attorney J. Klebanoff and Y. Rosenfeld, Chief Engineer of PICA. The four professors testified before the committee, and on that occasion made serious accusations against Pewsner which the latter said were completely "fabricated".

The committee submitted its report within a month. They found that Badian's dismissal had been perfectly legal. They also noted that Badian had asked not only for severance pay but also for expenses for his wife and himself for a return trip to Poland, together with all their belongings. The conclusion: "Both parties to the dispute are filled with good intentions and have the interests of the institution at heart. Unfortunately there is a lack of confidence on both sides, and a readiness to suspect and to put the worst interpretation on anything which is done." Since the teachers were ready to resume work, the committee recommended that the directorate take "an ample dose of forgetfulness . . . ."

The Zionist Executive decided to prolong Badian's services at least until the end of the school year. The immediate crisis passed, but l'Affaire Badian was to keep the Technion in the headlines for some time. Associates have characterized Badian as a highly talented but egocentric engineer. He had held important positions in Lwow, and expected to be treated with respect in Haifa. He stood on his dignity. When the students brought him a pair of tickets to the annual student ball he refused to accept seats in the second row. In vain the students pointed out that Row One was reserved for governmental and Zionist Organization dignitaries. A rupture was avoided when the students renumbered the rows. The VIP's sat in Row Zero and the senior academic staff, including Badian, occupied Row One, immediately behind.

Detailed reports of the scandal which appeared in the press both in Palestine and abroad seriously affected the prestige of the Technion. The students were of course well aware of what was going on and feared for the integrity of their studies. As a result, they too were drawn into the controversy. Gradually peace returned to the campus, at least on the surface. New by-laws were adopted for the Teachers' Council, broadening its composition and more clearly defining its functions and operations.

From time to time there were further eruptions. A group of the teachers went to the press with new complaints. Members of the Zionist Executive expressed themselves in terms critical of the directorate, but the latter was forbidden to make any reply or issue any statement lest this only aggravate the controversy. The Technion was constantly in the news, but the publicity was certainly not helpful.

The struggling young institution was in the meantime again called upon to defend its university status. At its eighth annual convention in 1927 the Engineers' Association repeated its opinion that the country required only technical workers and foremen, and there was therefore no need for an institution of higher learning. The same arguments that had been advanced previously were repeated, both in convention resolution and in the press. At the same time the Association requested that it be afforded official representation on the Technion Board.

The Institute drew up a formal statement setting forth the views of the directorate. The arguments were familiar, but perhaps there was a new emphasis. The point was made that on the basis of the immediate needs of the country there appeared to be no present need for a technical university, or even a technical secondary school. Such needs could be met by bringing skilled personnel from overseas. From a broader perspective the following additional points had to be taken into consideration:

1. Dozens of graduates of Palestine high schools left the country annually to study abroad. This was a loss to the country.
2. The country must absorb the new immigrants from Russia and Poland whose only opportunity for education was in Palestine.
3. Should there be a surplus of engineers at a given moment, the situation could quickly change, because many of them often moved on to other occupations.
4. The country required a technological center with testing laboratories and a research station. Only a university could provide the reservoir of skilled scientific and technological manpower necessary to man such a center.
5. Prestige was an element. The neighboring Arab countries would undoubtedly build universities, and they could afford to bring in expert teachers from abroad who would teach in foreign languages.
6. The standards of the Technion, thanks to the insistence of the students, were already very high and close to that of university level.

7. If financial resources were required, they would have to be raised, just as in the case of the Hebrew University in Jerusalem and other institutions. The important thing was to reach conclusive determination with respect to Technion's future.

This was by no means the last word. The big question mark continued to cast its shadow over the Technion, and the issue was raised anew on the eve of each of the biennial Zionist Congresses, the supreme authority in the Zionist movement.

Under conditions of internal disharmony, lack of clarity with respect to its status and ever-decreasing allocations despite growth in the number of students, it is surprising that the Technion managed to survive, much less make some progress.

First Graduating Class

The academic year 1927/28 opened with a new freshman class which now gave the Institute a full complement of four years. Obviously this necessitated more teachers and more equipment, but the net budget was reduced rather than increased. The problem was met only because the teaching staff agreed to accept overtime hours of instruction without extra pay, and in some courses, one teacher directed exercises in two classes simultaneously.

The burden was great upon the students as well. Part of their difficulties they were creating for themselves, knowingly. In their insistence on higher standards, leading to academic degrees, they were making the curriculum more difficult and raising the demands made upon them. Under the circumstances it became a matter of honor for them to study hard and to show that they were capable of carrying the academic burden they had themselves insisted upon.

Daily personal sustenance was a problem for some students. During the summer vacation the students sought work to earn money, usually in construction jobs. In 1927 a member of the teaching staff, Ariel Avigdor, went to Egypt and called upon a large number of private and public offices, soliciting summer employment for Technion's students. Hecker dispatched tens of letters, all with excellent results. The Director of Public Works in Cairo, the Chief Engineer of Alexandria and many others were happy to have the services (for a very small salary) of the talented and eager young students from Haifa. The program got off to a bad start when the first batch of students was unable to leave for the jobs promised them because, having been born in Russia, they were at the last moment not given entry visas. Thereafter things went more smoothly, and the young people gained both experience and money on their summer jobs. Some of

First graduating class of engineers. Left to right (seated): Leib Uretzky, Eliezer Rinkoff, Nahum Levin, Abba Kramer, David Yitzhaki. (Standing): Yaakov Alperovitz, Zvi Vizansky, Alexander Hassin, Shlomo Zakai. (Missing, Emanuel Friedman.) 1929.

them were employed on the first Aswan Dam. Similar opportunities were sought in Syria and Cyprus.

The first graduating class took its written and oral diploma examinations in Hebrew at the end of November and early December of 1928. There were no precedents for the administration of the tests, but every effort was made to conduct them on an unquestionably high level. The examination committee was composed of representatives of the teachers, the Zionist Executive, the Engineers' and Architects' Association and the Public Works Department of the Mandatory Government. The inclusion of the latter was considered a great achievement since it assured official recognition of the engineering diploma.

The oral examinations were conducted with great formality to impress on the students the seriousness of the occasion. Hence they were required to wear jackets when they appeared before the examining board. Not every student had a jacket, and it was common procedure when a student left the examination chamber, for him to doff his jacket and pass it on to the next in line.

The seventeen successful graduates in Technion's first class were: Engineers, Yaakov Alperovitz, Emanuel Friedman, Alexander Hassin, Abba Kramer, Nahum Levin, Leib Uretzky, Eliezer Rinkoff, Zvi Vizansky, David Yitzhaki and Shlomo Zakai; Architects, Zvi Gregory Frenkel, Zeev Wolf Gasko, Shlomo Ginsburg, Yehuda Leshtziner, Zipporah Neufeld, Shlomo Spektor and Moshe Tiegerman. Shimon Wasserman, a member of the first class, completed his diploma requirements at a later date.

Conferment of diplomas took place on February 27, 1929. Most of the addresses were in Hebrew, but representatives of the Mandatory Govern-

First graduating class of architects. Left to right: Zvi Fraenkel, Shlomo Ginsburg, Moshe Tiegerman, Prof. Alexander Baerwald, Zipporah Neufeld, Zeev Gasko, Yehuda Leshtziner, Shlomo Spector. 1929.

ment spoke in English and the Mayor of Haifa spoke in Arabic. Aviezer Yellin, on behalf of the Education Department of the Government, told the graduates that they were following in a great tradition of the engineers of the past: Noah had designed and built the Ark, and Solomon had built the Temple in Jerusalem.

There was immediate work for the graduates, though perhaps not in the lofty fields cited by Mr. Yellin. For example, in June the Municipality of Haifa announced a competition open to the seven architects who had just graduated, to design the municipal stables. First prize was to be £6, and carried with it the right of employment to draw the final plans and supervise the construction. Second prize was to be £4.

The years of struggle had been worthwhile. Students and faculty alike experienced a triumphant sense of achievement. Forgotten for the moment were petty squabbles, the unpaid salaries and the ever-present threat to Technion's continued existence which still hovered over the Institute as a result of the unwillingness or inability of the Zionist authorities to give the institution unqualified backing. To them the Technion appeared to be a nuisance, and since it was undeniably a necessary nuisance, constant efforts were made to find an easy solution. Thus the idea was once again raised of folding the Institute into the Hebrew University. In June, 1928, the executive body of the University (*Vaad Menahalim*) took cognizance of the pressures upon it and formally decided "to explore the relations between the University and the Technion" and to submit a full report.

Five Engineers a Year

From London Dr. Eder sent a fresh memorandum to Dr. Magnes in which he noted that the Haifa committee and the teachers had been defying the policies determined by the Board of Governors and thereafter by the Zionist Executive, and had proceeded to implement an academic program of university level. Wrote Eder: "An English University professor, to whom the syllabus of the mathematical course at the Haifa Technical Institute was recently submitted, observed that with his most advanced 'Honours' students he would not pursue such studies."

Eder addressed himself also to the financial problems and noted the difficulties of seeking funds. "The inauguration of an open campaign on behalf of the Technical Institute is everywhere vetoed by the local Keren Hayesod organizations, while the sympathies of individual donors, of whom a special interest in higher education in Palestine might be expected, have been monopolized exclusively by the Hebrew University ...." He repeated the suggestion he had advanced a few years earlier, together with Blok, that the University assume administrative and financial control of the Institute, but without giving it university rank. Indeed, such control would "tend to prevent the premature growth of the Haifa Technical Institute into a college of university standing."

His final argument pointed to a valuable public relations benefit for the University: "It should not be impossible for the Hebrew University, with its budget of over £42,000 per annum, to find another £3,000 or £4,000 for an institution which, while not pursuing University aims, is yet doing important educational work in Palestine, the immediate importance of which for the younger generation of the country can in no way be gainsaid. It may even be that the latter consideration may prove a not invaluable element in the propaganda on behalf of the University, which is frequently exposed to the criticism that it has no relationship to the actual requirements of Palestine."

The authorities in Jerusalem had in the meantime commissioned a special study. The result was a 20-page memorandum which reflected the views that were dominant in Zionist and educational circles at the time. The author added several original observations of his own. The document, entitled "Notes on Technical Education in Palestine and on the Future of the Haifa Technical School", was written by Louis Green, a civil engineer who had served with great distinction in the Colonial Service. He had been chief engineer of the South Indian Railways and had pioneered in the use of pre-stressed concrete for railway ties. Following retirement he settled in Jerusalem and was frequently consulted by the Hebrew University on engineering matters.

Green handled his assignment systematically. "The Hebrew University has been approached with a suggestion to take over the Haifa Technical School and make itself responsible for the future organization and management of that institution," he wrote. He noted at once that conditions of the transfer had not been clarified, and the whole proposal was vague and general. At the very outset he dealt with the question of Technion's scope, and the desire on the part of the Institute to train highly qualified engineers and architects. "On the other hand," he went on, "Dr. Weizmann and his colleagues are said to favor the view that Palestine already possesses a superfluity of engineers and architects and needs only trained foremen . . . ."

The Chancellor of the University had proposed that the matter be studied by an enquiry committee. Green was not a "committee man". "Committees in Palestine are not always very helpful in deciding conflicting issues," he said, and thereupon took upon himself the examination of the whole question.

There followed a full review of the character of skilled and unskilled labor in the country, an analysis of the national needs, a survey of available statistics with respect to the number of engineers then practicing in Palestine and even a frank review of their professional shortcomings. His conclusions follow.

With respect to training, he decided that it was impossible to combine both university level and lower level technical training in one institution. The manpower requirements? Considering the expected annual turnover, he estimated that there would be need for no more than four or five new engineers a year. There was no possibility of a sudden outburst of industrial activity, hence no provision had to be made for that. Therefore, until such time as technical education could be completely reorganized, he proposed that a program of scholarships be set up to enable four or five students to go to Cambridge every year, to remain abroad for practical training in England or even the U.S., and then return to Palestine to meet the local needs. This program could be followed for the ensuing five or six years.

From a long-range point of view he recommended that the Haifa Technical School should be reduced to the level of a training school for foremen. The need for engineers in the distant future could be met by careful organization of a new engineering school in Jerusalem as part of the Hebrew University. There was no need to hurry.

One of his novel ideas was this: "Candidates (for admission) should be unmarried, and a student should be considered ineligible for a scholarship if he has entered a state of matrimony before he has completed his practical training."

After a review of these recommendations the Hebrew University decided to withdraw. Another solution had to be found, and the question of the Technion's scope was again subjected to public debate.

In May, 1929, the Zionist Executive sent out a new questionnaire to a list of distinguished personalities, asking their opinions. The Executive attached a memorandum drawn up by Henrietta Szold, which reviewed the history of the controversy. It noted that the original London Board had been for a low level. The Palestine Committee and the Technion students and staff had favored university level. The Engineers' Association had been strongly against a high level, but it appeared there was beginning to be a split in its ranks on this issue.

The questions asked were pointed and direct:

1. Do you support the idea that there is need for a technical institute of higher learning?
2. If so, do you think the Technion should continue to develop as it is in that direction, or should it become a faculty of the Hebrew University?
3. What is your opinion about training of craftsmen and foremen?

As in previous opinion polls on the subject, the replies ran the gamut of views. Most outspoken was Ussishkin who contended that Technion should remain an institution of higher learning and the very suggestion that it be transferred to Jerusalem was "absurd".

Once again the matter was put before the Zionist Congress, this time the 16th Session, at Zurich in August, 1929. The Jerusalem Executive sent in a recommendation that the Technion be closed down altogether until there was clarification with respect to status and budget. Haifa sent Prof. Tcherniavsky to Zurich to rally all possible support to save the institution. His own service as co-director, together with Pewsner, was now approaching the end of its second year, and he determinedly pressed for engaging an experienced, professional Managing Director.

The Congress was almost exclusively occupied with debates over enlargement of the Jewish Agency with the addition of non-Zionist members, and the best Tcherniavsky and his supporters could do was to have the Technion matter referred to the Zionist General Council with the following terms of reference:

1. A committee of experts should be sent to Haifa to study the problems there. In consultation with the new Director, to be appointed, they were to examine the curriculum, study the possibilities of further development and determine in detail the educational aims of the institution with regard to the practical needs of the country.

2. Until the completion of this investigation, the institution was to maintain its character as a polytechnicum, with improvements in the practical instruction.

3. Immediate steps should be taken to appoint a new Director. Responsibility for the management and supervision of the Institute was to remain in the hands of the Zionist Executive.

Within the hierarchy of Zionist administration the General Council duly took up this resolution. It changed a key word in the terms of reference, so that the committee of experts would merely "propose" its solutions, and not "determine" Technion's scope. Need for a Director was reaffirmed.

The committee was appointed. Its chairman was Dr. Shlomo Kaplansky, a prominent Zionist labor leader, who had previously displayed an interest in the Technion. The committee never met. Years later, when he was the head of the Technion, Dr. Kaplansky confessed that he had deliberately not summoned the committee into session, apparently fearing that the opponents of the Technion would have the upper hand.

The text of a proposed advertisement seeking candidates for the post of Director/Principal gave an indication of the nature of the individual sought. "The Hebrew Technion in Haifa announces an opening for Principal beginning 1 October, 1929. Candidates with academic technical edu-

Prof. Ilioff supervises a class in the chemistry laboratory. 1929.

cation (preferably in civil engineering) and practical experience in teaching or administration are requested to submit documentation and curriculum vitae together with their (salary) requirements. Preference will be given to candidates who can instruct in technical mechanics or bridge-building."

There was dispute with respect to these criteria. A revised text asked for information on the candidates' knowledge of Hebrew.

The advertisement was never published. The Zionist Executive decided that since the committee of experts would "shortly" be making recommendations with respect to the structure and future functions of the Technion, it would be better to wait, and see what kind of Director might be required.

In the meantime there was need for management on the spot. Tcherniavsky and Pewsner, who had several times resigned during their two-year stint and had been compelled to withdraw their resignations, now made it clear that they would under no circumstances continue. In their second and final year their problems had multiplied, even as the institution grew.

Objectively, the achievements were impressive. Total student enrollment, which had reached 85 in the year 1926–27, had mounted to 100 in 1927–28, and to 125 in 1928–29. Superficially there was an atmosphere of growth and progress. The Technion was gradually being involved in the daily life and problems of the country. After the 1927 earthquake Professors Badian and Neumann went out to the field, on both sides of the Jordan, to study demolished or damaged buildings and brought back practical recommendations with respect to construction lessons to be learned.

The *Vaad Menahel* decided to set up a laboratory for testing building materials, and accepted Prof. Tcherniavsky's offer to finance it by a contribution of about 25% of his salary for the ensuing twelve months. Despite the generous offer, the Zionist Executive vetoed the opening of the new laboratory, apparently for fear that it would increase expenses.

The work of the Hebrew terminology committee, which had been suspended, was renewed and Saadia Goldberg was engaged as its secretary. The trade courses in the evening continued unabated.

Again America

The budget fluctuated at between £6,000 and £8,000 annually, but the grant allocated from the Zionist funds was systematically reduced from £6,000 in 1926–27 to £3,000 in 1927–28, to £1500 in 1928–29. Again desperate attempts were made to raise funds abroad independently. This time the good offices were sought of Dr. Markus Reiner, an engineer with the Government Public Works Department, who was to spend three months in the fall of 1928 at Lafayette College in Easton, Pa. He agreed with alacrity. In Bologna he had met with Professor Theodore von Karman, then already a distinguished scientist, and von Karman had promised to take an interest in the Technion during his stay in the U.S. Reiner decided to set up a top-level reception for von Karman and went after an American Committee on Science and Technology which had been established in America to advance the interests of the Hebrew University and the Technion, only to discover that a short time before, the Committee had decided to drop the Technion and devote itself to University affairs only. He then called on Leon S. Moisseiff, one of America's major bridge builders, who was surprised to learn that von Karman was Jewish and agreed to help. Von Karman had in the meantime gone to Japan, and was unavailable.

Reiner went back to the American Committee with a proposal that it encourage the establishment of such research institutes as would be of primary interest in the development of the country and could form a nucleus for an engineering department in the University. Institutes such as a material testing laboratory and a hydrological laboratory should naturally be housed at the Technion which would thus benefit from them.

Hebrew University records show that Dr. F. Julius Fohs, head of the American Committee, wrote to Dr. Magnes regarding the technological side of the University program and Magnes replied, suggesting that Committee members draw up tentative plans for an agricultural college at the University. "We are also considering an engineering college; perhaps some of your group could draw up tentative plans for this as well," Magnes added. He sent Fohs a copy of the Louis Green memorandum, a document hardly calculated to promote support for the Technion.

Getting little satisfaction from the Fohs Committee, Reiner sought out the members of the original Zionist Engineers' Society. Dr. Selig Brodetsky, on a trip to New York, had regathered the nucleus of its members and was trying to encourage them from London. Reiner met with them, but the results were discouraging. They felt the university-level Technion was "too high" for them and they were going to concentrate on vocational education. "It therefore appeared," wrote Reiner, "that the poor Technion was bound to fall through between the Scientific Committee for which it was 'too low' and the others for which it was 'too high'."

Reiner decided to activate the committee at least for vocational guidance, hoping "they could be moved to do *something* instead of promising many things." He was encouraged by the interest taken by Dr. Ferdinand Sonnneborn, described as an influential person. Promises for the future could not help meet the Technion's immediate budgetary crisis, however, and Reiner met several times with Laurence N. Levine, a wealthy man and president of the American Engineering Society. He and a number of his friends ultimately agreed to form a group of 50 people who would pledge $100 each for a period of four years toward the Technion budget.

He met also with Bernard Flexner and Julius Simon, of the Palestine Economic Corporation, and when he learned that they were planning to give $5,000 to the University for the setting up of a materials testing laboratory there, he argued that the proper place for such a laboratory would be at the Technion, rather than in the Physics Institute of the University, which, he was told, did not want it.

There were few immediate results from Reiner's efforts, but it cannot be said that his efforts were in vain. Von Karman rendered great help to the Technion much later. Moissseiff gave the Technion a major collection of his bridge plans. Ferdinand Sonneborn became the first President of the

American Technion Society. The PEC testing equipment eventually found its way to Technion's laboratory.

A bequest from the estate of Barnett Machanik, of South Africa, was evidence that the message of the Technion was getting through in other countries as well. Kisch had spoken to Machanik during a visit to South Africa.

In February, 1929, Kisch went to New York in connection with Jewish Agency matters, but he also found time to seek support for the Technion. Weizmann, too, helped. Felix Warburg gave $10,000. Max Shoolman, and a friend from Chicago, gave $20,000. Henrietta Szold was able to report to the Technion that there was now an endowment fund of $30,000. Haifa lost no time in asking for the money, but the equivocal answer was that the Technion would get only the annual income. It did, but only within the framework of the normal Jewish Agency allocation to the Institute.

Under these circumstances it was easy to understand why salaries were not paid on time. Following the Badian controversy all employment contracts had been discontinued, and there was no written undertaking to pay the teachers any fixed amounts except what was agreed upon from year to year.

As noted earlier, the students had made their own academic burdens heavier. Their economic problems were not diminished either. They set up their own cooperative enterprises to operate a canteen, a supply store and an employment office. They discussed dreams for construction of a student dormitory on the campus, but were no less concerned about the fate of the institution and decided to try their hand at fund-raising. During Weizmann's visit in 1928 they had suggested that he ask the Mandatory Government for moral and material help. He was pessimistic about the chances of success and declined to make the approach. The students thereupon decided to take the matter in their own hands and arranged to call upon the heads of the Education and Public Works Departments of the Government. The students made the point that the Technion was performing a service of national value in training highly skilled manpower, and therefore was entitled to help. When a small Government subvention was eventually received, and the Government representatives agreed to serve on the examination board, the students were convinced that it was due to their intercession.

The Government was not altogether unaware of the existence of the Technion. In 1929 the District Commissioner wrote to the Institute that on June 3, 1929, pupils of all Government schools were to be given a holiday on the occasion of the birthday of His Majesty, King George V, and he hoped Technion pupils would have similar privilege.

A meeting of the student representatives with the Keren Hayesod, seeking an increase in the budgetary allocation, ended unsuccessfully. The students, who sought to enlist the help of Weizmann, Ussishkin and Sokolow, did everything they could to help ease Technion's financial burden, except to enforce the payment of tuition fees.

Pewsner Under Fire

Few thought the situation could get worse, but it did. The opposition to the Technion administration broke out in new directions and in new forums. This time it was aimed at Pewsner, the Treasurer. In a flurry of news items appearing in April, 1929, in local and European Jewish papers it was alleged that he had misused or misappropriated Technion funds, and besides had adopted a haughty and condescending manner toward Technion's professors. The tone and content of the accusations were similar to those voiced by the four dissidents on the teaching staff, and it was generally believed that they had instigated the attack. The articles further charged that the Technion Board had been "packed" with incompetent people who interfered in academic matters and denied the teachers any voice in management or even in determining curriculum. The Board was also said to incite the students against one unfortunate teacher who had been improperly dismissed because of prejudice against him. The Badian affair was being kept alive.

The unpleasant publicity poisoned the atmosphere and still further undermined Technion's already low prestige. Replies, in print and at public meetings, never quite caught up with the allegations, which were recalled and repeated for some time by people who had no personal knowledge of the Technion.

The formal replies to the press took two forms: the official response from the Board, signed by two of its members, M. Aleinikoff and S. Itzkovich, and the personal reply by Pewsner.

The Board categorically denied all the charges made against it. The autonomy of the Teachers' Council in academic matters had been preserved, but the Board was duty bound to protect the institution against the destructive activities of a small group of men, one of whom had been discharged by the Zionist Executive for good reason, but continued to fight his dismissal. Further, the Board had not inflamed the students against the individual; exactly the contrary was true.

Pewsner called attention to the fact that the charges made against him had already been voiced before the investigating commission which had not only rejected them, but had dismissed one of the accusers and had reprimanded the other three. He denied as a "mean lie" the charge that he had ever misappropriated Technion money. To the contrary, he said, during the seven years he had been Treasurer he had frequently guaranteed Technion bills on the basis of his personal credit, and had helped the Institute through many financial pinches. He had been associated with the Technion Board for a total of 17 years without remuneration and had always received the confidence and cooperation of all concerned, except this group of four dissidents.

In the tense atmosphere that prevailed, Prof. Tcherniavsky declared that his good name had been maligned as well, and he demanded that the Executive appoint a committee to ascertain the facts. Messrs. M. Ladyjensky, Y. Ratner and N. Het were named, but the four professors who were asked to give testimony (Badian, Grossman, Baerwald and Neumann) refused to appear. The committee thereupon concluded it could not function and returned its mandate to the Executive.

The atmosphere had to be cleared, and the insistent resignations of Tcherniavsky and Pewsner were at last accepted. The latter withdrew completely from Technion affairs, deeply hurt by the attacks upon him and by the failure of the Zionist Executive to give him sufficient backing. He died in May, 1930, and the meeting of the Board held immediately thereafter determined to set up a memorial plaque for him at the Technion, though no public notice of his services was taken at the Technion until his picture was hung in the President's office in 1979. Tcherniavsky continued as a respected and senior member of the Physics Department until he became Professor Emeritus in 1954. He died in 1966.

In the absence of a Director, and in view of the ruling by the last Zionist Congress that no major changes be made in the Technion, two professors were given management authority. At first Prof. Alexander Ilioff was Acting Director, and then, beginning in February, 1930, Prof. Joseph Breuer was asked to administer the institution. This leadership, subject to constant carping and criticism from Jerusalem, continued until after the 17th Zionist Congress in Basle in July, 1931.

Those two years were relatively free of scandal and internal conflict, but all the other old problems, and a few new ones, arose to plague the leadership.

Central supervision of Technion affairs had been placed in the hands of the Education Department of the Jewish Agency in Jerusalem, headed by Dr. Isaac B. Berkson. He sought to introduce authority and responsibility into Technion affairs. When two teachers went to Jerusalem to consult the Government regarding participation of its representatives in the final examination, he regarded this as a bypassing of the Zionist authority. He noted that the students, too, were in the habit of conducting negotiations with the Government, and he asked acidly if this indicated a state of anarchy at the Technion.

Certificate testifying to participation of Architect S. Bernstein of the Technion in an exhibition in Vilna. 1930.

The *Vaad Menahel* members were indignant. They replied that in the face of indifference in Jerusalem and the resulting vacuum, those connected directly with the Technion were at least doing something. Berkson ruled firmly that his office, as the agency of the Zionist Organization, was the supreme authority. Even the Director of the Technion was only an employee, subject to authority.

For a while peace returned to the campus. The institution was operating on a Zionist-approved budget of £6,750, of which £6,003 was for salaries. The Haifa Board noted bitterly that they were asked to train engineers on a budget of less than £7,000 while the Hebrew University, with practically no students, had been given a budget of £70,000.

There was a possibility of some income, but on "terms". The Hilfsverein had not lost interest in its Technikum and Max Warburg, of Hamburg, suggested that $10,000–$15,000 could be made available annually if the school gave up its university ambitions. "There must be a solid foundation before one can think of a University in any form," he wrote to Magnes.

The University Chancellor replied that it was not possible to give the assurances which Warburg requested. "The Technikum is struggling along in the old way and with the old difficulties. Nobody in the Palestine Executive seems to give it much attention .... I am afraid that until the University gets ready to take the matter in hand it will not be done properly. But the University, at the present time, has not the money with which to take the necessary steps ...."

Is Technion Necessary?

At this stage Berkson reopened the question of the need for a Technion at all. According to Louis Green's memorandum, he said, the country required no more than five new engineers and five architects a year. If that were so, why was the Technion accepting a freshman class of forty? This seemed to be a waste of money, and he suggested merging the Technion with the Reali School, both to operate under the leadership of Dr. Biram. In a rare display of unanimity, the teachers, the students, the Board and the administrative employees all rejected the suggestion.

Berkson persisted. He thought it was a great mistake to accept more students. Many institutions had been destroyed by too large an enrollment, he said, but none had ever been harmed by limiting the number of students. Breuer, Itzkovich and other members of the Board pleaded that the Technion had a special mission to fulfill: to provide a place for students from overseas who were being denied admission to technical universities in the countries in which they lived. Was the Technion, too, now to put into practice a *numerus clausus?*

The issue was removed from the field of theoretical discussion, and practical implementation was begun. The Zionist Executive decided, on the basis of Berkson's recommendations, to save £800 by not taking in any first-year students for the forthcoming academic year. The reaction could have been anticipated. Pressed for further explanation, Dr. Lurie revealed that the decision was more than budgetary in nature. There were many who seriously questioned the need for a technical university at all, but if Haifa could find other ways of reducing the budget by £800 perhaps it would not be necessary to close the first year. In any event, he added, the internal discussions were "nauseating" and turned the stomach. Why all this combativeness?

There was no doubt that the authority in Jerusalem was in the hands of individuals who were, at best, unfamiliar with the role of technology and engineering in modern society, and at worst, hostile to the very existence of the Technion which they saw as competition for the Hebrew University in which they were themselves involved.

Several times during 1930, in January and again in June, there were sharp confrontations between Berkson and the Teachers' Council. He warned them that if Technion was unpopular and not receiving the support which it felt it should have, part of the blame was theirs. The constant bickering and the internal conflicts within the Technion had antagonized many. Unless they could create internal harmony and peace, they should not expect the Jewish Agency to support them since the whole matter of the need for a Technion was still open to question. And if no funds were forthcoming the result would be the gradual liquidation of the Institute.

The Board in Haifa appealed to the Zionist Actions Committee, due to meet in Berlin in August, 1930. A detailed memorandum reviewed the history of the Technion since it had been taken over by the Zionist Organization and the many unhappy experiences due to lack of understanding and clarity. The main reason for this situation, the memorandum concluded, was the absence of any agency which would take systematic responsibility for the development and maintenance of the Institute. It was treated like a stepchild, shunted around and put under the jurisdiction of persons who had no interest in it.

Once again all the arguments in favor of a high-level institution were repeated. At the same time, it was noted, vocational training was not being neglected. At the beginning of January, 1929, the first experimental classes had begun for a trade school. A second course had begun in October, 1929, and in October, 1930, a third class, for metal and electrical training, was due to open.

Since the British Mandatory Government had granted no entry visas, there were no students from overseas. This resulted in a saving. Furthermore, Berkson noted coldly that since Prof. Baerwald had passed away, they should make no appointment in his place, and that would save more money. If they insisted on engaging an assistant, it would have to be at the expense of the money otherwise earmarked as compensation to the Baerwald family!

The situation was becoming tense once again. Berkson made it clear that he was going to press for a clear-cut decision on the Technion's future, and he asked the Board to draw up an analytical statement presenting its case. The statement should be logical and objective, not a public propaganda piece. Matters were moving toward a climax.

Breuer carried on with the day-to-day business of the Institute. He acknowledged receipt of complaints by students that they did not understand Badian's lectures. He dealt with a formal complaint of the Senior Medical Officer of the District Health Office that examination of the water in the well (which was the drinking supply of much of the neighborhood) revealed that it constituted "a menace to the health of the public", and the Technion was asked to take steps to remedy the situation.

He waited patiently for the meeting of the investigating committee which the Zionist Congress had voted for. Mild and gentle-mannered, he was able to maintain an appearance of organizational stability even while the fate and future of the school were in question. He encouraged plans for the setting up of the Building Materials Testing Laboratory, which was to be headed by Prof. Badian. There the latter would be able to concentrate on scientific work. Badian was honest and sincere, Neumann testified, but a hard man to work with. There was talk also of getting the Herzl Club in Vienna to give a contribution for the new laboratory.

Berkson's unmistakable hostility found expression in a series of recommendations which he sent to the Jewish Agency Executive meeting in February, 1931. That body then decided on a reorganization of the administration and program of studies at the Technion as follows:

1. The main course in engineering and architecture should concentrate more on practical subjects, and should seek to turn out foremen and assistants, rather than engineers and architects. The lower level studies in the various trades should be expanded and the evening program should be extended also to Tel Aviv and Jerusalem.

2. A new Board of Governors (*Hever Neemanim*) should be set up as an independent body to administer the school. The Jewish Agency would continue some financial help for at least three years, but the Board would be responsible for day-to-day administration. The Jewish Agency would have to approve the composition of the Board, and would also retain final approval of the budget, appointments, program of studies, etc.

3. The new Board would assume responsibility also for the Reali School, following which consideration would be given to naming Biram executive head of both.

Berkson went to Haifa on February 25 to present the reorganization plan before a joint meeting of the Board and the Teachers' Council. Yellin came with him from Jerusalem. The new program, he said, had been drawn up only after careful study of all the available material on the subject. The facts were simple: there was no need for more engineers in Palestine; indeed, there was already a surplus. He repeated the old refrain about sending a selected few students for training overseas each year. On the other hand, he maintained, there was a need for skilled personnel who could work under engineers, perhaps to be called Responsible Assistants.

The next major problem was with respect to higher responsibility for the Technion. The Jewish Agency had no funds and could not continue its subsidy. Hence the recommendation for a new Board of Governors which, he hoped, would actually function, unlike the one in London that had failed.

There was need to economize. The reduced scope of the Technion would save money. Reduction in the number of years of study, reduction in faculty members, less hours of instruction — all these, he said, would also save money. The decision taken in Jerusalem was not final. It was in the nature of a recommendation to the Jewish Agency Council due to meet during the summer, and from there it would go to the Zionist Congress. It was now the turn of the Technion people to state their views.

The teachers present were particularly harsh in their replies. Prof. Ettingen expressed a sense of shock that the leader of a Zionist organization could speak as Berkson had about the future of the country, without vision and without hopes. Was this Zionism? That was how the Alliance talked when it considered putting up another school in a place like Damascus. The original differences with the Hilfsverein had not been about language alone; that was only a symptom. There were basic differences on the question of Zionism.

Itzkovich wanted to know why a change was being advocated at all. The Technion was already in existence, and flourishing academically. Its graduates were first rate. Change should be made only after years of careful study. As for Dr. Biram, proposed as joint head, he had no technical training. He already had his hands full with the Reali School. There was nothing in common between Technion and Reali except that they were neighbors. For that matter, the Haifa Hadassah hospital was even closer, but nobody had suggested that Technion amalgamate with it.

Aleinikoff bristled at Berkson's critique that emphasis at the Technion

In the workshop. 1931?

should be on the practical, with less of theory. What about the Hebrew University, he asked. Was that practical? They were building an academic hothouse on Mount Scopus catering very little to the practical needs of the country, and this at great cost. He became ironical: On Mount Scopus they were encouraged to educate intellectuals, but on Mount Carmel they were asked to produce intellectual cripples, as cheaply as possible.

Berkson weathered the storm and repeated that he was going ahead with the plan as proposed. He expected objections. "I know where those objections will come from. I shall try to mobilize friends in favor of the program, and we shall see who will win."

The threat was clear, and the battle lines were drawn.

Shmuel Itzkovich distributed a memorandum in which he reviewed the last developments and restated the Technion's case:

"If we look at the technical schools in Egypt and Syria we shall see that they are doing exactly what Berkson wants us to do, train low-level technicians without intellectual level. Perhaps what they are doing is suitable to the general intellectual level of their population, but that is not the situation here nor among our youth. To the contrary, the difference in cultural level, which is in our favor, is a tremendous weapon in our hands in the battle for our existence in this land ...." Wise and prophetic words indeed.

Who was against a university-level Technion? he asked. He charged that they were the circles around the Hebrew University in Jerusalem. They were engaged in setting up esoteric departments, remote from the needs and realities of the country, while at the same time preaching to the Technion the need to be "practical". If the University is said to have great cultural and political value, would not the same be said about the Technion, he asked. There were some who thought and spoke of European culture in lofty terms, but referred to Jewish youth in Palestine as if they were "natives". "The Technion is not a dying body. It is alive and flourishing. It is successful. Let it live and develop. Don't throttle it," he concluded.

Others joined in the campaign. As was to be expected, the student body rose up in protest and called upon the entire population to resist a step which in effect meant the closing down of the Technion.

A very welcome ally was the Palestine Engineers' and Architects' Association which had gradually undergone a change in its attitude, to a large extent at the urging of Ing. Uriel Friedland (Shalon). By now there was also a Technion Alumni Association to voice its "astonishment and grief" at the Berkson proposals.

The Teachers' Council was particularly outspoken in its criticism of the plan, and of the failure of the Jerusalem authorities to measure up to the needs of the Technion. A delegation from Technion went to Jerusalem to meet again with Berkson and the confrontation was almost violent. Berkson had before him a copy of the minutes of the teachers' meeting on March 28, and he termed the discussion reported therein a *hutzpah*. He resented the implication that he, Berkson, was failing in his duties. He charged that the teachers were interfering in administrative matters that were none of their concern. The trouble in this country, he said, was that people were under pressure, got excited, and had no consideration for others.

The conclusive decision was to be made at the 17th Zionist Congress in Basle in July, 1931, and all concerned prepared memos, brochures and reports to express the various points of view. To those close to the Technion there was only one issue before the Congress: the attitude it would adopt toward the institution.

The Congress, however, had other weighty problems facing it. Just as Herzl, three decades earlier, had been faced with a revolt against his leadership, so Weizmann was now battling a vigorous opposition. It became clear that there might be major changes in the political constellation of the movement.

In this atmosphere, those charged with Zionist budgeting preferred to avoid any commitments until after the Congress had spoken. Accordingly, on June 9, 1931, notices of termination of employment (as of September 30, 1931) were sent to all employees of the Technion, together with all other employees of educational institutions supported by the Jewish Agency. This measure was justified on the grounds that a sweeping reorganization was expected in the educational field, and there was a desire not to tie the hands of whatever body would in the future take over responsibility for education. It was noted that recommendation would be made to such new body, when it proceeded to set up its program, to give priority to the re-employment of former teachers, with the hope that the great majority would be able to continue their teaching assignments without interruption.

Technion's formal report to the Congress contained an objective accounting of what had happened since the last Congress two years earlier. A beginning had at last been made on the laboratory for materials testing. The number of students had declined because of the uncertainties about the institution's future. The investigating committee that was supposed to look into Technion's problems had not been set up, and instead the Zionist Executive had asked Dr. B. Cohn and Dr. Berkson to make recommendations, in consultation with Prof. Brodetsky.

The Technion budget for 1930–31 was £6,816, of which the Jewish Agency provided £4,600; tuition fees, £1,050; miscellaneous local income,

Prof. Aharon Tcherniavsky (1887–1966).

£ 1,166. This amount was not sufficient to maintain the Technion on any level, let alone take into account expansion.

The Congress Decides — What?

The Technion designated Prof. Grossman to represent it at the Congress, and the Student Association financed the trip for its chairman. This was Dov Haimovitsh (Givon), older than most of the other students, and already regarded as an influential person in the Labor Party. Givon began his lobbying in Palestine and lined up support from members of his party. En route to Basle he stopped off at Vienna, which was world headquarters of the Jewish Student movement, and arranged for cables to be sent from the World Union as well as from groups of Jewish students at various European universities, calling on the Congress to affirm university status for the Technion, so that Jewish students would come there from overseas.

Grossman and Haimovitsh lobbied constantly among the delegates, testified before the sub-committee which had been appointed and sought to make the Congress Technion-conscious. The campaign was difficult, for the delegates were more interested in other exciting political conflicts. Restrictive British policies in Palestine aroused more interest than the fate of the Technion. The Revisionists led the attack on Weizmann's policies, and succeeded in defeating him. Nahum Sokolow was elected a compromise president of the Zionist Organization. The Congress passed a vote of no confidence in Dr. Magnes of the Hebrew University. The Executive Committee was reshuffled, and the Labor Party rotated its members. Sprinzak and Kaplansky were dropped; Chaim Arlosoroff and Berl Locker stepped in.

The resolution on the Technion was relatively brief, and was equivocal. It took into consideration that the Jewish Agency had decided to hand over the control and direction of the Zionist school system in Palestine to the *Vaad Leumi* (General Council of the Jewish Community of Palestine). This body was recognized by the Mandatory Government as the representative body of the Jewish community, and enjoyed quasi-taxation rights to provide funds for its communal activities. Indeed it served for many years as the government-in-training for the State of Israel. In these circumstances the Congress decided that any changes to be effected in the Technion by the Jewish Agency Executive would require the agreement of the *Vaad Leumi* after clarification by a committee of experts proposed by that body. The Technion should develop various levels of technical education according to the needs of the country, but no changes should reduce the standing of the existing institution as a school of higher learning.

Prof. Jeremias Grossman (1884–1964).

This was a masterpiece of double talk, and both sides found cause for both distress and joy. A budget of £7,000 was approved. After the decision, Leo Motzkin congratulated the Technion lobbyists on their success, but asked how they expected to run the Institute with so little funds. With a straight face Grossman replied that he would go to the U.S. and stand in the middle of Times Square, carrying placards announcing that he was collecting money for the technical university in Haifa.

Nevertheless, the Congress marked a turning point in the history of the Technion. There was at last movement in the direction of engaging a responsible Director. Breuer, who had steered the Technion through stormy waters, had had enough. He had given four months advance notice that as of September 20 he would vacate the post of Director, thus giving the authorities sufficient time to choose his successor. The political tides of the Congress cast up a candidate, Dr. Shlomo Kaplansky. He was one of the leaders of the Labor movement who had been dropped from the Zionist Executive, and some way was sought of retaining the services of this talented and capable man. Berl Katznelson is said to have been the first to propose that he be offered the Technion post. He suggested to the two-man Technion delegation that if Kaplansky were to agree, this would solve Technion's managerial problems and also help the Labor Party solve its own problem of what to do with him. Grossman approached the candidate and was later joined by Givon. Though the two had no authority whatever, they made the proposal.

Prof. Joseph Breuer (1882–1962).

Kaplansky had not been a stranger to Technion affairs. While still a student at the Technical School in Vienna, it was said, he had at times expressed a desire to be affiliated in some way with the institution in Haifa, whose construction was then widely publicized. He immigrated to Palestine in 1912, and recalled later that one of his first tasks in the labor exchange office to which he was attached, was to send Jewish building workers to the Technion.

He had for a short while been a member of the Palestine Management Committee, and later, it will be recalled, had been asked by the Zionist Congress to look into the problems of the Technion. Kaplansky was one of the few engineers in the Zionist movement.

He had played a leading role in Zionist affairs and his experience had been considerable. From 1913 to 1919 he had been principal Secretary for Colonization of the J.N.F. He had directed the affairs of the Finance and Economic Committee of the Zionist Organization from 1921 to 1924 and in the three years thereafter had been Treasurer. He had been a member of the Palestine Executive in charge of agricultural administration.

Faced with the unofficial offer, he at first deferred decision. When he asked Brodetsky what was really going on at the Technion, the latter replied that it was only a matter of reaching decision with respect to its level and scope. Kaplansky was one of those who believed in an engineering university, and Ussishkin therefore urged him to accept.

There was much behind-the-scenes discussion, and politics was not absent from the argumentation. Labor leaders were anxious to have Kaplansky well placed. Members of other political parties objected. A well-reasoned presentation of the case in his favor was written by Yitzhak Wilkensky-Elazari (Volcani) to Brodetsky:

"My friends Arlosoroff and Berl Katznelson have told me of your doubts regarding Kaplansky and the Technion. I do not agree with you and should like to present my reasons. There are two different periods in the establishment of a research or teaching institution: the organizational period and the operational period. Possibly there may be grounds for your doubts with respect to the second period, but as for the first, in which stage the institution is at present, and in which the Technion will be for many years, what is required is a man with organizational abilities, with personal connections and influence and with long-range vision. For these Kaplansky is unusually qualified. Even if his technical background is not sufficient for the second period, because he did not follow his studies with practical work, it is certainly enough for the first period." He pointed out that Kaplansky had the confidence of the teachers and the Board of the Technion. If he were to go to America he would be able to enlist the support of Jewish engineers there. And above all, he wanted to make the Technion an institution of higher learning, which for economic and political reasons, was all the more important at that time.

In the meantime, Prof. Breuer again reminded the Jewish Agency that his term was over. One week before expiration date he formally asked Berkson to inform him to whom he should transfer the affairs of the Technion. Temporarily, Prof. Grossman was named Acting Director.

The office in Haifa announced to the press that registration of candidates for admission was continuing and classes would resume in the fall, but no date was given. Now the presssure was on the Jewish Agency to do something about a new Director.

In the interim period conditions became more chaotic. Aleinikoff rushed a complaint to the *Vaad Leumi* that despite the Congress decision not to make changes in the structure of the Technion, Berkson had called for institution of reforms, lowering the admission requirements and reducing the course from four to three years. Financial pressure was put on the Technion. Though most of the teachers in the country had been paid after the Congress, the Technion staff complained as of mid-September that they had not yet received their salaries for the months of April and May.

160

The *Vaad Leumi* was moved by overriding patriotic considerations, and its committee, composed of Eliahu Berligne, Eliezer Kaplan and David Zvi Pinkas, came to firm conclusions: It was considered both desirable and essential that the Technion be stabilized as a school for training engineers and work superintendents, and not be transformed into an institution for overseers and foremen. The committee pinpointed the need in Palestine for an institution of higher education both for local youth who would otherwise go abroad, and for immigrants. Amalgamation with the Reali School was rejected as harmful to both institutions, though it was recommended that Biram become a member of the Technion Board.

The *Vaad Menahel* in Haifa seized the initiative, and on October 31, 1931, Eliahu Berligne, then serving as chairman of the Board, wrote to the Jewish Agency Executive in Jerusalem that Kaplansky was their choice to be head of the Technion for a period of one year. His task would be not only to manage, but also to procure the necessary funds, and they asked that the appointment be ratified. The *Vaad Leumi* gave its consent, in accordance with the Congress resolution.

Approval did not come easily. Hexter, a non-Zionist member of the Jewish Agency, objected. Berkson wanted to restrict the task to administrative duties only. Other opinions in Jerusalem were divided. Toward the end of November word came from the central office in London that Sokolow, Brodetsky and Locker agreed to a one-year appointment, without commitment for the future.

The *Vaad Menahel* held its meeting on November 29, 1931, in the presence of Kaplansky. Announcement was made that the Jewish Agency had approved the naming of Kaplansky to the post which in English would be called Executive Director. The appointment was for one year only, and the title would be qualified as "Acting".

Kaplansky expressed thanks for the confidence which the local Board had placed in him. However, he feared that the "Acting" was an indication that the Jewish Agency did not share that confidence. Titles in themselves were unimportant, he said, but this meant that his administration would be starting under a cloud. It raised grave uncertainties with respect to the future of the institution.

He hesitated, and the members of the Board divided. Some agreed that he was right, and called for a battle against the Jewish Agency, even to the extent of a mass resignation of all the members to indicate their protest. Others urged that he accept. Kaplansky wavered. On the one hand, it appeared that he was not really wanted. On the other hand, as a disciplined Zionist, he accepted the supreme authority of the authorized bodies.

One member brought the sense of the meeting to a point: "Technion stands before a catastrophe. He is our last hope. Resignation of this Board would destroy the institution, and that's what 'they' want. We must go on. Technion is more important than a title."

It was decided that Kaplansky should go to Haifa on December 1, two days later, and take up his new duties. The final word of the Zionist Executive was not helpful. They informed the Technion that his major function would be to set up a group of Technion supporters, to collect the funds necessary for maintenance and to see to the Technion's future. At the same time they expressed serious doubts whether he could succeed.

With this "encouragement" Kaplansky assumed the post which he filled with distinction for 19 years until his death at the end of 1950.

*　　*　　*

Among the signatures of many distinguished personalities who signed Technion's Visitors' Register: Martin Buber (1927) and Marc Chagall (1931).

Voluntary Waiver of Salaries

Dr. Shlomo Kaplansky entered upon his duties as Director* of the Technion under trying circumstances that gave little promise of the nineteen years of growth and development which were to follow under his leadership. He undertook to manage an institution which was wracked by personality conflicts and ideological dissension and on the very verge of dissolution for lack of funds. If a miracle was required to save the Technion it happened twice: once in the determination and strength of its Director, the first true leader the Technion had had since its creation, and a second time in the devotion and self-sacrifice of a staff which, at a time of stress, rose above all personal interests and saved the institution from collapse.

The miracles did not occur all at once, and they were accompanied by much agony and thunder, but they emerge unmistakably as historical peaks in the chronicles of the Technion.

Formal notice to the staff that their employment was to be terminated as of September 30, 1931, was technically still in effect. As that date approached the staff met and came to the conclusion that if the responsible authorities in Jerusalem and London would agree to certain basic requests in the interest of the Technion, they would be prepared to continue work even without immediate assurance of payment of salaries, thus making it possible for the school to open its new academic year on time. On September 11 they drafted their conditions: 1. Visas should be procured for overseas students waiting to come to Palestine, and these would constitute the bulk of the new incoming class; 2. The composition of the new Board of Governors of the Technion should be determined in consultation with the teachers; 3. The new Director should be appointed as quickly as possible, with the consent of the teachers; 4. The £2,000 approved by the Congress and the Jewish Agency for the year 1931–32 should be paid on time.

News of their offer had little impact in the Jewish Agency offices which were undergoing a reorganization of their own following the results of the Zionist Congress. Even the issuance of Technion's announcement that the school year would open as usual drew no comment. More than ever before, there simply was no one who cared. When Nahum Sokolow, the President of the Zionist Organization and the Jewish Agency, visited Palestine at the end of the year he spent time in Haifa. Not only did he not visit the Technion, but in a press conference held in that city went out of his

Dr. Shlomo Kaplansky (1884–1950).

* The title was officially Director, but Dr. Kaplansky for many years identified his post, in English, as Principal.

way to express great pleasure at the development of the Hebrew University. To him the Technion did not exist.

A week before Kaplansky was officially engaged and took over, the staff formally signed and presented its offer. Typed on the letterhead of the Hebrew Technical Institute, dated November 22, 1931, and addressed to the *Vaad Menahel*, the statement read:

"We the undersigned employees of the Hebrew Technion in Haifa hereby express our willingness to work at the Technion during the year 1931–32 (with the understanding) that no one will bear legal responsibility for paying salaries for 1931–32. That portion of the salaries which will not actually be paid during 1931–32 will be carried on the books of the Technion to our credit. The time and the manner of payment of this portion shall be determined in the future by such body as shall assume the responsibility for the Technion, and in accordance with the financial possibilities of the Technion."

Even the four conditions previously set had been dropped. The offer was clear, unqualified, generous, self-sacrificing. There were thirty signatures of teachers, workshop instructors, librarians, technicians, clerks and maintenance staff. One absentee member joined later.

It was understood that whatever funds were available to the Technion treasury for the purpose would be distributed to the staff as partial payments. Local storekeepers joined in the spirit and agreed to carry Technion staff members on credit. The students, not to be outdone, volunteered to pay higher tuition fees.

Technion's operations continued, though the staff offer was not formally accepted in writing until April 28, 1932. Even the acceptance letter was not transmitted to the staff until June 22, due to "technical reasons". During the year only partial salaries were paid. The maximum was about 60%, and some payments were as low as 25% of salary, depending on circumstances. Some of the junior staff members received no pay at all. Many of the teachers also had outside jobs, either at the Reali School or in engineering work, and were therefore not fully dependent on income from the Technion. The accumulated back salaries were eventually paid off over a period of about five years, without interest.

The immediate emergency was over and the Technion was still in business, but the treasury was empty. There were outstanding debts. One, which hung heavy on the Director's conscience, was the obligation to pay a pension to Baerwald's widow, who had been left penniless.

Kaplansky sought to change the focus and to create a new image for Technion. In his first address, on January 11, 1932, he reviewed the Technion's achievements, which had been obscured by the scandals and crises. Sixty-five young people had already graduated from the institution

Prof. Avner Badian in the Building Research Station. 1931/2.

and were at work. There were nine in the Government Public Works Department; five engaged in the construction of the port of Haifa; four in the planning of the Haifa-Baghdad railway; others in the employ of municipalities and private firms. Some were working successfully in Egypt and Syria, thus vindicating the vision of those who had foreseen a regional activity beyond the borders of Palestine. More than a hundred workers were enrolled in the advanced evening classes. The trade school had an enrollment of sixty. A modest beginning had been made with the opening of the Building Materials Testing Laboratory.

Under these circumstances, the new Director was highly critical of the Jewish Agency which had reduced even further its contribution to the budget on the theory that the Board of Governors should provide the needed funds. The basic problem was still money. For the first time a fund-raising campaign within Palestine was tried, and letters went out to a number of notables seeking their support. Societies of Friends of the Technion were organized in Haifa, Jerusalem and Tel Aviv. Receipt of a check for £50 from the Anglo-Palestine Bank was seen as a good omen.

Dedicated to Esmeralda Muggia, of Italy. Plaque is in Lecture Hall 12 of the old building.

Kaplansky's First Mistake

It became clear that big money could be obtained only from overseas, and the new Director decided to go abroad and raise the funds without which even the most devoted and idealistic administration could not survive. His departure so soon after taking over was a serious mistake, but this did not become clear until much later.

Before leaving he tried to set the house in order. Wanting to forestall any problems with Badian, he gave him almost free rein as operator of the Building Materials Testing Laboratory, and ignored charges that Badian had been operating the unit almost like a private business venture. There was no budgetary control by the *Vaad Menahel*. Kaplansky was anxious to avoid a rupture and he asked Grossman, who was to be Acting Director in his absence, not to take any action on the matter until his return. He gave Badian signature rights on all the Laboratory's affairs, but warned him that a set of by-laws and regulations was required to regularize the operations of the unit. He begged Badian not to do anything which would cause disturbances in his absence.

He dealt with charges that Prof. Breuer had withheld funds during his term as Director, and tried to settle claims for severance compensation and suits for back salary from a few dissident employees.

Kaplansky left for abroad in March and returned five months later. He was subjected to severe criticism for abandoning his post at the very beginning of his stewardship, but he had come to the conclusion that it was absolutely necessary to secure a firm financial basis.

Technion's Director discovered that there was a great difference between coming to a community as a distinguished member of the Zionist Executive, in which capacity he had previously travelled, and visiting as the head of an educational institution seeking funds. His tour took him to Italy, Switzerland, France and England, with mixed results. His talks with old Zionist contacts in Naples, Rome and Milan brought a pledge of 50,000 Italian lira from the Muggia family in memory of their mother. Kaplansky recommended that a plaque be set up in a lecture hall, and this was duly done. He left behind him a Technion Committee, composed of Prof. Mario Giacomo Levy, Comm. Ing. Norsa, Comm. Ing. Finzi, Sig. Roberto Adler, Comm. Ing. Giuseppe Muggia, Ing. Giacomo Hirsch, Comm. Ing. Roberto Almagia, Comm. Pardo-Roquez and Ing. Ernesto Recanati, of Milan, Rome, Parma, Ferrara and Pisa.

He had less luck in Switzerland. There he found other fund-raising causes already well-established. As a result, the local communities had set up fund-raising control mechanisms, and without their approval it was almost impossible to launch a drive. To top it all, the worldwide economic depression was now beginning to be felt in that country as well. Under the circumstances he favored a suggestion that the existing Hebrew University committee be broadened to include the Technion as well.

His most discouraging experiences were in Paris. He did not know enough French to talk to the French Jews, so he concentrated on the colony of Russian Jews there. He knocked on many doors and saw many people, including Rothschild, Citroen, and the head of the Galeries Lafayette, all without results. In France, as in Switzerland and Italy, he visited technical schools, and observed both their administrative procedures and academic programs, so his trips there were not a complete loss.

He hurried on to London, where he felt more at home. Here, too, he found fund-raising as difficult as "parting the Red Sea", but he had many contacts. His meeting with the Federation of Synagogues did indeed lead to ultimate presentation of equipment. He met with leaders of Bnai Brith and with the Friends of the Hebrew University. Small contributions were available, but the big money which he sought was not to be had. "The big, generous contributor is disappearing from a world reeling in economic crisis," he reported.

He spoke to the American Zionist leaders who were in London, and got a pledge that they would raise $5,000. He placed little reliance on the pledge and wondered if it were worth sending a man to the U.S. to collect. Perhaps it would be cheaper to engage a fund-raiser in America. He met with South African Jews, with a view to organizing fund-raising in their country.

The Zionist Actions Committee was due to meet in July and Kaplansky decided to stay on for that meeting, at which Technion budgetary affairs were also due to be discussed. Weizmann was cordial and helpful. He enlisted friends of the University to help Kaplansky, and himself agreed to head the Technion Society in Great Britain. Kaplansky felt encouraged and became optimistic again. However, when the Zionist sessions got under way he realized how little anyone really cared about the Technion. The session of the General Zionist Council devoted to education took place on August 7 and lasted from 9 P.M. to 4 A.M., but the minutes do not show that the Technion was more than casually mentioned. The major issue was the transfer of the entire Zionist educational system in Palestine from the Zionist Organization to the *Vaad Leumi*, and the need to provide funds. As for the Technion, it sometimes appeared that as between the Zionist Organization (Jewish Agency) and the *Vaad Leumi*, each wished to divest itself of the responsibility for the Institute so that the other would have to bear the budgetary burden.

Action on the budget was not what Kaplansky had expected. The Actions Committee in principle approved his proposed operating budget of £7,000 for the year 1932–33 but offered an allocation of only £1,500 toward that budget. He had hopes of obtaining an additional grant of £1,000 toward the outstanding deficit which totalled twice that amount. He felt that the Technion itself could raise the balance required for the budget, so that the teachers would not have to forgo part of their salaries again.

The long trip had been very trying. The expenses were to have been covered from the Technion budget in Haifa, and more than once he had sent back urgent pleas to provide funds since it was a matter of principle not to use money collected. At times he had borrowed money while on the road.

He sent back full reports on his activities, expecting that they would be read to the *Vaad Menahel*. He had some doubts with respect to his prolonged absence and sought guidance from Haifa. Early in May he received a cable signed by Berligne, chairman of the *Vaad Menahel*: "Don't hurry to return; complete your work." In August he returned to Haifa and found a hornet's nest.

Communications had not been all that they should have been. Kaplansky's reports reached the attention of the *Vaad Menahel* only when that body met, and delays were inevitable. The pessimistic note which pervaded his letters from Italy, Switzerland and France sounded even more gloomy when read in Haifa where staff members were working at reduced salaries. To the teachers it seemed that their new Director had failed, and that even he seemed to acknowledge he could not continue. Some who had been ardently in favor of Kaplansky spoke up and admitted that they had been wrong. The question was asked if the Technion needed a Director at all. So far he had only been an expensive item in the budget and had contributed nothing. In his absence it appeared that the Technion was running like a cooperative, managed by its own staff who were making all the sacrifices. They did not want to hurt Kaplansky, but sought some way to inform him discreetly that he could not continue.

There were others who defended Kaplansky. It was disgraceful, they said, to talk about dismissing him, when they were the very ones who had sent him abroad. Judgment as to his success or failure should be suspended at least until he had returned and had given his full report.

The bitterness grew out of a feeling that only they were making sacrifices and nobody else was interested. Some thought the Jewish Agency was only waiting for the Institute to fall apart. In that case perhaps it would be better for the Board to quit and hand back the keys at once. Others warned against giving ultimatums. The Agency would be happy to receive the keys back; it would make destruction of the institution easier. The year had been difficult financially, but successful academically. No students had dropped out. The level was high. At least the public was interested.

During the months of Kaplansky's absence the debate raged on. The feeling continued to grow that the Technion was a stepchild. Kaplansky had been an experiment that had not succeeded, and his salary was being paid at the expense of the teachers, though he had taken the voluntary cut as well. The money promised from Muggia in Italy had not yet arrived. The picture was bleak.

The Building Materials Testing Laboratory was again a bone of contention. Badian was running it alone on the basis of the confidence put in him by Kaplansky but it lost money, whereas a similar laboratory operated by the Engineers' Association in Tel Aviv was returning a profit. Badian must go, was the call.

Such was the mood and the atmosphere when Kaplansky returned to Haifa. He was shocked. If things were that bad, why had he been told to remain away longer, he asked, and indeed to this there was no reply.

The attacks on Kaplansky spread outside the Technion and began to assume a political coloration, reflecting the Zionist politics of the day. The Revisionist press charged that he had been a failure and should be replaced. His trip overseas had cost £7,000, and he had raised only £1,200, of which half was a contribution that would have come anyhow. He had been a political appointee to begin with and he had not proved his worth.

L'Affaire Grossman

Kaplansky was accustomed to political attacks, but he found it difficult to understand the opposition of his colleagues in the Technion. He read the minutes of all the meetings that had taken place in his absence, and they contained frank, accurate coverage on what had been said. Most of all he felt that Grossman had been disloyal and there was a confrontation between the two. Grossman, as chairman of the Teachers' Council, refused to call a meeting of that body, and a rump session was held with only eight attending. They voted, seven in favor of retaining Kaplansky, and one abstaining. Personal relations within the Institute were again becoming tense, and some staff members did not talk to others.

In November Kaplansky formally informed the Central Committee of the Teachers' Union that the *Vaad Menahel* had decided not to renew Grossman's employment because he had been disloyal to the Institute, had acted in a manner hostile to his colleagues, and had plotted against the Director in his absence. His remaining would only lead to endless argumentation and controversy. Another storm broke out.

A group of Haifa notables joined in a protest against the dismissal. The

Engineers' Association called for the closing down of the Technion. The Teachers' Union rallied to Grossman's defense and when they heard that Prof. Tcherniavsky had been asked to teach Grossman's mathematics classes they forbade him from doing so. He refused to heed, on the grounds that the Histadrut was not acquainted with all the facts in the case. This confrontation could have led to the closure of the Technion completely, but the Histadrut did not wish to take such a drastic action and therefore obtained from Tcherniavsky a pledge that he would yield up those classes if Grossman were returned to the staff.

The *Vaad Menahel* responded to the public criticism by maintaining that in actuality it had not fired Grossman. All the teachers had been dismissed on the eve of the 1930–31 year and the new Board was free to re-engage whomever it wished. This had been clearly understood by all. When Grossman was Acting Director he had used this very argument against the re-engaging of other teachers, among them Badian.

The Histadrut suggested that a truce be declared for a year, but Kaplansky refused and proposed arbitration instead. For four weeks the air was thick with hostility until the *Vaad Menahel* and the Teachers' Union agreed to the composition of a neutral committee to pass judgment in L'affaire Grossman.

The committee was composed of Dr. Moshe Smoira (later to be first President of the Israel Supreme Court), Haim Salomon and A. Krauze. They held six sessions, heard testimony, and on February 23, 1933, came to a unanimous decision:

1. On legal grounds the dismissal was not justified.
2. On moral and public grounds, it was true that Prof. Grossman did make life difficult for the directorate and did unjustly criticize Kaplansky, but the committee members were of the opinion that he did so in the honest belief that he was helping the Technion. There were therefore insufficient grounds to support the dismissal.

For the sake of peace they accepted Grossman's own offer to reduce his teaching program and to abstain from any non-academic activity in the institution.

Kaplansky considered this decision a criticism of his own actions, and he presented his resignation. The *Vaad Menahel* refused to accept it and only upon its insistence did he agree to remain, but on condition that internal labor regulations be reformed and some solution be found to the ever-present budgetary problem.

Parenthetically it must be noted that the tension between Grossman and Kaplansky was never fully dissipated. A few short years thereafter the Director again confronted his professor of mathematics, this time on an academic matter. In the final examinations in Math II in 1936 Grossman failed 85% of his students. A year later, only five students out of 57 passed Math I, and only 17 out of 79 in Math II. To Kaplansky these results were highly unreasonable and he politely but firmly insisted that Grossman review both his teaching program and his examination standards.

Despite the sympathetic hearing given Kaplansky at the Actions Committee meeting in London in the summer of 1932, the Jewish Agency Executive was still uncooperative, and there was a possibility that no appropriation at all would be made to the Technion. Kaplansky pleaded that he had applications for admission from 70 qualified candidates, 20 from Palestine and most of the remainder from Poland. If the Executive were to make it impossible to accept these students, he warned, it would be committing a serious crime. The actual sums involved were paltry: £1,000 on account of the year 1931–32, and £1,500 toward the 1932–33 budget.

If the hostility to the Technion was based on political grounds, because Kaplansky was a member of the left wing of the Labor Party, he was prepared to leave. In any event, he warned that he would not and could not continue unless a proper budget could be planned and adhered to. He protested that the Zionist Congress and the Zionist Actions Committee had both affirmed the continued existence of the Technion, but the Executive in Jerusalem was by its actions nullifying those decisions.

The Technion needed internal reorganization as well; two teachers should have been discharged, but the Institute could not let them go for lack of money to pay them their back wages. The Hadar Hacarmel Neighborhood Council owed the Technion some £3,000 for water and other services rendered, but no one had any great expectation that this debt would ever be collected.

The tensions and the pressures were undiminished. Life in Palestine and in the Zionist movement went on as before, oblivious of events in Europe that were to alter the course of history and change the face of the world. On January 30, 1933, Adolf Hitler became Chancellor of Germany; on February 27 the German Reichstag building went up in flames.

Growth and Progress

Though he had had far more difficulties and problems than he had anticipated, Kaplansky could also look back upon progress and achievements as he guided the destinies of the Technion into his second and third year. When he had assumed leadership in 1931, only 15 new students had enrolled because of the uncertainties of continuation. In the year 1932–33 there were 51 first-year students, and the following year 103 new applicants were received. The 1934–35 year saw 172 new students, and by 1935

Kaplansky could tell the Zionist Congress that he had in hand a thousand applications but could accept only 200. As a matter of fact the freshman class in that year numbered well over 300, of whom 230 were from overseas and 75 from Palestine. Some 60% of the new students were from Poland, and about 20% from other countries. Among them were 25 girls.

In 1936 the Polish Government, having examined the Technion's credentials, formally recognized the school as a professional institution of high standing, and qualified students from Poland who were accepted at the Technion were granted exemption from their compulsory military service at home. There was an immediate sharp rise in applications from that country. The Polish Consulate in Palestine proved cooperative both in assisting with the processing of applications and with the transfer of funds from the parents in Poland to their sons enrolled in Haifa.

The large number of applications from abroad may be explained in both political and educational terms. Entry visas into Palestine for students became easier to obtain when the British Government agreed to recognize Technion's guarantee on such visas, instead of putting them through Jewish Agency channels, as had been done previously. It was no secret, at least not in Zionist circles, that at this time and during a number of years to come, large numbers of young people escaped from Europe and obtained entry into Palestine under the guise of Technion students. When they reached Haifa they dropped out and were absorbed into the general community. There were occasional scandals about this which endangered the whole system, but there is no doubt that Technion's role in this glorious chapter resulted in the saving of many hundreds of lives.

The events in Germany were driving to Palestine's shores more than students. Soon after Hitler came to power the Technion began negotiations with Jewish scientists and academicians who expressed an interest in leaving Germany. Funds were required to enable absorption of the emigres, and in September, 1933, Kaplansky wrote to the Central British Fund for German Jewry in London, asking for its financial support. Dr. Albert Einstein suggested names of men who would be suitable and might be interested, and wrote a strong letter of support to the London Committee. "Some time ago I had an opportunity of discussing with Mr. Kaplansky the names of professors available for this task and ready to go to Palestine," Einstein wrote. "This is a great opportunity of creating in Palestine an important center of Jewish technical science. I therefore earnestly appeal to your committee to give its wholehearted support to the Technion and to enable it to save for Palestine some of the best Jewish technical brains."

London was interested and asked for names. Agreement was reached on Edwin Schwerin, Max Kurrein, Fritz Naphtali and others, who did come to the Technion. The first two remained with the Institute until their retirement many years later. Another name suggested was that of Dr. George Schlesinger who had helped Paul Nathan establish the school some 25 years earlier. However, there was no way of ascertaining if Schlesinger would be interested, it was noted, because "he is still detained in the Moabit prison in Berlin." Schlesinger spent nine months in the Nazi jail, was finally released and escaped to England where he built a new career. Erich Mendelsohn, the architect, whom the Technion wanted badly, decided to settle in London. Others who came later, from Germany and elsewhere, included Izhak Haberschaim, Hugo Heimann, Guido Tzerkowitz and Alexander Klein. Still another was Franz Ollendorff, who had visited the country as a tourist in 1926, had been hailed then as a brilliant young scientist, and had given several scientific lectures at the Technion. This time he came for good. There were others as well, and most of them found a warm home at the Technion where they could continue their creative careers. Enrichment of the academic staff with personnel of such high standing contributed considerably to the raising of the standards of the Institute as a whole, and made possible the introduction of new disciplines. There were sources other than Germany, as well. In 1934 Dr. Luisa Bonfiglioli came from Italy, and eventually became a full Professor, teaching Descriptive Geometry to successive generations of students until her retirement in 1973.

Prof. Max Kurrein (1878–1967).

Like their predecessors almost a decade earlier, the newcomers did not always know Hebrew, and some never did master the tongue. The tales were legion of the malapropisms uttered in class, and Prof. Max Kurrein was the central figure in many of them. His laboratory was in the basement of the main building, and he would sometimes send a short message in Hebrew to Dr. Kaplansky asking the latter to "come to my underside". Another classic was Kurrein's annual account to a new class of his engineering background. Before coming to Palestine he used to work in an armaments *(neshek)* plant in Germany, but what came out was a *neshikot* (kisses) factory. The tale was a tradition; it was awaited annually, and always brought a subdued chuckle. There was common agreement among the students never to reveal the error to the good professor, and so he repeated it year after year.

Enhancement of the academic stature of the Technion finally brought about belated formal recognition of the need to fix its status once and for all. The Jewish Agency Executive had objected even to use of the word "diploma" in the certificates given to the first graduates. What they received attested only to the fact that the graduate had satisfactorily completed a four-year course of studies and had successfully passed a set of examinations. There was no degree and no title, though the students had continued their battle unceasingly for formal academic recognition. Engineers practicing in the country were accepted as members of the Engineers' Association, sometimes even if they lacked academic qualifications.

Since 1930 Technion's final examinations had been conducted under the supervision of an appropriate representative designated by the Mandatory Government. The diploma requirements also called for a year of practical work after completion of the four years of study. Planning the tenth anniversary of the opening of the Technion, the *Vaad Menahel* in 1934 came to the conclusion that ten years was long enough to go on without definitive clarification of the status of the Technion, and application was made to the Jewish Agency to permit Technion to grant degrees, as the University was doing. Formal application was also made to the *Vaad Leumi*, and the latter sought the opinions of Eliahu Berligne and Yitzhak Ben Zvi. On the basis of their recommendations the Technion was at last authorized to grant the professional title of *Ingenieur*, and the graduates could add to their names the identification: Ing.

The graduating class of 1935 was the first to receive the new diplomas, and the degree was authorized retroactively to the previous graduates. This brought to an end at last the battle which had been waged so bitterly and for so long with respect to the level of the Technion's academic program. The granting of Bachelor, Master and Doctor degrees came much later.

The academic burgeoning of the Institute was matched by a flourishing physical development as well. Kaplansky, a consecrated, zealous Socialist, also had bourgeois financial acumen. As the Hadar Hacarmel neighborhood grew in population and business, he came to see the commercial possibilities in the long frontage of the Technion site facing on Herzl Street, which was emerging as the main shopping thoroughfare of the area. As early as 1933 it was proposed to seek permission from the Zionist authorities to put up a building on Herzl Street, which would contain stores and offices, and would become a source of income for the Institute. Since the plot was occupied jointly with the Reali School, the latter was made a partner in the enterprise as well. A year later approval was granted and plans were drawn up. An unexpected difficulty arose. In 1932 the Maccabi Sports Organization had been given permission to erect a gymnasium at the corner of the Hadar site, and Reali School pupils were given rights to use the facility. Maccabi had gained squatters' rights by its occupancy, and a protracted dispute ensued before the group could be bought out, and given funds to enable it to construct its own building elsewhere. It was a bitter lesson, but it was to be repeated some years later with respect to another squatter.

In May, 1935, the contract was approved with agreement of the Keren

Hayesod and the Jewish National Fund, and soon thereafter construction began on what has since become known as *Bet Hakranot*, a commercial and shopping center of considerable size. In the intervening years additional floors have been added, and the structure constitutes yet another monument to Kaplansky's vision and energy.

The pace quickened and the plans for the future developed simultaneously along three lines: academic expansion, physical development, and broadening of the technical training program.

Kaplansky's decision at this time to open a new department of industrial technology was a master stroke of vision. As he recalled it some years later, the act was a daring one under prevailing conditions. The new department was set up without laboratory or equipment, without teaching staff, and even with some doubt if the graduates would be absorbed into local industry. He explained that they were motivated by a sense of historic and economic needs. The immediate purpose was to provide an academic haven for the refugee scientists who were coming out of Germany and to seek to use their talents. In this there was no overriding

Physics Laboratory. 1938/39.

Zionist policy; each organization operated on its own, and the Technion leaders moved in accordance with what seemed to them necessary. Perhaps their actions were intuitive, growing out of the situation just beginning to unfold in Europe. As early as 1934 Kaplansky was already referring to events there as the "German catastrophe".

Industrial technology covered a broad field and included options in electro-mechanical and chemical engineering, all made possible by the availability of suitable teachers. A curriculum in town planning was added to the Department of Architecture. By 1935 the teaching staff had reached 30. A year later it went up by another 10.

The Materials Testing Laboratory was by then in full operation. Its work included research projects, mainly industrial, and systematic tests of materials used in building. A 200-ton Universal testing machine of the Amsler type, presented by the Palestine Economic Corp. of New York to the Hebrew University, could not be used there and with permission was passed on to the Technion. Several years later Shragga Irmay of Haifa found another Amsler machine under wraps at the Hebrew University Physics Laboratory. It had been received from Germany through the *Haavara* arrangement and had been intended for the Technion, but ended up in Jerusalem instead. It was not suitable for the research work there, and took up precious space. Irmay hastened to inform the Technion, and this piece of equipment, too, found its way to the Building Materials Laboratory. Both machines are in use to this day. The Herzl Club of Vienna presented a concrete mixer.

A laboratory for industrial chemistry was opened in 1936. Plans were drawn up for a laboratory to do research on soil for foundations. A laboratory for sanitary engineering was the basis for what later became a Department of Environmental Engineering.

Kaplansky also made plans to increase Technion's annual intake of new students to at least 100 per year, with a view to graduating some 60 engineers on the completion of each four-year program. He was convinced they would be absorbed into the economy. A field program was launched to provide technical guidance for kibbutzim and farm villages. Technion's roving instructor visited dozens of farms to provide the expert advice required. Rahel Shalon and Joseph Edelman, serving as volunteers, visited many scores of kibbutzim, conducting a year course to provide guidance and advice for those in charge of building construction in the settlements.

The trade school and evening school programs were likewise expanding under the competent direction of Yaakov Erlik. In November, 1932, Dr. Biram recommended Dr. Shlomo Bardin, who presented a detailed program for the setting up of a comprehensive vocational school. A trade

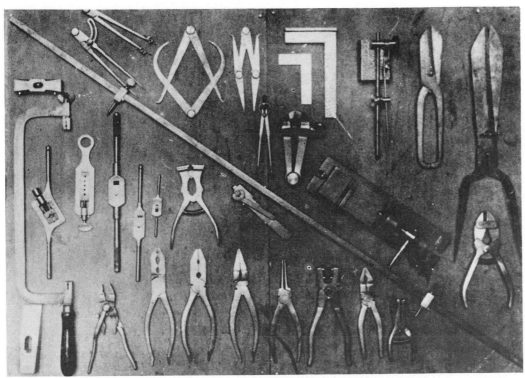

Tools made by the pupils of the Technical High School. 1935.

school had actually begun operations in January, 1929, with 14 pupils. In the fall of that year another class of 25 was added. By 1931 there were 55 pupils studying metal and electrical trades. Plans for a carpentry shop had to be deferred for lack of funds. The three-year program called for five hours daily in shop work and three hours in classroom studies. Bardin sought to elevate the standards, and called for a four-year Artisans' High School for pupils between the ages of 14 and 18.

Biram was extremely anxious to have his Reali School involved, and sought to have the new school operated jointly by Technion and the Reali. After some initial exploration in this direction, the Technion decided to run the school by itself. The relations with the Reali School were on the whole amicable, with occasional clashes. A written agreement had been reached with respect to division of the area on which both institutions were located, and there were sometimes differences of opinion on the precise boundary lines. For some years the Reali pupils were enabled to utilize the Technion laboratories, but after continuing disputes Dr. Biram finally (in 1946) set up his own laboratories and brought about a complete separation of the two schools.

The Artisans' High School began in 1933 in accordance with Bardin's program, and with Bardin as Director. New courses were introduced for mechanics, plumbers and electricians, and the academic program was broadened to include non-technical subjects as well. In 1937 the old program and the new were combined into what became the Technion's Technical High School (*Bosmat*).

As the number of German immigrants into the country increased there was ever-growing need for retraining of professionals and white collar workers in vocations providing employment. The Technion's courses in welding, sanitary installations, carpentry, painting, plastering, etc. helped make it possible for large numbers of the new arrivals to be absorbed into the economy. An early 1934 report put it statistically: 13 university graduates, 26 merchants, nine officials, 20 unskilled laborers and 12 of miscellaneous occupations were transformed into 39 concrete mold and form makers, 22 iron workers, 10 carpenters and nine plumbers. Because engineers associated with the building industry could always be absorbed, German engineers in other fields were retrained in Civil Engineering. Technion staff members volunteered to teach these classes, which were conducted in German. All these immigrants were able to earn a satisfactory livelihood.

Kaplansky looked ahead. He needed facilities for a hydraulics laboratory. He wanted a library building, and suitable quarters for a student center and restaurant. There was need for electronics and chemistry laboratories. The architecture department, long dormant, was expanding and needed space. He ventured to draw up an expansion and development budget, totalling £20,000 and believed that with joint participation of the Zionist Executive, the Mandatory Government and friends abroad, this budget could be attained.

Turning a Financial Corner

His vision was projected against a background of a still-existing deficit and a continued drop in the Jewish Agency grant, which now reached only about £1,000 a year. Nevertheless, the improvement in the financial situation was amazing. In 1931–32 the budget was covered 70% by the Jewish Agency grant, and 30% from funds provided by the Technion itself, principally tuition fees and direct contributions. Within three years the grant constituted only 20% of the budget, and the Technion itself covered the remaining 80%. How was this miracle accomplished?

For one thing tuition fees were raised from £18 to £24 per year, and for the first time there was systematic enforcement of payment of full tuition fees by all who could afford to do so. The increase in number of students further boosted income from this source. The willingness of the staff members to defer payment of their salaries, in whole or in part, for the 1931–32 year should also not be forgotten.

Contributions from overseas also began to increase, in large part a delayed result of Kaplansky's near-fatal extended visit to Europe in 1932, and in part because of the resources available for aid to German refugees. Technion's role in providing a place for German scientists and engineers was recognized and the Central British Fund for Help to Refugees from Germany, headed by Herbert Samuel, gave funds that exceeded considerably the grant from the Jewish Agency. The Union of Jewish Communities in Italy, Bnei Brith of Czechoslovakia and others made generous donations. The large stone gravity saw given in 1934 by the Federation of Synagogues in London is to this day still in use in the Building Materials Testing Laboratory.

American Jewry still stood aloof. A Brooklyn engineer, Simon Cooper, who visited the Technion in 1932, offered his help. He received a letter from Grossman suggesting that he contact the "Provisional Stirring [sic] Committee" in New York. There are references in several letters to this same committee, in which M.H. Sugarman and B.H. Halpern were active, but apparently they did not succeed in stirring up enough interest in the Technion. The one exception was an old friend, Mrs. Frieda Schiff-Warburg, who in 1934 gave £3,000 for the new Vocational High School. Later she added £3,000 on her 60th birthday, which made possible the opening of a department of automobile mechanics.

The outstanding debt from the Hadar Hacarmel Neighborhood Council was renegotiated and opportunity given to pay it off in small installments so that at least some of it was salvaged.

In 1934 the High Commissioner for Palestine announced that the Government would make a contribution of £10,000, which later made possible construction of new physical facilities for the vocational school and for the new department of hydraulic engineering. Engineers in the civil government were well aware of the Technion's functions and appreciated what it meant to the development of the country's economy. There was an affinity to the military as well. Kisch reported a dinner in June, 1931, with members of the Royal Corps of Engineers. Because the relations between the Jews and the Government were strained at the time, every effort was made to avoid political talk, but one of the speakers could not avoid quoting the lines from one of Kipling's poems:

"When the Jews had a fight at the foot of a hill
Young Joshua ordered the sun to stand still
For he was a Captain of Engineers.
When the Children of Israel made bricks without straw

They were learnin' the regular work of our Corps,
The work of the Royal Engineers."

Appreciation of the Technion could flourish in such an atmosphere.

The gradually improving situation did not deter Kaplansky from exploring every possible avenue which might lead to a long-term solution of the Technion's perennial financial problems. In December, 1933, he once again looked into the possibility of some kind of organic relationship with the Hebrew University. A committee of three, Dr. Redcliffe Salaman, Sir Philip Hartog and Prof. Louis Ginzberg, which had been charged with a general investigation into the whole structure of the University, visited the Technion, made a careful inspection of its facilities and met with Kaplansky, Aleinikoff and Breuer. The visitors made it clear that they had no authority to negotiate regarding coordination, but were happy to hear Haifa's views.

Kaplansky reviewed the case carefully. Technion had in the past been favorably disposed to becoming the engineering branch of the University, and he believed that the *Vaad Menahel* would be willing to consider this again. Because of the difference in the character of the two institutions, and their separate geographical locations, it should be clear that a degree of autonomy for the Technion would be required. He saw two organizational possibilities.

If organic union were found advisable, the Technion could become the University's Faculty of Engineering, perhaps along the lines of Imperial College and the University of London. If it were preferable to maintain complete separation, then there were many areas for joint coordination, for example, agreement to avoid duplication, such as the University's opening a competing department of engineering; University participation in the Technion's diploma examinations; joint invitation to distinguished visiting professors from abroad and even joint operation of certain departments, such as agricultural engineering; and finally, joint fund-raising.

The Jerusalem visitors heard the proposals with interest, promised to give the matter consideration, and they parted in "warm friendship". Technion's *Vaad Menahel* received a report on the visit and authorized the Director to continue the negotiations leading to either union or closer cooperation with the University.

In July, 1934, Kaplansky and the University's Judah Magnes met and continued the discussion along similar lines. Other areas of close cooperation which were mentioned, in addition to agricultural engineering, were chemistry, physics and hydraulics. The two university heads seemed to agree on many aspects of the outline and they looked forward to continuing the discussions, but the subject did not again reach a serious stage of consideration until 1938.

The *Vaad Menahel*, Technion's administrative committee, was renewed annually, with occasional resignations and additions. In 1935 the membership was E. Berligne, chairman; M. Aleinikoff, treasurer; Dr. Berkson, S.D. Sourasky, Z.D. Pinkas, E. Kaplan, B. Rosenblatt, I. Reiser, Prof. J. Breuer and Dr. Kaplansky.

Para-Military Activities

With all their resources strained to provide home and haven for the immigrants from Europe, the Jews of Palestine were subjected to new tensions growing out of the Arab unrest and riots beginning in April, 1936. From the very days of its opening the Technion had come to play a central role in the defense program of the Jewish community. As early as 1924, at the initiative of Yaakov Pat, two rooms had been set aside in the workshop building where members of the Haganah, the defense organization, could store their supplies and practice with weapons. Much later the main room was part of the toilet and washroom and was known as *Hadar Habrazim* (the Faucet Room). The "slick" (hidden cache of weapons) was in the wall behind a dirty cabinet. Pressure on a board behind the cabinet opened up the wall. When training was going on, or the slick was open, there was need for careful watch through the still sparsely settled area of Hadar Hacarmel, and guards were posted at the corners of the Technion buildings. At first each guard had an electric button which he could press to give alarm if a suspicious stranger approached. However, Yosef Baratz, student, teacher and eventually Director of the Technical High School workshops, early pointed out that in case of a surprise the guard might bolt, leaving the telltale button, and British soldiers could quickly follow the wire, like Ariadne's thread, and it would lead them straight to the hidden chamber. Baratz changed the system. First he ran the wire from the alarm buzzer up through the ceiling to the roof, then around the building, inside the walls and out of sight, and finally after circuitous loops brought the wire to several points on the outside walls where he made unobtrusive plug sockets. The guard carried only his button with a short stretch of wire attached to a plug. When he went on duty he plugged in to one of the sockets. When he left his post he unplugged. Some of these sockets, serving no earthly purpose today, are still to be seen outside the old building.

Perhaps the first "public" appearance of Technion's Haganah members was in 1926 when Ze'ev Jabotinsky visited Haifa. His private, unobtrusive bodyguard was composed of three Technion students.

In 1929, when big British warships were in the Bay, their powerful searchlights would from time to time brightly illuminate the front win-

Ze'ev Jabotinsky speaking from the steps of the Technion main building; in the foreground his student bodyguard. 1926.

dows of the Technion, and personnel inside, engaged in illegal activity, had to make sure they stooped every time they approached a window, lest the searchlights pick them out. The area Haganah headquarters was in the Technion's main building. During the 1929 Arab riots Nahum Levin was in charge of the Technion unit, and Rahel Shalon was his deputy. Shlomo Ginsburg (Shaag) also became an early ranking officer.

For more than two decades Technion personnel was engaged in Haganah activity at every level, and Technion buildings were utilized for these purposes. The activity expanded during periods of stress such as the Arab riots, or protests against British policies. The following pages record only some selected reports on these activities, for the most part prior to World War II; the period embracing the war years will be dealt with in a later chapter.

The Technion workshops also engaged in manufacture of vital items. Cartridge clips were frequently lost in action and the workshop made clips for the parabellum then in common use. They also produced many parts for guns, except the barrel. Many dozens of signalling keys for use on both lamps and wireless sets were also made.

For years the workshops produced heliographs for signalling. The Lucas light was adapted for signalling purposes, using the Morse code. A place on the Hadar, just outside the Technion campus, was the main signalling point to Hanita up in the Galilee hills during the early days of that colony. In the frequent morning fogs the flashes from the Hadar were not visible at Hanita, so Baratz and his associates built a shutter on a powerful Bausch lamp which was able to penetrate the fog, and more than once provided the Hanita settlers with life-saving information.

Ratner permitted the use of the balcony of his home on Panorama Road for the signalling. Vision was clear and unobstructed from there, across Haifa Bay to the top of the Galilee hills. There was nothing illegal about such signalling, and the British obviously were aware it was going on right under their noses, but they apparently never cracked the code to enable them to determine the texts of the messages.

No hidden corner was overlooked. The well in the central courtyard, it will be recalled, went down some 100 meters. There was a narrow staircase inside leading down. Sometimes they would test repaired weapons by descending into the well and shooting bullets into the water. Not a

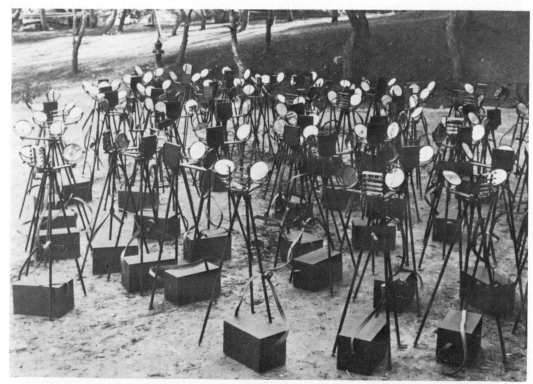

Heliograph signalling instruments produced in the Technical High School. 1947

Prof. Alexander Ilioff (1860–1942).

sound could be heard above. At other times pistols were sometimes fired in the basement, with a long corridor serving as the firing range. The noise of the machinery in the workshops above drowned out the sound of the shots.

Prof. Ilioff produced a hand-thrown bottle bomb which was known as the Ilioff Cocktail. It was mass-produced in the War of Independence, and the Syrian tank halted at the perimeter of Kibbutz Degania is said to have been a victim of the device. Later came a demand for an instrument that could throw the Ilioff Cocktail and other small bombs a considerable distance. Baratz went to the Technion library and perused drawings of the Roman catapults as they appeared in various encyclopedias. They were ingenious in those days, too, and he adapted the Roman devices, substituting springs for their supple wooden limbs. As matters turned out, Technion's Roman catapult, though produced, was never used. In still another field, Shimshon Turetzky (Ben-Tur), of the Chemistry Department, created mixtures which when ignited in a small *leben* cup, produced a smoke screen.

Services to the Haganah took many and varied forms. Avraham Akavia, while still a student, did much of the translating into Hebrew and the editing of the military texts used by the Haganah. Saadia Goldberg both translated, and where necessary coined new military terms in Hebrew.

There were services of a stranger but no less valuable nature. A welding instructor in the technical school workshop, Yaakov Zacks, had been assigned the task of cultivating the friendship of the British police. Zacks loved beer, and he used to invite the policeman on duty to drink with him. They would sit for hours, and in the meantime the Haganah members went about their assigned tasks in the neighborhood, knowing they would be undisturbed. A very considerable amount of beer was consumed.

Students were of course the ideal human material for the underground defense movement, and almost everyone enrolled at the Technion had his place in the Haganah. Shirkers were rare.

Two full companies were mobilized on the campus and were subject to training and call as necessary. As has been seen, the girls too were full-fledged members. During the 1930's Rahel Shalon instructed classes of mature German Jews in the handling of weapons. It came as a shock to most of them to find that their weapons instructor was a pretty slip of a girl.

Not all of the Haganah members were beginners. One day Rahel Shalon was surprised to find that one of her Technion teachers, Prof. Yohanan Ratner, had also been invited to the instructors' course. It quickly became obvious that he seemed to know more about military matters than his instructors, but he said nothing and completed the course. He had agreed to join the Haganah on condition that he be permitted to learn local methods from the bottom up. Only later did he reveal that he had been an officer in the Russian Army, first under the Czar, then in the Kerensky forces, where he was a division head, and later in the Red Army. He was quickly advanced and before long joined the top leadership of the Haganah in the country.

For years the Technion was one of the major mobilization points for the Haganah in Haifa, but it was more than just a meeting place. Shortly after the 1929 riots, when Arabs began throwing stones at Jewish cars passing along the main road through the Arab quarter in east Haifa, some members of the underground were assigned to prepare in their Technion hideout a small bomb made of water pipes stuffed with explosives. This they placed at the door of the home of one of the Arab ringleaders, as a warning that the Jews could strike back if necessary. The young men were officially reprimanded, not so much for their action, as for letting it be known that it had been the work of the Haganah because at this period an earnest effort was being made to restore peaceful relations between Jew and Arab.

With the departure of Pat for Jerusalem, Yaakov Dostrovsky (Dori) became head of the underground. He intensified the training, not only in the Technion chambers, but in the field as well. At first he went to the hills of nearby Naveh Shaanan, where approximately twenty years later, as President of the Technion, he supervised the training of the students in more peaceful pursuits. Training was later extended into nearby kibbutzim.

In 1936 the Haganah unit in Haifa was the largest in the country, numbering some 2,400, a good many of them Technion students. Selected recruiting of likely candidates went on all the time, and as more and more of the German immigrants joined the movement a new and occasionally surprising atmosphere was introduced. Once Dori summoned a meeting of the Haganah officers, who included a number of German Jews, to his headquarters at the Technion. The time set conflicted with a music concert in Haifa which, of course, all the local German Jews had planned to attend. They protested the timing. Dori was furious at first, but relented and deferred the proposed meeting until much later in the evening. The moment the concert had finished, the Germans sent their wives home and at once reported in at the Technion, still wearing their formal dress from the concert. At the command of 'tenshun!' they froze in their places. They made an incongruous picture and even the usually grim Dori burst out laughing.

The Technion provided a cover of another nature; Haganah men on missions would often pass themselves off as Technion students.

April of 1936 saw the beginning of the prolonged disturbances known as the Arab riots. Jewish refugees from downtown Haifa were temporarily housed in the new Hydraulics Laboratory. The same place also became the center for another course in weapons training and a Haganah workshop was opened under the basement floor. Armed confrontations took place throughout the country and tension heightened. The students were called upon for missions, many of which took them far out into the countryside. The service they gave was at the sacrifice of their studies. Night drilling left them sleepy the following day.

The students served with devotion and energy, but the circumstances took a toll. The 1936 graduation exercises had to be cancelled, and no plans were made for the 1937 graduation. As a matter of fact in both years only 19 graduates had managed to complete their courses and pass the examinations. The remainder had missed major parts of the program, or had even dropped out altogether. The somber mood in the country also made it difficult for young people to devote years to study for a future which was so uncertain.

Informed that the Jewish Agency was again to reduce its grant, Kaplansky in June, 1936, sent a desperate appeal for help to the Joint Distribution Committee in New York, the Anglo-Jewish Association in London, and the Jewish Colonization Association in Paris. To each he wrote that "... the collapse of the Haifa Technical Institute, which seems inevitable if no immediate and lasting assistance is rendered, would be an irreparable loss to Jewish Palestine and a serious blow to the Jewish economic prospects in this key position of the Near East ...." There were no encouraging replies.

Morale Rises

Kaplansky determined that he had to do something to bolster sagging morale and restore hope, faith and optimism. The year 1937 marked twenty-five years since the laying of the Technion cornerstone. After approval of the Zionist Executive and the *Vaad Leumi* he decided to make the occasion festive, even dramatic.

It was a gala event. The British High Commissioner, Sir Arthur Wauchope, was the distinguished guest of honor and formally unveiled a bust of Schmaryahu Levin, executed by the sculptor Moshe Dykar, and presented to the Technion by Morris Eisenman. Levin had died two years earlier. The engineering diplomas were presented to the small class of graduates. The Chairman of the Jewish Agency, David Ben Gurion, also came. Appropriate greetings were extended by the Mayor of Haifa, Hassan Shukri, to a large audience which included Jews, Arabs and British offi-

In celebration of the 25th anniversary. 1937.

cials. Among those present were the widow of Prof. Baerwald, Shabbetai Levi, then Vice Mayor of Haifa, Gedalyahu Wilbushevitz, Yosef Barsky and Nathan Kaiserman, all people who had had a share in the creation of the Technion.

When the Technion had in 1934 marked the tenth year of its opening Kaiserman had not been invited to the party. It had been an oversight, but he had been hurt. He had at the time written a note to the Technion, reviewing his role in acquiring the land, obtaining the permits and completing the construction of the building. "That shows how history is written among us," he wrote sadly. This time he was one of the distinguished and honored guests of the celebration.

Kaplansky reviewed the events during the fateful years 1912—37, from the World War to the rise of Hitler, and the rapid growth of Jewish Palestine. He paid special tribute to Baerwald on the occasion of an exhibit of his major architectural plans.

Technion's Director was beginning to overcome the initial difficulties and problems. Though many serious problems still lay ahead, he was becoming more the master of the situation and could provide that degree of leadership which had for so long been absent. He could even afford the luxury of looking forward to a more distant future: "The time will come

Sir Arthur Wauchope, British High Commissioner, addresses the 25th anniversary celebration. To his right, Dr. Kaplansky and David Ben Gurion. 1937.

when we shall be free of materialistic problems and I can go back to my first love, when I was a teacher in a small suburb of Jaffa, in the Gymnasium, teaching Descriptive Geometry. Then I can teach something I know, and become a teacher among teachers," he wrote on one occasion.

He had also learned to be diplomatic. At a time when the Jewish Agency was cutting and withholding budgetary allocations he could be gracious enough in a public statement to pay tribute to the Zionist agencies which had rescued the Technion, seen to its opening and had made possible its continuance as a Zionist project.

Though personally far removed from religious feelings, he was mindful of the sensibilities of others, and of the status of the Technion as a national institution. The Maccabi sports organization maintained its headquarters on the Technion campus, and its parallel group in the labor movement, Hapoel, asked for similar privileges. Kaplansky objected, but in any event he warned that the sports organizations must not violate the Sabbath on Technion premises, or they would have to leave.

He dealt not only with the prosaic, mundane matters of administration and finance but also with underlying, basic academic philosophies. In an address to the Palestine Society of Friends of the Technion in 1934 he spoke of the public discussion then going on with respect to teaching and research; which should be given priority? He did not want to argue with such authorities as Ussishkin and Bialik, he said, but in technical science such a conflict did not exist. Theory and application, learning and life, could not be separated in the technical sphere, he declared.

And in response to the anxiety expressed by some that Palestine might be producing too many engineers and other academically trained professionals, he granted that the influx of intellectuals from Germany might indeed create a problem. However, he urged that questions of principle should not be considered under pressure of catastrophe. Such matters should be viewed in the light of what the future might hold in store in the ensuing five or ten years, he stressed.

Kaplansky had emerged as a major personality in the Haifa community, and was increasingly called upon to take a role in public affairs not immediately associated with the Technion. Thus in early 1936 he took up the cudgels for another cause — the movement to have Theodor Herzl's remains brought to Palestine for reinterment. There was almost general agreement, on the basis of incontrovertible evidence, both written and oral, that Herzl himself had desired to be buried on Mt. Carmel. On February 18, 1936, Kaplansky appeared as spokesman for a Haifa committee at a meeting of the Zionist Actions Committee in Jerusalem, at which this subject was taken up. The local committee had already selected a suitable location for the burial shrine, not far from Ahuza. The Zionists were in no position to take action at the time, and when Herzl's remains were finally brought to Israel in 1949, national-political grounds were found for overruling his own wishes, and he was reinterred in Israel's capital, Jerusalem.

Technion leadership was called upon to study major political matters following the report of the Peel Royal Commission in 1937. That report, it may be recalled, proposed partitioning of Palestine and establishment of a truncated Jewish State which would not include Haifa, Tiberias, Acre and Safed. Although rejected out of hand by the Zionists, the report was nevertheless subjected to careful study, and the Jewish Agency set up brain trust groups to analyze it. A special Haifa committee was asked to document every aspect of the separation of Haifa from the Jewish State. Members of the committee were Kaplansky, Aleinikoff, Frederick Kisch, David Hacohen, Abba Khoushy, Dr. M. Soloweitchik and Dr. Aaron Barth.

The first meeting took place in Kaplansky's office at the Technion.

More than a dozen meetings were held, and full details were presented showing what the exclusion of Haifa would mean. Since the British viewed everything in terms of their interests, the committee went to great lengths to show that making Haifa part of the Jewish State would in no way endanger British naval interests. The point was made that it would be unfair to the large Jewish population in the city to place them outside the borders of the State. Suggestions were advanced for protecting Arab interests. In the event that decision went against the Zionists, the committee called for detailed assurances as to how Jewish interests would be protected. The discussions continued from November, 1937, until well into May, 1938.

In the cloudy months of 1938 and 1939 no one could have predicted what was to come. Even against mounting, ominous signs, perhaps in unconscious preparation for vital tests ahead, the Technion continued to grow. The new hydraulics laboratory building, constructed with the help of a Government grant, was completed and put into use. The Institute's academic program was organized in three departments: Civil Engineering, Architecture and Industrial Engineering, with the latter providing specialization in electro-mechanical and chemical engineering. The student exercises as well as research were carried out in 11 laboratories: Building Materials Testing, Hydraulics, Electrical, Heat Engines, Metallography, Metrology, Industrial Chemistry, General and Inorganic Chemistry, Physics, Physical Chemistry and Soil Mechanics.

The library boasted a collection of about 6,000 volumes. The workshops, used by both Technion students and the pupils of the Technical High School, comprised a well-equipped mechanical shop, a forge shop, a welding department, a motorcar repair shop and a carpentry shop.

By 1938 the number of students had reached 499, two-thirds of them from Poland. The teaching staff consisted of 43 individuals, 26 of them listed in senior rank. Of the latter, 13 came from Germany, four from Russia, three from England, and one each from the U.S., Czechoslovakia, France, Hungary, Switzerland and Holland. There was an impressive list of research papers and scientific publications by staff members. The research projects in which staff members were engaged covered a wide range of subjects and included, among others, building materials, irrigation problems, chemical fertilizers and — reflecting the times — projects in development of landmine detectors and production of activated carbon (for use in gas masks) from raw materials locally available.

A new academic constitution was formally approved on January 24, 1939, and the first Council of Professors was set up (later to be known as the Senate) composed of Professors Ilioff, Tcherniavsky, Kurrein and

Schwerin. In July three more professors were added to the elite list, Ollendorff, Breuer and Ratner. Neumann also became a member. The Deans of the three Departments were: Civil Engineering, Prof. Breuer; Industrial Engineering, Prof. Kurrein; Architecture, Prof. Ratner.

The Technion was beginning to feel and act like a university, and one day in 1939 it occurred to the academic authorities that they could also confer honorary degrees. Two names were proposed: Menahem Ussishkin and Pinhas Rutenberg. It seemed presumptuous to think of granting two such honorary degrees at one time, and the decision fell on Rutenberg. The Council of Professors formally recommended that on his 60th birthday Rutenberg be named an Honorary Engineer of Technion in recognition of his activity in advancing the technical development of the country. The proposal was discussed with his brother, Abraham, and then with Pinhas Rutenberg himself. The idea got no further. It appears that the very idea of such an honor was repugnant to Rutenberg and he asked to be excused. In writing his will some two years later he again expressed his opposition to projects or memorials in his name. Crestfallen, the Technion academicians waited seven years before they again suggested a name for an honorary degree.

As a university the Institute was beginning to experience some of the problems familiar to institutions of higher learning. Despite construction of the new hydraulics building there was still dire need for more space. The basements of the workshop were adapted for the electrical laboratories. A Technion leaflet in 1938 described them as being housed in "rooms hewn into the living rock of Mount Carmel." It sounded exotic, but what it really meant was that Prof. Ollendorff was in a dingy cellar without proper light or ventilation, as an American visitor, Harry Fischbach, observed to his horror when he visited him later. The American was so moved that within a few years he made possible construction of a magnificent new building to house the studies and research in electrical engineering. Until then, Ollendorff continued in his basement rooms which had originally been part of the military wartime stables, and were still divided up into stalls. The partitions were not destroyed, but the stalls were cleaned up, painted and transformed into offices and study cubicles, with appropriate electrical connections and panels. Ollendorff wryly observed that in effect very little had changed. Originally there were horses there; now they had horse power. The first five students did the remodelling themselves, working long hours late into the nights, mixing study with physical installation work.

After the place was tidied up, Prof. Ollendorff invited a representative of the Electric Corporation to visit the laboratory. Proudly he showed the visitor the miniatures and models and equipment on which the students were trained. The Electric Company man was disdainful and told the professor that at the power plant they had huge machinery, many times bigger than what he was being shown. Replied Ollendorff: "The difference is that you have a *Tahanat Koach* (Power Station), whereas we have a *Tahanat Moach* (Brain Station)."

In an entirely different, but no less important area, agriculture, the Technion was extending and expanding its operations. Professional advice was given on agricultural machinery of all kinds, including proper maintenance, repairs and adaptation to special conditions. Field courses were also conducted on both practical and theoretical levels. An instructor visited about twenty settlements a month, reaching some fifty colonies in all. In addition the Technion conducted intensive one-month winter courses in Haifa, attended by a representative of each settlement. Aid was given in the development of a new form of food production a few years later when the Hydraulics Laboratory took part in studies that led to the widespread and profitable establishment of fish ponds.

The status and role of the lower-level technical training underwent clarification. The Trade School first set up in 1928 and the Technical Secondary School established alongside it in 1933 were by 1938 combined in one institution known as the Technical High School, with a total enrollment then of 220.

In October, 1938, the Technion ventured into a new field and opened a Nautical School with the participation also of the Palestine Maritime League and the Jewish Agency. Forty pupils were enrolled in the four departments: Navigation, Marine Engineering, Radio and Boatbuilding. A 300-ton sailing vessel, the *Cap Pilar*, was presented as a gift through the Palestine Maritime League in England, from Dr. Richard Seligman and Adrian Seligman. A British Naval Reserve Officer, Commander R. Stevenson Miller, was named head of navigation. The program was conducted under the aegis of the Technical High School.

Occasionally questions arose which recalled problems of the past, but they were no longer major issues. During the 1930's there was a spasm of public agitation for the uncompromising use of the Hebrew language on every occasion. Protests were directed to the Technion when a visiting lecturer from abroad addressed the Engineers' Association, in German, in a Technion lecture hall. In keeping with the spirit of the times, when the Technion received a letter from a customs clearance firm, asking for signature on certain documents to enable clearance of a shipment, the letter was indignantly returned to the sender because it had been written in English. The zeal was sometimes even more excessive. A foreign embassy, seeking to be cooperative, asked that certain academic documents which they were asked to submit to their Foreign Ministry in Europe, should be translated into English. They were told, in effect, that the language of the Technion was Hebrew, and that should be good enough for them.

Admission requirements were also sometimes called into question. Students at two technical schools in Tel Aviv complained that they could not get into the Technion because the Institute did not recognize the diplomas of their schools, and they went to the press with two pointed questions: 1. Why does Technion accept so many students from overseas, at the expense of places for Palestinians? 2. Technion demands a matriculation certificate for entry, but accepts graduates of the Vilna Jewish Technical School without such certificate; why not us?

To this the Technion replied that every qualified Palestinian who had applied had been accepted, so the students from abroad were not taking places away from local youth. With respect to the second question, Technion said it required not just a technical background but a good general education as well from candidates for admission. The Tel Aviv schools in question were told to raise their standards before they rushed into the press with complaints. The Technion would not lower its standards.

Despite the growth of the Institute and the concomitant increase in budget, Kaplansky had no major financial problems during this period. The annual operations cost between £20,000 and £23,000 per year. Toward this the Jewish Agency contributed no more than £1,000 a year. Yet in the year 1934–35 there was a small operating surplus, followed by several years with small deficits. By 1938 revenue from tuition fees, workshops and testing laboratories covered from 75% to 80% of the expenditures. A review of income and expenditures for the eight-year period from 1931 to 1939 showed the following, in pounds:

			INCOME			EXPENSES
	Jew. Agency	Contribs.	Tuition Fees	Misc.	Total Inc.	
1931–32	—	1,360	1,463	1,243	4,066	6,907
1932–33	750	548	1,956	1,265	4,519	6,112
1933–34	750	1,480	4,326	1,878	8,434	8,890
1934–35	1,250	1,874	8,274	3,643	15,041	14,425
1935–36	750	6,103	11,023	2,195	20,071	21,131
1936–37	1,000	5,593	10,740	2,845	20,180	20,610
1937–38	1,000	2,086	10,487	2,190	15,763	16,297
1938–39	1,000	5,210	11,477	3,825	21,512	23,590

The annual budget included both operating and capital items lumped together in both the expenditure and the income columns. In the above table, the large contributions received in the years 1935–36, 1936–37 and 1938–39 reflect the installments paid by the Mandatory Government on account of its pledge of £10,000. Similarly the increased expenses in the middle of the decade reflect the construction program which was taking place. The sharp increase in receipts from tuition fees is clear.

In 1938 the Technion published an honor roll of the major contributions received from overseas during the twenty-year period since the War. The list follows, as originally published. There had been other contributions, but the funds had apparently sifted through the Keren Hayesod and had been applied to previously allocated grants for Technion.

The Late Mrs. Jacob Schiff endowment	$10,000
Mr. M. Berliner, Pittsburgh	$25,000
The late Mr. Felix Warburg	$10,000
The late Lord Melchett	£ 2,500
Mr. Sigfried Hirsch	£ 2,400
Comm. Ing. G. Muggia	Lire 50,000
Anglo-Jewish Association	£ 100 (Annually)
Grossloge Bnei Brith (Prague)	Kc. 50,000
Central British Fund	£ 1,250 (For scholars from Germany)
Union of Jewish Community of Italy	Lire 50,000 (For the same purpose)
Mrs. Frieda Warburg-Schiff (New York)	$25,000 (For Technical High School)
Mr. G. Benenson	$12,500
Messrs. Schulman (*Shoolman?*) and Gordon	£510

The figures were by no means extraordinary, but Kaplansky recognized the potential, and he had great plans for the future. In 1937 he rebelled against the restrictions which the Zionist Organization had placed on independent fund-raising by the Technion, and uttered a *cri du coeur*:

"At a time when the famous Oxford University embarks on a drive for £500,000 and the University of Manchester issues an appeal for £300,000 for improvement and enlargement, a young and budding institute like the Technion will certainly be permitted to appeal to the Jewish public and the Zionist movement to share our anxieties for the near future."

New Expansion Program

A year later he drew up and published his ambitious program. First, he called for the establishment of three new departments: 1. A Department of Economics in which students could qualify for administrative and executive posts in business, industry and the civil service. 2. A Department of Agriculture (or Reclamation) Engineering to operate in conjunction with the agricultural college which the Hebrew University proposed to open. The new department would deal with machinery, irrigation, drainage and water supply. He noted that the Jewish National Fund was ready to provide land for field work as soon as the Technion could acquire machinery and equipment. 3. A Department of Navigation and Marine Engineering.

This was quite distinct from the new Nautical School; it would train higher qualified navigation officers and marine engineers.

He laid down a program for increase in the research activities, emphasizing the need for more space for those activities which served industry, like building materials, chemistry and electrical studies.

The third element in the proposed extension of the Institute was the distressing situation of Jewish students in so many countries of the Diaspora. He noted again that of 1,000 applications from abroad in one recent year, they had been able to accept only 200. Furthermore, dozens of students could be seen standing in every lecture hall because they could not find a seat.

He added a call for new buildings to provide social services for the students, and for expansion of the existing departments and the library. He did not overlook the need for equipment. "If the Jewish community wants the Hebrew Technical Institute to develop into a first-class college and to meet the demands of education and industry, it must be prepared to make a generous response to an appeal for funds," he concluded.

A major fund-raising operation requires an imaginative and devoted director. This Kaplansky found in Prof. Rudolf Samuel, the very one who had advised Max Hecker when the latter was seeking support in Germany in 1923. Samuel had come to Haifa in 1937 from India, where he had been Head of the Physics Department at the Muslim University in Aligarh. His field was physical chemistry, but it was not easy to absorb him academically. It was agreed that he should proceed to England as Technion's representative, seek to reorganize the Friends of the Technion which Kaplansky had initiated in 1932, and engage in some limited fund-raising among selected people. News of his coming produced negative reactions in London. The official bodies all requested that he delay his trip because there were too many campaigns going on, the time was not right, etc. A simultaneous inquiry to New York produced a cable from Henry Montor to the effect that he "must not come" because the multiplicity of appeals from Palestine were undermining the UPA (United Palestine Appeal) structure. Benjamin Halpern, who had brought together the Zionist engineers in 1917, tried to revive the New York group. He got encouragement and help from Israel Brodie, Robert Szold, Jehiel R. Elyachar, M.H. Sugarman, Dr. Ferdinand Sonneborn and others, but the attempt petered out.

Despite the warnings, Samuel arrived in London in January, 1938. The general reaction was: "You could not have come at a worse time. The Hebrew University has just had a campaign; go on to America."

He remained and began the slow, laborious work of gathering friends and supporters of Technion around whom he could build an organization. The master target was £20,000, of which they hoped to raise one

Looking out on the gardens from the main entrance of the old building.

third, or around £7,000 in Britain, this in two separate drives for £3,500 each. At the end of April he was joined there by Dr. Bardin, but the latter's visit was devoted almost exclusively to efforts for the Nautical School. It was difficult to obtain cooperation. Kaplansky wrote to him: "Everything points to the same moral. We must help ourselves, then God and other people, too, will come to our assistance."

Samuel discovered that the constitution of the Friends of the Hebrew University in England provided also for help to the Technion. On this basis he procured from them a pledge to give the Technion £500 from their last campaign, but the agreement was cancelled on instructions from the University in Jerusalem. He then explored the possibility of a joint fund-raising drive for the two institutions, while Kaplansky, at the Palestine end, discussed matters with Dr. Werner Senator of the University. Again the matter of possible unification arose and Kaplansky once more voiced his willingness to having the Technion become the Engineering Faculty of the University. Jerusalem's reaction again appeared to be negative, largely out of fear that it would mean a financial burden.

Against great odds Prof. Samuel succeeded in raising about £3,000. The largest single contribution, £700, was from the Federation of Synagogues, and it was used to construct a large assembly hall *(Ulam Haknesset)* over the workshops. The hall was dedicated in the name of the Federation. Lord Hirst agreed to assume the chairmanship of the organization; the £3,500 goal was within reach.

Into the Cauldron

After five months Samuel returned to Haifa. He had hoped to go to Germany where, despite the worsening situation in July, 1938, he believed he could raise funds from old friends and associates. The Zionist authorities would not approve his trip, but Kaplansky urged him to launch the Technion drive in Germany and direct it by correspondence from Haifa. Samuel asked a non-Zionist friend in Berlin, Hugo Herzer, to take matters in hand, and again applied to the Zionist Organization for cooperation, pointing out that through Herzer there would be access to many Jews who would otherwise not give to Zionist funds. Since Prof. Ollendorff was planning a trip to Germany to visit his elderly mother, it was hoped that his presence could be utilized to further the Technion collection.

Seen in retrospect, and with the knowledge that the Munich conference was to take place on September 30, the German occupation of the Sudetenland within a week thereafter, and *Kristallnacht* on November 9, the Technion campaign during the late months of 1938 and early months of 1939 can be regarded as almost Kafkaesque.

The Zionist Organization in Germany informed Prof. Samuel on July 28, 1938, that its fund-raising committee had approved his request for permission to collect contributions for the Technion, subject to approval by the foreign currency department of the Government, and the Organization was prepared to cooperate in a campaign for up to 12,000 marks. The consent was given in view of the great help extended by the Technion to German Jewry. Both the Jewish National Fund and the Keren Hayesod were asked to receive the contributions and look after the transfer.

Herzer himself could not do very much; he was fully occupied trying to arrange for the emigration of members of his family, and he therefore engaged a Dr. Suse Michael to serve as a professional fund-raiser. The hope was that she could obtain contributions which could thereafter be sent to Palestine in the form of scientific and technical equipment within the framework of the *Haavara* arrangement whereby the Nazi authorities permitted German Jews to take some of their possessions out of the country in goods. In the list of approved merchandise Prof. Samuel had found particular items which he would have liked to have for his own new laboratory at the Technion, such as optical measuring instruments, chemistry equipment and materials for instruction in physics.

Suse Michael made contact with the Zionist office and asked for a list of prospects whom she could approach. This was not given to her; she was expected to locate entirely new prospects. In the meantime she married a Dr. Bach and left on a short honeymoon. It was felt that when Ollendorff reached Berlin his presence could be utilized for a fund-raising meeting.

In his letters to Samuel, Herzer referred also to difficulties with the police, who had frozen permits for fund-raising. He inferred that this was due to the current registration of all Jewish capital. His letters contained only indirect reference to what was going on in such terms as, "there is no need to say anything about our situation."

One week before the Munich conference Samuel sent Suse Bach full guidance on how to solicit contributions. He approved the arrangements made with her by Herzer whereby she was to receive 5% of everything she collected.

In the meantime, the first contribution had already been recorded, 3,000 marks given by Prof. Samuel's aunt, Mrs. Anna Hornthal. But Prof. Ollendorff, whose arrival had been anticipated, had not appeared and his colleagues in Haifa became worried. Much later Ollendorff revealed that when he reached Berlin he discovered that the Gestapo was looking for his brother, and might well arrest him instead. Friends urged him and his wife to catch the first express train to Switzerland, but the scientist objected. The express would be the very train that would be checked carefully, he pointed out. He and his wife therefore boarded a slow, local train which stopped at every village, and took many hours en route. As he surmised, it was waved through at the Swiss border with only a cursory check, and they returned to Palestine.

The correspondence with Germany, and then with Austria, continued on through the end of 1938 and into 1939. Books and equipment were sent. Students received permission to take out limited resources to pay for their school expenses overseas, and this was expected to make some 30,000 marks available. Some 200 German and Austrian students received entry certificates from the British Government. Contractual arrangements were made with the various bodies involved in the rescue of Jews from Europe, chief among them the Central Bureau for Settlement of German Jews, an arm of the Jewish Agency, the Association of Settlers from Germany and Austria in Palestine, and eventually with the Keren Hayesod. The agreements regularized the distribution of immigration certificates and the allocation of funds for absorption of the personnel involved.

By the spring of 1939 Technion's own correspondence with individuals in the Nazi-dominated countries petered out.

The heavy drama of the times was reflected in an incident of reverse fund-raising. Early in 1939 the Technion received a letter written from Detroit by a German Jewish refugee, and forwarded by a refugee aid organization. The writer told that as a prosperous industrialist in pre-Hitler Germany he had made generous contributions of electrical equipment to the Technion. Forced to flee the country he had left everything behind and now, at the age of 60 and almost penniless, he had no prospects for a livelihood. Could the Technion perhaps see its way clear to making some payment now by way of compensation for the equipment? Investigation in Haifa confirmed the truth of his story, but the Institute was reluctantly compelled to reply that with all its appreciation for his previous generosity it had no resources from which to make even a gesture of payment at this time. No doubt there were many others in a similar position.

Ferment at Home

The intensification of the tragedy in Europe was accompanied by a fermentation in the Jewish community of Palestine. British attempts to restrict immigration were fought with every means available. "Illegal" ships sought to run the British blockade. Some got through; others were intercepted.

The center for community demonstrations in Haifa was the Technion courtyard fronting on the main building. The stairway at the entry led up to an open vestibule which formed a natural raised platform overlooking the space where thousands could gather. The courtyard was also used as the departure point for the funerals of those who had been killed by Arab action. On April 23, 1939, the community gathered there to protest a British attempt to deport a sailboat full of refugees that had been brought into the port. After the heated speeches the meeting erupted into the adjoining streets. Leading the crowd was a band of Technion students who paraded in the neighborhood, chanting and carrying protest banners. There was a confrontation with the British police and their armored cars. The parade was dispersed and regathered in the courtyard. At this point the police burst onto the Technion grounds, scattered the crowd, and pursued students within the building from room to room. There were some physical encounters and several students were taken to the hospital. But the protest made its point, and the hapless refugees from Europe's terrors were permitted ashore, interned for a while and then allowed to remain in the country. The occurrence was but one of many similar confrontations being held all over the country, but in this case the drama was heightened by the successful outcome of the protest.

In the following month the MacDonald White Paper was published, calling for further sharp limitation in immigration of Jews, and restriction on sale of land to Jews. This too brought a sharp reaction. Three hundred Technion students proclaimed a sit-down strike at the Institute, this time protesting the failure of the Zionist leadership to take decisive action against the restrictions.

In this atmosphere it was decided not to hold the diploma ceremony for the tenth graduating class. More and more students — and their teachers — were absent from class. It was difficult to maintain normal academic operations.

Kaplansky's responsibilities grew. In addition to administering the Technion, he was temporarily burdened also with the Technical High School and the Nautical School, since their Principal, Dr. Shlomo Bardin, was absent on a mission to the U.S. He continued to lecture on economics. He was named by the Jewish Agency as chairman of a committee set up by the Zionist Congress to go into the problem of Arab-Jewish relations in Palestine. The subject was of great interest to him. His days and nights were filled with activity.

The dark clouds of the gathering storm cast a deep shadow over the world. The 21st Zionist Congress met in Geneva on August 16 to reject the White Paper. One week later, and while the Congress was in session, von Ribbentrop signed the German non-aggression pact with the U.S.S.R. The Congress dissolved, as the delegates hastened to leave Geneva and get to their homes. On September 1, 1939, Germany invaded Poland. The Second World War had started.

CHAPTER SEVEN *The War Years and After*

Crisis

There could certainly be no doubt regarding the policy of the Jewish community in Palestine, the *Yishuv*, on the conduct of the war against Nazi Germany. At the same time, the *Yishuv* and the entire Zionist movement were also locked in battle against the British Government which had over the years been whittling away at the Balfour Declaration of 1917. The Passfield White Paper of 1930, the Peel Commission Report of 1937 and the Woodhead Commission of 1938, among others, had systematically undermined British obligations to establish a Jewish National Homeland in Palestine. The MacDonald White Paper of May, 1939, went to further extremes. It called for limitation of Jewish immigration to Palestine to 75,000 spread over a five-year period, this in the face of the refugee pressures from Europe; it banned purchase of land by Jews in much of the country; and finally, it called for establishment of a government by the assured Arab majority at the end of ten years.

The feeling among the Jews was best expressed in David Ben Gurion's much quoted policy statement: "We will fight the war as if there were no White Paper, and we will fight the White Paper as if there were no war."

The Technion found itself seriously affected and deeply involved. Immediately affected were more than 200 students from Poland who were dependent on the financial help which they received periodically from home. This was now cut off. Indeed, a fairly large sum which had already been paid by the parents to the Zionist office in Warsaw for dispatch to Palestine was swallowed up and lost in the overwhelming developments of the war. Not just tuition fees were involved. Most of these students relied on the remittances from home for their actual subsistence, and Kaplansky feared genuine emergencies. The crisis was not slow in coming as 50 students reported immediately that they were destitute. In response, a number of Technion's teachers assumed a voluntary tax, deducted from their salaries, to provide scholarships for such cases. A special contribution for the purpose was also received from Solel Boneh, the Histadrut construction company, in appreciation of Technion's importance to the upbuilding of the country.

Against the background of the overrunning of Poland and the beginning of the establishment of the Warsaw Ghetto, the Technion held its graduation exercises on May 25, 1940. It was a double graduation; the ceremony, which was to have been held a year earlier for the 20 successful candidates in the tenth graduation class, had been postponed because of the mood induced by the White Paper. Though there was certainly no cause for celebration in 1940, the formal handing out of diplomas had to take place, and the 31 graduates in Technion's eleventh class joined their colleagues of the year before at the restrained, internal affair. Kaplansky

200

soberly addressed himself to the emergencies of the hour. In the midst of the maelstrom of blood and agony, he said, the Jews of Palestine must stand firm and strong, confident that they were preparing the means for preservation of the remnants of the Jewish people. He reported highlights of growth and progress. In financial affairs, the Technion was carrying on with a budget based in large part on funds which it had procured itself. He told of the establishment of a new Technion Society in the United States. Whereas there had been two laboratories in 1931, there were now ten. The number of students had risen to about 500, with an additional 275 in the Technical High School and Nautical School.

The graduation ceremony included also an eloquent address by Ussishkin and a concert by the student orchestra. There could be no jubilation, but the event certainly helped raise morale in the institution.

The atmosphere of crisis was brought home to Haifa on July 15, 1940, when the city had its first air raid; more followed. Enrollment at the Technion dipped substantially because of three factors: the cessation of immigration, the dropout of students who had lost their financial support from abroad, and military enlistments.

It was a matter of Zionist policy to encourage voluntary enlistment into the British forces, and the program drew students away from their studies. By 1941 some 150 Technion students had joined the British army, and the recruiting program was given further stimulation with formation of the Jewish Palestinian units. Curiously, the Jews were formally known as the "Palestinians" in those days. There was practically no Arab enlistment. The Technion encouraged its students to come forward, and even offered inducements. Those who volunteered for military duty at the end of the academic year 1940–41 were automatically promoted without examination. In 1942 no freshman students were accepted unless they could produce military exemption certificates from the Enlistment Committee. On the other hand, students in their last year were permitted to complete their studies and sit for their final examinations. The national defense effort was supreme, but it was policy not to permit the complete dismantling of the institutions of higher learning.

Opinion on the campus ran ahead of the official Zionist policies and programs. At a rally on May 9, 1941, speakers launched attacks on the Zionist leadership to such an extent that Kaplansky walked out in protest. The students adopted a resolution demanding immediate establishment of a Hebrew army with its own flag and its own commanders, to take part in the war as partner and ally. While extending greetings and best wishes to all students who left to join the army, they withheld endorsement of recruiting (for British units) under the existing conditions. In reply Kaplansky rebuked the students for their political immaturity and for

Prof. Heinrich Neumann (1888–1955).

their violation of university and Zionist discipline. He termed the demand for a Jewish army unrealistic.

There was still an element in the country that seized the opportunity to call for closing down the Technion entirely, and Kaplansky at once rebutted the argument that engineering training was a luxury. The demand made by some that all Technion students who were not in the army should be sent into agriculture or industry drew his fury. He spelled out the contributions which technical training could make in augmenting the war effort, and noted that reduction in such training would have a reverse effect, undermining defense industry and the military alike.

The pressures on Kaplansky were enormous, and in mid-1941 fatigue and the poor condition of his health forced him to take a vacation. Ratner served as Acting Director.

Technion's role in fighting the war as if there were no White Paper was part of the institution's official policy. The Technion's share in fighting the White Paper as if there were no war was completely unofficial and under cover. There were some extremely unusual cases in which the operations on one level were linked to the effort on the other, but on the whole the two activities were at opposite poles, and must be reviewed separately.

To Fight the War

To obtain maximum mobilization of all the resources of the area for its war effort, the British Government set up a Middle East Supply Centre. It was described as a kind of "economic directorate" to plan and control the economic life of the region with an eye to contributing to the successful prosecution of the war. Kaplansky and Professors Ollendorff, Neumann, and Heimann headed important sub-committees on the Scientific Advisory Committee of the War Supply Board or filled other important functions for that body.

The industrial and technological potential of Palestine in a vast region where other sources of such assistance were notoriously lacking was not lost upon the British. At the Eastern Group Supply Conference held in New Delhi in December, 1940, British representatives from eleven Empire points east of Suez drew up a long list of articles that Palestine could produce for the common war effort. The modern industries which already existed provided the basis for the great expansion which was required, thanks to the cadre of professionally trained personnel. Perhaps some recalled that a dozen years earlier efforts had been made to limit the production of engineers to no more than four or five a year. The Technion had its own vision of the needs, and while such vision certainly did not include any prophecy of a great world war, the maximalist program had prevailed. By 1943 about 600 students had gone through Technion's 15 graduating classes. Although only 291 had received engineering diplomas, the others too had been equipped for technological and scientific service to local industry. Together with graduates of the Technical High School, a total of more than 1,000 qualified engineers and skilled technicians had been made available to the armed forces and to the war industry. It was for this reason that Palestine, more than any other country in the Middle East, was a solid and dependable source for the supplies the British needed for their war effort.

The value of the manpower could be measured qualitatively as well. The Technion was proud when a building contractor wrote in 1945 that five graduate engineers from the Technion had taken part in the construction of the Gut Bridge across the Euphrates River in Syria, and their high degree of professional qualification had been matched only by their conscientiousness.

Prof. Yohanan Ratner (1891–1965).

The list of new industries which developed as a result of the immediate needs was long. Capital was no problem; in wartime money is always available. But only in Palestine, the British found, was there a supply of well-trained professionals in addition to a highly intelligent population which could quickly acquire even the most complicated skills. There were broad developments in the textile field. The first wool spinning plants were opened, as well as the first steel foundry and plants for the construction of machinery and machine tools. New chemical industries included pharmaceuticals, paints and dyes, and glue. The leather industry was broadened. Wherever there was a military need in Africa or Asia which could not be met from elsewhere, Palestine was called upon to provide. The list of industrial products was extensive.

An objective review of the operations of the Middle East Supply Centre by historian Martin W. Wilmington was not sparing in its tribute to what had been done. "The new manufactured goods were a blessing not only to the civilian population but to the Allied forces. Two minesweepers were built and many other warships repaired in Palestine shipyards. The Vulcan Foundries in Haifa delivered considerable quantities of heavy armaments .... A Tel Aviv workshop switched overnight to the output of bolts, the lack of which had stalled numerous tanks in the Libyan desert. Another succeeded in developing a compass that could be used by tank crews during severe desert storms, and manufactured it from local raw materials."

Practically all of the land mines planted by the British troops in the Northern Desert, and which contributed so much to breaking the back of the Nazi legions, were produced in Palestine. The production was a tribute to the hundreds of Jewish engineers and technicians whose talents and skills were involved.

The Technion's role in the war effort was specific as well as general. There was the closest of cooperation with the Royal Air Force, whose planes used Pratt and Whitney engines. Parts were hard to get, and at least once a precious cargo of replacement parts was sunk by German submarines before it reached port. The RAF turned to the Technion in desperation to ascertain if the Institute could do anything. Yosef Baratz, of the Technion workshops, recalled that blueprints were flown in from England and the Jewish technicians applied themselves. Working for a year or more, on two shifts daily, they turned out thousands of parts and precision tools. Baratz was proud of the Technion role and he did not let the contribution go unrecorded. As every piece came off the production line, bearing its part number carefully engraved on it with electric stencil, he saw to it that the symbol of the Technion *Bosmat* workshop also appeared. Years later he heard from Jewish members of the armed forces, serving as technical crews with the RAF in such widely scattered places as Egypt, Eritrea and the Western Desert of Africa, what a thrill it had given them, in making their repairs, to discover the little tell-tale mark that showed the source of the precious pieces required to keep the RAF planes in the air.

The fine desert sands were taking a toll of the British tanks, and there was need for frequent renewal of the steel pins in the track links. The Cluson Steel Works in Haifa, the only steel foundry in the country, produced the pins in the raw, and the workshops of the Technion did the fine finishing by means of special gauges produced for the purpose. The first emergency order was received early in 1941, and the students of the Technical High School worked three shifts a day to produce the 3500 pins that were flown at once to the Western Desert. Many other orders followed. The same workshops made other parts for British tanks, as well as specialized equipment.

As early as June, 1941, in response to inquiry from the Chairman of the Jewish Agency Executive, David Ben Gurion, the Technion could list specific work being done in its laboratories for the war effort. The report singled out: Building Materials Laboratory: working with the Royal Engineers in the testing of building materials and helping to select suitable materials to replace those not available in the market. Chemistry Laboratories: producing active carbon from local materials, to be used in gas masks; material for pharmaceuticals; improving the quality of flour made from domestic wheat; study of Huleh peat for fertilizer or fuel; studying methods of substituting sodium sulphate for sulphuric acid, which was unavailable. Electrical Laboratories: repairing of electric equipment and

Repairs on British army equipment in the workshops. 1946–1947.

motors both for the army and for companies working for the army, such as the cable company, the Iraq Petroleum Co., the Potash Works, the Electric Company and others; equipment for hospitals, calibration of equipment used in the war effort; development of instruments to detect the sound of planes (Gamma Wave). Lack of funds slowed down the latter work. Metallography Laboratory: checking valves in the pumps of the Iraq Petroleum pipe lines, which frequently went out of commission. Workshops: miscellaneous work in welding, lathe operations and the production of heliographs for the police.

By 1944 the Technion could report a further variety of important contributions coming directly from its laboratories. The Metallography Laboratory carried out investigations in the process of manufacturing cartridges. It helped in the development of a water container which could be thrown to troops fighting in the desert. The Electrical Laboratories were deeply involved in an area which more than three decades later was to become a major field of activity at the Technion, the development of medical instrumentation. A crystal stethophone was created. Other medical apparatus constructed or improved upon in the Laboratory included optical amplifiers, cardiographs, and thermal sterilizers. For more general use the Laboratory engaged in the construction and winding of A.C. synchronous generators and the construction of automatic regulators for pressure regulation of synchronous machines, transformers for testing purposes, recording apparatus for determining the temperature inside buildings, fine measuring bridges, crystal microphones, oscillographs and pyrometers. The High Tension section was able to test insulators and insulation material up to a tension of 100,000 volts.

Radar had not yet been brought to Haifa, and the first air raid had caught the city completely unprepared. It was Pinhas Rutenberg who stormed into the Technion and insisted that the scientists do something at least to provide an early warning. Prof. Ollendorff set up a small task force which worked day and night to produce a giant "ear" with a sensitive microphone which could pick up sounds from a great distance and magnify them. A filter was developed to eliminate the sound of the sea, or the swishing of the branches of trees. The efficacy of the "ear" could be tested only by exposing it to the sound of an approaching plane. The British had only a Piper available, and would not send it up for test purposes. Leo Schaudinischky made a sound recording of the roar of the Piper's engine, and set up the recording with an amplifier on the roof of the Technion workshops. The detector was taken up to the top of Mount Carmel and tuned in to the sound coming from below. The device was not actually used, .and soon thereafter the British brought in a radar installation.

The Chemical Laboratories, too, were deeply involved. Of the long and varied list of activities, special mention was made of the following: development of a substitute for sulphuric acid for accumulators; preparation of industrial and cosmetic emulsions; research on aniline dyes and dyeing; development of tanning materials from Palestine herbs and trees; solidification of benzine; routine test work on food, metals, agricultural supplies; investigations on the corrosion of iron tubes in water supply lines and the methods for prevention; electrolytic production in a pilot plant of persulphates for both flour and textile industries. Other investigations included methods for electrolytic production of copper tubes and wire; methods for making aromatic compounds by treating water of the local superphosphate industry (using acid sludge from the oil refinery). The Laboratory helped make possible the production of carbide, without which no welding could have been done in Palestine.

The Building Materials Laboratory worked on the use of natural fibers, such as reeds, instead of steel as reinforcing in concrete structures. It developed pre-stressed concrete sleepers for the Palestine Railways. Its

Electrical regulator for generators, built at the Technion for the Iraq Petroleum Co. in 1940. It was sent to the oil fields in Kirkuk. Following the coup by Rashid Ali in 1941, he took the equipment to his tent where he would entertain his guests by alternately illuminating and switching off the bulbs.

research for the Iraq Petroleum Co., to combat sulphate attack on concrete, was used in Iraq. In the Hydraulics Laboratory there was much work connected with irrigation, water supply, hydro-electric power, etc.

On the educational level, too, the Technion provided services. Beginning in 1942, at the request of the British army, special evening courses were opened in technical subjects for soldiers. In 1943 there were 50 students. A year later the number had risen to 260. Teachers from the regular faculty of the Technion taught such subjects as elementary mathematics, physics, radio technology, electrical engineering, chemistry, construction, Diesel engines, etc.

The British Intelligence, too, called upon Technion's brain power. In the early years of the war, with French forces no longer available in the area, the British sought new sources of reliable information, and Prof. Ratner was drafted to head a special unit. He commandeered the services of a number of Technion scientists and others who gathered and analyzed information on technical installations and transport arrangements in various Middle Eastern countries. Many of these reports were so helpful that they were at once sent on to General Wavell in Cairo.

There were sacrifices, too. Many Jewish boys, "the Palestinians", lost their lives while fighting in British uniform against the common enemy. In 1943 Frederick Kisch, then a Brigadier with the British army in North Africa, was killed by a German mine in Tunisia. Kisch had become a member of the Zionist Executive as early as 1921, and was said to have been the first British Jew to take out Palestinian nationality. He had for many years been a member of the Technion's *Vaad Menahel*, and in his latter years had been a resident of Haifa. As Chief Engineer of the British 8th Army, his last military contribution was the building of the water pipe line to El Alamein. The line, not quite finished, was tested with sea water. At this point a portion of the pipe line was overrun by a German unit which had been in the desert for some time and had run out of water. In great excitement they began to drink, but the salt water only increased their thirst enormously. Their morale collapsed, and more than a thousand of them surrendered to the British. The Technion friends in the U.S. later undertook a major memorial project at the Technion bearing the name of Kisch.

While the military authorities were only too eager and willing to encourage industrial development in the country and to foster the growth of a technological society, the colonial administrators had their own political doubts. They could advance no arguments in wartime against anything that would help make the region self-sufficient and a strong military base, but they feared that the long-range effect would be to strengthen the hands of the Jews in their eventual goal of setting up a Jewish State.

Frederick H. Kisch (1888–1943).

The same political motivation called for minimizing the Jewish participation in the war effort as much as possible. The full story of the Technion's role has never been told before, and many of the records are no
longer available. In 1943 British censors killed news dispatches reporting
on campus activities in support of the military.

Technion engineers were to be found on many fronts, but the exploits
of one large group merited special attention. Early in the war Solel Boneh
offered the British a complete company of its employees, a self-contained
unit of engineers, draftsmen and technicians with all the other skills
required. The 745th Palestine Artisan Works Company of the Royal Engineers, or Plugat Solel Boneh as it was more commonly known, was
headed by Major Dov Givon who, as a student, had lobbied for the Technion at the 1931 Zionist Congress. The company built ports, airfields,
roads and military buildings throughout the Middle East, ranging from
Syria, Egypt and North Africa, and after the invasion also in Italy. Givon
was awarded the MBE for his services and the company received the
highest of commendations for its distinguished engineering performance.
One of its many projects was the result of Egypt's insistence on maintaining neutrality despite, or perhaps because of the threatening approach of
the Nazis in North Africa. The British were thus denied use of Egyptian
ports for loading military supplies, and Plugat Solel Boneh was asked to
construct pier facilities near the Suez Canal. Some 2,000 Egyptian laborers
were employed in the task. When asked how he would get work out of
them, Givon replied that some thousands of years earlier the Egyptians
had been taskmasters over his ancestors, and he would know how to
reverse the roles now.

Jewish determination to do everything possible to help in the struggle
against Hitler often overcame hostility to Britain's anti-Zionist policies.
Sometimes methods developed by the Haganah to fight the Mandatory
Government were diverted and placed at the disposal of the British
instead. "Jenka" Ratner, a Technion graduate who later headed the Israel
Defense Forces scientific research program, was brought to England by
the British during the war to help them in their work on anti-submarine
devices. Did they know that he had helped to design the limpet mines
used by the Haganah against the British patrol craft intercepting Jewish
immigrants into Palestine? The answer is not known.

A device that Ratner did tell the British about was his "umbrella mine",
and there was even public exhibition of it in the Technion courtyard.
Munya Mardor, a key figure in Haganah activity, tells the story. The Special
Operations Unit of the Haganah (p'ulot meyuhadot), POM, had devised a
number of projects to sabotage the British when the fight against the
White Paper was at its height, before the outbreak of war. Ratner had

created what he called an "umbrella mine" which could be inserted into an oil pipe line. It was so designed that it would flow with the oil and not detonate until it had entered one of the big storage tanks. It was an ingenious piece of mechanism and everything had been set to use it. The plan was called off at the last moment when Britain declared war on Germany. Cooperation between the Jews and the British became the order of the day, and Mardor was ordered to reveal the secret to the British and even demonstrate the mine, which perhaps could be used by Britain against the Nazis. Mardor set up a miniature plant in the Technion courtyard. It was built to scale, and involved precise calculations of oil pressure, speed of flow, etc. The students stood by and watched curiously as the Haganah demonstrated to the British exactly how it could be done.

To Fight the White Paper

There was another side to the coin, and the Haganah expressed its opposition to the British whenever matters of Jewish immigration or Zionist policies came into open conflict with the Mandatory Government. The Technion continued to serve as the central rallying point for Haifa's Jewish community, and covertly as headquarters for the Haganah.

In March, 1940, a physical confrontation took place on the campus. A women's demonstration in the streets of the city was dispersed by the British, and the Haganah authorities, assembled in their Technion headquarters, decided on an immediate counter-rally. Before they could go into action the lookout reported that British soldiers were entering the courtyard. Hurriedly the Jews destroyed or hid incriminating papers and rehearsed their alibi that they were a group of students who had been studying in the building and had been trapped there when curfew had been proclaimed. The act was easy and the performance convincing since most of them really were students. Nevertheless, the British arrested more than 50 of them including Haim Laskov, Munya Mardor, Meir Zorea, Nehemiah Brosh and Israel Libertovsky. They were held in detention for two days. In the face of their defiant refusal to cooperate and the absence of any evidence against them they were released.

Libertovsky was one of the more fiery of the students, and a source of original ideas to make life difficult for the British. Among other things, he proposed burning down the main post office in Hadar Hacarmel, confident this would cripple the police and the soldiers for a long time. Mardor vetoed the idea on the grounds that the post office also served the Jewish community, and the latter would suffer perhaps more than the British. Libertovsky felt that his initiative was not appreciated. He completed his studies at the Technion where he served also as chairman of the Student

Abigail Weinbrand (1927–1946).

Association, became one of the founders of the *Palmah*, and went on to a distinguished career as an engineer, capped by his appointment as managing director of the Israel Shipyards. The small arms "slick" under the workshops was still in use, and an even larger one, a veritable arsenal, was maintained in the new Technical High School building. High school pupils who had been accepted as trustworthy members of the Haganah engaged in frequent oiling and cleaning of the weapons. The members also filled cartridge cases, made fuses and prepared explosives for hand grenades.

One accident occurred. On April 26, 1943, one of the dedicated Haganah volunteers, Yaakov Ziniuk, was engaged in moving what were supposed to be empty Italian grenades. One fell. He leaned over to pick it up and it exploded in his hand. Severely wounded, Ziniuk seized a piece of canvas and flailed away until he had extinguished the fire that had started. Then he carefully locked the door behind him. He went at once to the nearby firefighters' station, which was also a cover for the Haganah. He told them what had happened, gave them the keys, and asked them to go back to eradicate the tell-tale traces, including the trail of blood which he had left. Having completed his self-imposed assignment, he fainted away. At the hospital it was found necessary to amputate his right hand.

Several years later (August 13, 1946), during a street demonstration against the deportation of Jewish immigrants, a 19-year-old Technion secretary, Abigail Weinbrand, was victim of a confrontation with the British troops; she was shot in the chest as she marched in the front ranks of the demonstrators.

The same workshops and laboratories which were engaged in the war service also found time to produce equipment and various devices ordered by the Haganah. Things were done on an arbitrary, personal basis and few records were kept. The important thing was to produce and to meet immediate needs. Yosef Baratz recalls when the underground asked him to build mine-trippers that could clear the dirt roads for Haganah vehicles. The British did not need mechanical devices, Baratz said grimly. They would simply load a car with Arabs and have it drive ahead of them. If there was a mine in the road, it would detonate under the first car. That was their mine-tripper. The Haganah needed more humane devices, and the Technion workshops produced them.

The Building Materials Laboratory sought an improved material for the walls of the Haganah armored cars, as well as for the walls and ceilings of large "slicks" in kibbutzim and elsewhere. Experiments were made with an asphalt-cement mixture which could stop bullets. In the Electrical Laboratories infra-red equipment was developed for night-time signalling.

The secret "slicks" at the Technion multiplied, and not everyone in the

Haganah knew of all of them. Baratz, who was close to these matters, tells that the secrecy was deliberate. Nobody knew everything. The Haganah's emergency hospital was located in the Building Research Station in the basement of the main building.

Why didn't the British police ever search the Technion? They would have found considerable evidence of the underground activity. The Haganah men have asked themselves this question many times. It was inconceivable that the British did not have a pretty good idea of what was going on there. Perhaps there was an unspoken gentleman's agreement. There were too many times when they needed the Technion badly themselves.

There was little doubt that the actual Haganah activities on the campus interfered many times with the normal operations of the Technion as an educational institution. Kaplansky sought to steer a middle path, but it was not easy. The underground forces wanted more and more rooms placed at their disposal. They sought greater use of the Technion's equipment. More than once Yaakov Dori had to smooth down the relations in the inevitable conflicts that developed. Dori later testified that Kaplansky was always understanding and helpful, though naturally concerned for the proper operation of the institution.

Again in 1941 the Technion skipped the annual graduation exercises. In 1942 a joint ceremony was held to distribute diplomas to the 31 students of the 12th class, and the 26 from the 13th class. Many were absent in the military services, and members of their families received the diplomas on their behalf.

The next ceremony was held in December, 1943, and again telescoped two graduating classes, a total of 72 students from the 14th and 15th classes. Special note was made of the fact that this time there was no representative of the British Government present and not even a message of greeting, though a number of the graduates appeared in British uniforms. Relations with the Mandatory Government were again reaching a breaking point.

In his message to the 1943 graduation exercises Dr. Magnes, head of the Hebrew University, made a familiar point: "I have always thought of the Technion as the Engineering School of the University, although the administration of each is separate . . . ." he wrote. If this was intended as a hint of interest in resuming negotiations, it got no response from Haifa.

Classes were still going on, but under the most difficult of circumstances. Students appeared irregularly, not for lack of desire to study, and sometimes it was not clear whether they had dropped out, or were engaged in duties which could not always be publicized. Teachers, too, would occasionally disappear for one or more days, and would return to

Quick leave from the army to take diploma examinations. 1943.

class, weary and unshaven, at about the time the press was reporting some Haganah exploit.

Examinations were difficult to schedule because of the prior claim of the military, whether British or Haganah. Students who were attached to British units stationed in Palestine or nearby countries, were often given short leaves by their Commanding Officers to get to Haifa to sit for examinations, but such leaves were sometimes approved only on the very eve of the exams. One student told that two days before the examination, he had not received his leave order and had already abandoned hope of reaching Haifa in time. He was on guard duty somewhere in the desert, far from his unit, when a dispatch rider passed by and asked to be directed to unit headquarters where he had to deliver a telegram. The guard undertook to deliver the telegram himself to his Commanding Officer, only to find that it was his leave order. He departed immediately and arrived in Haifa just in time for the examination. One is tempted to ask how well he could have been expected to do under such circumstances. On more than one occasion students were called out of the very middle of an examination to

Dr. Chaim Weizmann listens as Dr. Kaplansky reads the opening address at celebration of Technion's 20th anniversary. 1945.

report back to their units. There was no cause for surprise that the student enrollment had dropped sharply from about 500 to less than 200 in 1942.

As the Nazi troops advanced across North Africa toward Egypt the mood in the country stiffened. The Jews were under no illusion as to what would happen if the Nazis ever broke through and reached Palestine. Top secret plans were drawn up for every contingency. Prof. Ratner was in charge of the preparation of the Haganah's emergency program whereby the Jews would, under such circumstances, be withdrawn into a defensible area on Mount Carmel and would there make their fighting stand against the enemy.

The defeat of Rommel at El-Alamein at the end of 1942 lifted the threat. A year later student enrollment at the Technion showed an upturn.

Russia was bearing the brunt of the Nazi attacks, and Kaplansky headed a League to Provide Help to the U.S.S.R. He and others believed that a victorious Russia would in turn support the program for the Jewish National Homeland in Palestine.

The Valdora, *training vessel of the Nautical School. 1947.*

The Tide Turns

By 1945 the tide of the war had turned, and the Technion was able to mark its 20th anniversary with appropriate celebrations. The distinguished guest of honor was Dr. Weizmann, and Kaplansky expressed the fervent hope that when the Technion came to mark its 25th anniversary Weizmann would again be present, not as the President of a "Diaspora longing for Zion", but as the President of an independent people living its own national life. The wish became truth, and Weizmann did indeed attend the Technion's 25th anniversary celebration in 1950 as the President of the State of Israel.

The 20th anniversary of the opening of the Technion provided an appropriate platform for a summing up of what had been achieved and, at the same time, for a projection of Technion's plans for the future. The courtyard and corridors and rooms of the Institute were filled with an elaborate graphic, pictorial and physical display of the Technion's achievements in every field. In a sense 1945 marked a watershed in Technion's development. After all the Institute had gone through, it was encouraging to have a paper like the *Palestine Post* (now the *Jerusalem Post)* comment, after reviewing the history of the Technion, that "rarely has the vision of practical idealists been so strikingly vindicated by achievement as in this great pioneer effort of technical education."

The student enrollment, swinging sharply upward, reached close to 400. The number of graduates was approaching 700. Technion's role in absorbing immigration was obvious from the list of countries from which the 2,000 students had come during those 20 years. The breakdown was: Poland, 47.5%; Russia, 12%; Austria, 6%; Germany, 5.7%; Romania, 5%.

Important academic changes were instituted with the opening of the 1945–46 school year. The Department of Industrial Engineering was divided into two independent units, Mechanical-Electrical, and Chemical. Together with Civil Engineering and Architecture, the Technion now had four departments. Earlier, even in the midst of the war, a Research Institute for Town Planning and Housing had been set up, with Prof. Alexander Klein as its head.

The financial situation had improved. The most important item in the income budget was tuition fees, which in some years covered half of the operating budget. The fee had been £25 in 1939; it was £40 in 1945. In Technion's 20 years of existence the allocations from the Zionist Organization and Jewish Agency had totalled only little more than £50,000, but there was expectation that in this area, too, a corner had been turned. The Agency grant, which had been £4,200 in 1943–44, rose to £20,000 in the following year.

Technion's management committee, the *Vaad Menahel*, took its duties seriously and met regularly. In 1944 its members were E. Berligne, chairman, Z. D. Pinkas, Dr. U. Kopelowitz, Y. Gruenebaum, J. Reiser, Dr. M. Soloveitchik, E. Kaplan, Prof. Y. Ratner, M. Hecker and Dr. Kaplansky.

Changes in the administrative organization also took place. Krupnick died in March, 1942, and the secretariat was divided. Nessyahu was named head of the finance division. Two years later Dr. Mordecai Levy was appointed academic secretary. Eliyahu Oren was in charge of public relations until his death in 1950.

At various times in the early 1940's Josef Kastein and Raphael Patai were engaged to write promotional material about the Technion. For many years the legal Counsel of the Technion was David Bar Rav Hay. Years later his son, Meir Bar Rav Hay, held the same position.

The Technion's affiliated units for younger pupils, the Technical High School and the Nautical School, were also progressing despite financial problems. In the first 15 years of its existence the Technical High School, *Bosmat*, had graduated some 450 technicians who were serving industry, agriculture and the military. Enrollment in 1944 was 285, with an additional 90 pupils in the Nautical School. However, the financial support promised by the Jewish Agency and the Palestine Maritime League was not always forthcoming to the degree necessary. Promotion of a seafaring program attracted much attention, but not generous support. In 1940 the Hadassah organization made a small contribution. A few years later, with the help of the Maritime League, the Nautical School acquired a 106-ton training yacht, the *Valdora*.

Dr. Shlomo Bardin, who had been the first principal of the schools, had gone to the U.S. on a fund-raising mission, had been caught there by the war, and had not returned. In 1944, after he had been absent for six years, the *Vaad Menahel* considered that he had resigned, and Mr. Gershon Ahroni, who had for much of this time been acting principal, was formally declared headmaster.

Looking Ahead

As early as 1941 Kaplansky began to spell out his ambitious plans for the future, designed to meet the needs of the post-war period. He called for early planning for the establishment of two new departments: One, a Department of Economics and Administration, to provide a general education, as well as a technical and economics background for administrative personnel for industry, commerce and the civil service. The other was a Department of Agricultural Engineering, with specialties in drainage, irrigation, soil enrichment and rural construction. He made it clear that the purpose was not to train agronomists, who were being produced by the Faculty of Agriculture at the Hebrew University, but to apply engineering and technology to farming.

Not long thereafter he called attention to the possibilities of large-scale production of chemical fertilizers in Palestine, especially nitrogenous fertilizers. He pointed to desalting of saline water as a challenge to technology. Implementation of these plans required suitable physical facilities. Even before the outbreak of the war the Technion had begun negotiations with the Jewish National Fund regarding additional land, near Kiryat Eliyahu, where it was hoped to build student dormitories. The negotiations dragged on for several years, and the site went to a building company to put up housing for port workers. In 1944 the J.N.F. informed the Technion that land could be had if buildings were put up within two or three years. In the midst of war this was impossible, and the opportunity was lost.

Kaplansky recognized that buildings, while important, were not in themselves sufficient, and he called for the funding of Chairs and Research Fellowships. In view of the fact that the Institute was filling a national function, he asked the Government and the national agencies to

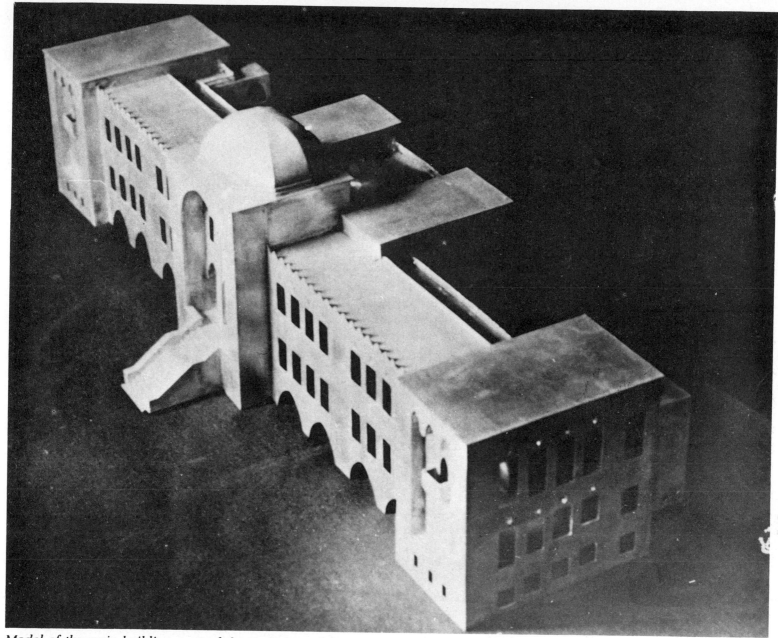

Model of the main building, part of the Technion exhibit at the New York World's Fair. 1939.

provide the Technion with the means. This theme recurred again and again in his speeches and reports, and was emphasized even more strongly in the decades that followed.

He also made the point that friends overseas would gauge their generosity by the degree of support received from within Palestine and he called on local industrial and financial institutions to come to the aid of the Technion. There had been efforts to set up an organized committee of friends of the Technion. The group in Haifa was reorganized in 1941, and Shabbetai Levi became chairman. Two years later a Technion Society was established in Tel Aviv with an executive composed of Messrs. A. Echtman, M. Ladyjensky, A. Sinyaver, D. Remez and A. Shenkar. The latter was chairman. For almost ten years an intensive fund-raising program took place, part of which included collection of debts from Technion graduates. Several attempts were made to organize the alumni, but the war atmosphere made organization of any kind difficult. Dr. Felix Rosenbluth, and thereafter Dr. Shlomo Ravikovitz, were professionally engaged in these activities. Early in 1944 the Government tax authorities ruled that contributions for specific research would be recognized as a business expense.

The major development in the procurement of financial help took place in the United States. The Technion had neglected an opportunity to place its case before American Jews at the Century of Progress Fair in Chicago, in 1933, but Kaplansky insisted that the Palestine Pavilion at the New York World's Fair in 1939 should have a Technion exhibit, and appropriate material was gathered and shipped. The organization of the American Technion Society in the following year took place so precisely on the threshold of Technion's emergence as a vital factor in the establishment of the State of Israel that one is tempted to speculate on the strange and mysterious ways in which history provides solutions to the problems which it creates.

Mission to America

Encouraged by his relative success in England, and despite the pessimistic warnings of the Zionist fund-raising agencies, Prof. Rudolf Samuel offered to go to the United States to seek help for the Technion. He left Palestine late in 1939, travelling east by ship. He took advantage of every stop to explore the possibilities of finding friends and supporters, including such places as Bombay, Singapore, Hong Kong and even Java. In Hong Kong he met with Lawrence Kadoorie who told him the family had been sympathetic to the Technion and had made an initial contribution, but interest had waned because the Institute failed to keep in touch with them. Nothing daunted, Samuel signed up members in the Crown Colony.

He arrived in New York in mid-January and at once ran into problems. Zionist officialdom was of no help whatsoever. He reported back to Haifa that Henry Montor was "hostile", Stephen Wise "unfriendly", Louis Lipsky, "human". Samuel decided not to seek help from the Zionists and to operate with a free hand. Other fund-raisers in New York predicted that he was doomed to failure and hoped that he had brought enough money at least to keep himself alive.

Samuel went about his work systematically. His arrival had been preceded by a number of letters from Haifa, so that he did not come unannounced. He decided that his best chances were with people not

Prof. Rudolf Samuel (1897–1949).

normally reached by the usual Jewish fund-raising drives — people who would understand the importance of engineering and technology. He lost no time in calling on Einstein at Princeton and received a warm welcome. Einstein sent out letters introducing Samuel to his friends, including M.I.T. President Karl Compton, and the latter introduced him to Gerard Swope. Louis D. Brandeis told him that he had always been under the impression that the Technion did not need money.

He shared his observations on American fund-raising with Kaplansky in a letter soon after his arrival: "Everything our Zionist friends say about raising money in the U.S.A. is wrong. They have built up a bureaucracy which commercializes Jewish charity and they consider everything from the viewpoint of this machinery, their own influence, and sometimes their own jobs . . . ."

It was not long, however, before he came to realize the need for a bureaucracy, at least in the form of a permanent organization with a full-time professional director, which could prosecute an ongoing campaign of public relations and promotional activity.

He located some of the original members of the old Society of Zionist Engineers, and with their help a new organization was created with the imposing name of American Society for the Advancement of the Hebrew Institute of Technology in Haifa, Palestine, Inc. Much later this was shortened to American Technion Society. The first officers were Dr. Ferdinand Sonneborn, President; M. Henry Sugarman, Vice President; B. M. Halpern, Executive Vice President; William Ginsberg, Treasurer; J. R. Elyachar, Associate Treasurer. Among members of the Board who later became active were Dr. William Fondiller, Jacob Mark, Joseph W. Wunsch, Philip Sporn, and Bernard Rosenblatt. Patrons of the Society included Dr. Compton, Judge Julian W. Mack, and Samuel Zemurray, President of the United Fruit Company.

Provisional headquarters was located in the private office of Elyachar, and the Society then moved to its own premises, first at 55 West 42nd Street, and then to 154 Nassau Street where it remained until it occupied its own building at 1000 Fifth Avenue in 1952. Judah Wattenberg was the first Executive Director.

"The new Society is still a bit shaky in that so far the number of Nobel Prize winners is greater than the millionaires," Samuel recorded, but at least it had a letterhead of its own.

He had a surprise call from Weizmann, then in New York, asking him to cable Haifa and ascertain what tools and equipment would be necessary to enlarge Technion's workshops and prepare them for any emergency such as repairing airplane motors, arms, etc.

The Society launched its career with a festive dinner in New York in honor of Albert Einstein, on May 8, 1940. The distinguished scientist spoke about Palestine and emphasized that its development was of importance to all Jews. As for the Technion, he called upon Jewish engineers, architects, scientists and industrialists to support it "with your advice and your funds." Prof. Samuel told the assemblage the goal: to provide an annual income of $30,000 to $40,000 for the Technion to replace income lost because of the war, and to enable construction of new facilities made necessary because of the Technion's increased responsibilities.

He hammered away at the unique contribution Technion could make to the development of the country. He pointed out that chemical fertilizers were imported from Belgium, where they were made from materials imported from Cyprus and Tunis. Research at the Technion could develop local sources of such fertilizers. Sulphuric acid was also brought in from Belgium, and the shipping cost more than the material itself. Millions of cases of oranges were shipped annually, each orange wrapped in a piece of tissue paper which came from Finland. A little study and research could find a way to produce it in Palestine. He made his points effectively. Contributions pledged at the dinner totalled $10,000.

Two days later came the news that the Germans had invaded Belgium and the Netherlands, the Western Front had collapsed, and the Nazi threat had become very real to the rest of the world. Samuel felt that all he had created in America was now about to be destroyed. Who would be interested in helping a university in Palestine when civilization itself was in danger? He chanced to meet David Rose, a New York builder, who had already pledged $250. As if anticipating Rose's reaction, Samuel, with a heavy heart, offered to waive the pledge. If the world was collapsing, he would not hold the American to his promise. Rose countered in a different mood. In view of what had happened, and in demonstration of his faith in the Technion, he undertook on the spot to double his pledge. Samuel realized that in America he was dealing with a different type of person, and his experiences in the months that followed confirmed the revelation.

By the end of June it was announced that new personalities had joined the Board of the Society, including Abraham Tulin and Sol Pincus. Philip Sporn assumed the chairmanship. Announcement was also made of the first large contribution, given by Joseph W., Harry and Samuel Wunsch to establish a Mechanical Engineering Laboratory in memory of their brother, David, who at the time of his death in 1935 had been a senior engineering student at Purdue. The significance of the Wunsch gift was in the high standard which it set, ultimately reaching well over $100,000. In the following decades Joseph Wunsch sponsored other vital projects at the Technion. The circle of friends began to widen even further. Soon after

220

the first Wunsch contribution, a group of friends of the late Charles Hardy raised a generous sum to set up a memorial to him at the Technion in the form of a Machine and Tool Shop. One of the leaders in this effort was industrialist Alexander Konoff.

During all this period Samuel had only peripheral contact with Dr. Bardin, who was then serving as organizer and fund-raiser for the Palestine Maritime League. Though he retained his nominal association with the Technical High School, Bardin was asked by Kaplansky to discontinue his separate fund-raising for the School since this would fragmentize the Technion's major effort.

Samuel did not remain in New York. He travelled tirelessly around the country, and before the year was up support groups had been established in Pittsburgh, Chicago, Indianapolis, Milwaukee, Cincinnati and Detroit.

By 1941 there were Technion groups in many more communities. Benjamin Cooper initiated the activity in the New Jersey area. Abba Hillel Silver and A. M. Luntz headed the chapter in Cleveland. In Baltimore Dr. Abel Wolman, Dr. Jonas Friedenwald and Dr. Alvin Thalheimer assumed the leadership. Lionel F. Levy became the first chairman in Philadelphia, backed up by Lester M. Goldsmith, Samuel S. Fels, Dr. Louis Gershenfeld, Morris Newmark, Howard Levy and others. Dayton and Washington, D.C., also reported activity. Membership and fund-raising drives were reported from every group. It appeared that American Jews, who even in those days complained about the surfeit of fund-raising, had responded to a new and unique appeal, that of science and technology.

In June, 1941, the new Society received a tempting offer from Edward Norman, who had set up the American Palestine Fund to serve as an umbrella organization for various institutions in Palestine. He promised a fixed allocation from the joint proceeds if the Society would join, and give up its separate fund-raising with all its bother and risks. The committee in New York refused, but Norman turned directly to Haifa. Kaplansky was tempted, especially by the guarantee of $4,800 a year, come what may, and he urged the Society to seek to bridge the differences of opinion. The Technion friends insisted on retaining their independence of action. Within the first year they remitted $9,600, and began to think in terms of endowment funds. Establishment of a Chair was priced at $40,000.

The big dinners in New York became an annual event. In 1942 Boston, St. Louis, Providence and Buffalo joined the list of Technion cities. Lazarus White succeeded to the presidency of the ASAHITECH, and first efforts were made to organize a Women's League, headed by Mrs. Miriam Cooper, who had been a former student of Biram in Haifa. The work was continued by Mrs. Eva Frost, but it was not until more than ten years later that Mrs. Belle Bernstein, in Chicago, took the initiative in organizing her friends and neighbors in support of the Technion, and soon thereafter the American Women's Division took on national form in New York.

Technion activity continued to spread. Some of the leading Jews in the Boston community affiliated with the local group, among them Joseph H. Cohen, chairman; Jacob J. Kaplan, co-chairman; Max Mydans, James D. Glunts, B. L. Landers, Louis E. Kirstein, Julius Daniels, Alexander Brin, Milton Kahn, Harry Levine, Yoland Markson, Sidney Rabinowitz, Max Shoolman, Mark Linenthal, Frank Vorenberg, E. M. Loew, Albert H. Wechsler and Dr. Charles F. Wilinsky. Dr. Joshua Loth Liebman addressed the first meeting.

Detroit leaders were Harvey H. Goldman, Karl Segall, Fred M. Butzel, Leon Kay, Benjamin Wilk, Philip Slomovitz, Isidore Sobeloff, Morris Mendelson, Louis Redstone and Louis Gelfand. In Chicago Dr. Max Woldenberg headed an impressive Board which included A. K. Epstein, Samuel A. Goldsmith, Ben R. Harris, Joseph Tumpeer, Harris Perlstein, Joseph Wertheimer and Samuel Wolberg. Providence, R.I., was headed by Milton Sapinsley and Irving Jay Fain. The Pittsburgh committee included Charles J. Rosenbloom, Dr. J. E. Rosenberg, Louis Caplan, Prof. Max M. Frocht, Edgar J. Kaufmann, Sam M. Levinson and Morris Neaman. In St. Louis Dean Alexander Langsdorf presided over the committee.

On the fourth anniversary of the Society in 1944, Dr. William Fondiller became national president, and a drive was launched to raise $500,000 for a project at the Technion in memory of Brigadier Kisch. General Montgomery agreed to serve as Honorary Chairman of the Kisch Fund. Speakers at the dinner that year included Walter Lowdermilk, who told how impressed he had been with the way Jewish settlers and colonists were making use of old land in Palestine.

Dr. Einstein, who had retained his interest in the Technion ever since he had been chairman of Technion's first society of friends in Germany more than 20 years earlier, sent frequent messages of encouragement. In 1944 he wrote the Society: "It gives me real satisfaction to know that I took part in the founding of the Society and to see that your work in the past four years has amply fulfilled our expectations . . . . I want to stress again that the Technion, as the only technological training and research institution in Palestine, is indispensable for the development of industry and every kind of practical work in that country. Its importance for the rehabilitation of our people after the war is obvious . . . ." A year later his message to the American Technion Society dinner paid further tribute to the Technion: "You may also be justly proud that during the war it helped to bring victory by supplying the armed forces of the United Nations with technically trained personnel and modern research facilities . . . ."

There was competition, too, at this stage, chiefly from the Weizmann

Institute. In December, 1944, Technion's friends in New York sent an urgent cable of protest to Kaplansky informing him that the new fund-raising project for the Rehovot institution had been launched in the name of the "Weizmann Institute of Science and Technology". They felt this gave a misleading impression of overlapping. Kaplansky made inquiries, and a few days later was able to inform the Society in New York that Weizmann himself declared the word "Technology" had been added without his approval. He had given instructions that the name should be only "Weizmann Institute of Science".

Prof. Samuel's original mission to the United States had been intended for a short period, but when his time was up the Americans asked Kaplansky to extend his leave for another year and still another. Mrs. Samuel joined him in New York in 1941.

In the fall of 1942 he accepted a post as Visiting Professor at the Illinois Institute of Technology, but again at the urgent insistence of the Americans he returned to the Society for the year from June, 1944 to July, 1945.

His last task was to set up the Kisch Memorial Fund, and in November, 1945, he returned to Haifa. He discovered that the family to whom he had sublet two rooms of his flat had in the meantime taken over the whole apartment, together with all its contents. Legally they also acquired squatters' rights and could not be evicted. Samuel and his family had housing problems. Furthermore, the years spent in the hectic atmosphere of fund-raising and public organization had affected his ability to adapt to the more sedate atmosphere of Haifa. His contacts with the wide world had also shown him the possibilities for promotion and public relations, and he wished to continue with that activity. However, his views brought him into conflict with the more conservative attitudes of both Kaplansky and his colleagues, who could not reconcile academic dignity with what they called showmanship. Samuel was asked to write occasional articles about the Technion for the overseas activities, but basically he was expected to return to academic life, with a reduction in salary. He was the head of the Laboratory for Physical Chemistry and Molecular Physics, which he had been responsible for setting up and equipping before the war. He enjoyed an international reputation in his field, but he was not happy, and relations between him and Kaplansky were not of the best. He became ill, and on February 7, 1949, he died. The American Technion Society had been his creation.

Success in America

That Samuel had built soundly and well was evident from the fact that the Society continued to flourish and expand even after he left America. The precedents which had been set for large contributions were followed by others. In 1946 Mr. and Mrs. Herman Lebeson of Chicago established a Radio Communications Laboratory in memory of their son, David. In 1947 the Houston oil geologist, F. Julius Fohs, announced a contribution of $50,000, over a period of five years, for engineering fellowships. In the spirit of enthusiasm that prevailed, Morris Schultz, a New York Board member, called for efforts which would assure the Technion an income of a million dollars a year to meet both development and operating budget, and for establishment of a Technion building in New York which would serve as headquarters for these efforts. Within the decade his "exaggerated" ambitions became reality.

The Americans found still another outlet for their interest. Technion was approved by the Veterans' Administration for the education of American war veterans under Public Law 347, the G.I. Bill of Rights, and a campaign was launched to encourage and assist American youth who wished to study in Haifa. A Technion admissions committee was set up in New York, composed of Benjamin Cooper, chairman, William Fondiller, Lazarus White, J. R. Elyachar, and J. W. Wunsch. The first group of G.I.'s arrived in Haifa on the *S.S. Vulcan* in the fall of 1946. A second group sailed on the *Marine Carp* in August of the following year. Cooper noted that it was easier to get the financing from the American Government than to obtain admission visas from the British Mandatory Government. As it turned out, not all those who came were bona-fide students. Some of them were idealistic young Zionists who used this as a cover to get to Palestine, and did not spend even one day at the Technion. A whole group of them formed the nucleus for a kibbutz in the south. In this sense they were following in the tradition of the "students" whose lives had been saved a decade earlier because that was the only way they could qualify for British immigration certificates and get out of Eastern and Central Europe while there was still time.

The genuine students among them were a great headache. The committee in New York had not been properly informed about the requirements, and the young people were inadequately prepared in mathematics and physics. Few of them knew Hebrew, and almost all required special tutoring. They were unhappy with the housing assigned to them in the Haifa suburb of Sabinia. Furthermore, the G.I. payments were usually late in arriving.

Despite all the problems and difficulties, many of them were absorbed into the life of the country and two, Harold Monasch and Benjamin Walter Stein, died in defense of the country during the War of Liberation.

In still another direction, the Society set up a Committee on Technological and Engineering Developments, headed by Moses Heyman, to mobi-

American students on arrival at the Technion to study under the G.I. Bill of Rights. June, 1946.

lize American engineering and scientific know-how for the needs of industry in Palestine. Similar efforts were made sporadically in the years that followed.

Lazarus White, President of the Society for four years, tried his hand at some bold engineering thinking. In an article in the journal published by the Society, he proposed construction of an aqueduct which would draw water from the ample flow of the Euphrates River, provide for the irrigation needs of vast areas of the Syrian desert and then, after bypassing Damascus, continue on into Palestine, bringing agricultural prosperity wherever it reached. His proposed Euphrates River Authority was calculated to have the joint participation of Palestine, Transjordan, Lebanon, Syria, Iraq and Turkey. Political factors seemed to be secondary to the engineering aspects of the project.

In 1946 the American Technion Society marked its sixth anniversary, with a membership of 1200. J. W. Wunsch, Vice President of the Society, came on a month's visit to Palestine, thus establishing a close personal relationship between the organization in New York and the Institute in Haifa. The Technion marked the occasion by conferring on Wunsch its first honorary degree, Honorary Engineer. A year later Wunsch became

President of the Society, serving for four years. At the sixth annual dinner in New York the guest of honor was Alexander Konoff. The Technical High School became his special interest, and through the years his financial support for it ran well over a million dollars. Much enthusiasm was generated by a fund set up by alumni of New York's Hebrew Technical Institute, an East Side school, to honor the memory of a former principal of the school, and an Edgar S. Barney Memorial Laboratory was established at the Technion.

At the end of that year Kaplansky went on a two-month visit to the United States, which took him to a number of major cities. He heeded Samuel's advice, reluctantly, and consented to have himself billed as "President" of the Technion, though back in Haifa he was still known as *Menahel*. His addresses were memorable, not only because of their length, in typical Zionist Congress style, but because of the bold challenge which he voiced.

Planning Ahead

Though conservative in his personal life and in his public relations, Kaplansky was a man of soaring vision. He had reached the conclusion that the end of the war would open up an entirely new and different era for Palestine and for the Technion. He brought to the United States an ambitious program of expansion that he had already begun to develop during the war years. It was a two-phase program, the first calling for maximum utilization of the existing campus, and the second looking ahead to expansion to an additional and larger campus located elsewhere.

The more immediate program called for suitable facilities on the old grounds to accommodate the increasing student enrollment, the Chemical Engineering Department, which had been separated from Industrial Engineering, and the Electrical Engineering Laboratories, which were cramped and crowded. Prof. Alexander Klein had already drawn up a construction program for the site. This called for the building of a three-story block to house Chemical Engineering, the addition of two floors to the existing Hydraulics building, which would constitute the Kisch Memorial, and the construction of Electrical Laboratories in the southeast section of the campus. The master plan also called for additional lecture and drafting halls in the center courtyard, the addition of a new two-story wing to the western side of the main building, and creation of a large assembly hall. In all there would be five new buildings in the front gardens, three on the east side along Balfour Street, and two on the west, facing Schmaryahu Levin Street. New quarters would thus be provided

Model of the old Technion campus, showing plans for construction of additional buildings. The program was abandoned in favor of the move to Technion City. 1945.

also for Physics, Architecture and Mechanical Engineering. The new floor space was to total about 80,000 square feet and the cost of construction and equipment was estimated at $1,000,000.

The second phase envisioned the growth of the Technion to meet the post-war needs. It became clear to Kaplansky that transportation was to play a vital role in the new world. Palestine, situated at the crossroads of international communication by land, sea and air, had to be mindful of its connections with the world, and he therefore proposed establishment of what he called a Transport Engineering Department, comprised of sections dealing with Automotive, Marine and Aeronautical Engineering. As early as 1943 Dr. Fondiller had also suggested the need for studies in the latter field, and a small sum of money had even been raised, which was sequestered and later made available to help establish the new department.

Kaplansky returned also to his earlier recommendation for the opening of a Department in Economics and Administration to train officials for industry, the civil service and the various national and public institutions. He again called for a Department of Agricultural Engineering. This would raise to six the number of departments in the Technion. There was also need for quarters for students, a recreation center, and adequate sports facilities as well as more space for the Technical High School.

These developments would require land area far exceeding that available to the Technion on its original site, and he cast his eyes in the direction of the Bay area, on the other side of the Kishon River. This seemed most suitable because there was ample space for airplane runways, for agricultural experimental fields and other needs of the future. He had already begun discussions with the Jewish National Fund for the allocation of the land. Funds required were estimated at an additional one million dollars, though he hastened to add that this was for future growth, perhaps after celebration of the 25th anniversary, in 1950.

He had also calculated the sources for these funds. The plan was to confront the Palestine Government and the Jewish authorities, show them that the Technion had already raised half a million dollars (for the Kisch Memorial) and demand their participation on the basis that each of them should match what the Technion had provided. That would then total a million and a half. It should then not be difficult, he felt, to obtain the additional half million from local industry and other friends abroad.

For an institution which had been operating on a shoestring budget, the concept was grandiose and overwhelming. The Americans were still a long way from completing their campaign for the $500,000 Kisch Memorial but they liked the new plans, which put the Technion in the category of a major institution.

At one stage the Americans came up with a proposal which they thought might solve all the Technion's financial problems at one fell swoop. They had observed that the old site in midtown Haifa, once located on the distant outskirts of the city, had now become prime property in the very heart of town. Its value as real estate was enormous and they suggested selling it. The proceeds would be more than enough to acquire larger areas outside the city and still leave ample funds for development. Kaplansky explained that Jewish National Fund land could not be sold.

On his return to Haifa Kaplansky reported to a joint meeting of the *Vaad Menahel* and the Council of Teachers. He said he had gone to America with many doubts and misgivings, fearful that his manner of speaking was not according to American tastes. He had been pleasantly surprised to find that the American Jewish public had had its surfeit of emotional propaganda, and had responded warmly to his talks on Technion's role against the background of Palestine's economic and technical needs.

Attendance had been good at his meetings, but he was disappointed at the slow pace of the fund-raising. Yet when he asked the Americans to double their contributions they were not frightened.

He had also learned much. He discovered that in America, too, the heads of institutions spent a great deal of time looking for money. He learned the difference between a contribution and a pledge. As a Socialist, he was heavy-hearted at the deproletarization of American Jewry.

He had taken the opportunity to visit some 20 universities and technical schools. And before his departure, in February, 1947, he had met with a group of people who were interested in promoting aeronautics at the Technion, among them Theodore von Karman. The suggestion had been advanced that money be raised for a wind tunnel.

Interlude of Growth

The end of the war provided the expected boost to enrollment, and the 1946–47 academic year opened with registration of 200 first-year students, bringing the student body up to a total of 670. There were 300 pupils in the Technical High School and 95 in the Nautical School.

There could no longer be any question about the value of a Technion education; the graduates had more than proven themselves. A survey made in 1945 showed that the alumni were occupied in the following sectors: military work, 10%; government services, 10%; municipality services, 12%; contracting firms, 27%; general industry, 16%; public institutions, 10%; private practice, 2%. The balance had presumably drifted out of technical employment.

The total of graduates rose by a further 99 when the 17th and 18th classes jointly received their diplomas in May, 1947.

The Technion bustled with activity and signs of growth. As a result of the increased demand from the various laboratories, the small gas generator which produced fuel for the Bunsen burners was rendered inadequate and obsolete. Word had reached Haifa of a new type of generator which had been developed at the Hebrew University and its designer, Eliyahu Sochazewer, was asked to install one at the Technion. After he had completed the equipment at his Jerusalem workshop in 1946 he was raided by a British searching party that had heard (with considerable justification) that he was secretly making armaments for the Haganah. Fortunately they found nothing, but one soldier stuck his bayonet through the tin shielding on the side and the Technion's gas generator had to be repaired. It was subsequently installed successfully in Haifa.

What Kaplansky had seen of technological universities in the United States made it clear to him that the Technion still had a long way to go. On

Members of the Anglo-American Committee of Inquiry on Palestine are received at the Technion by Dr. Kaplansky, right. 1946.

a visit to Cairo he had opportunity to make comparisons also with the Engineering Faculty of Fuad University. The department there was nearly 100 years old. It had ample space and buildings and equipment far in excess of what was available to the Technion; the language of instruction was mainly English, he noted.

The expansion of the Institute cost money. In 1945–46 the expenditures rose to £94,000. Income from tuition fees, laboratories, workshops and other internal sources covered about 70% of this, leaving a balance of £28,200, equivalent to around $140,000, which had to be covered by the Jewish Agency or by Technion societies overseas. Going into 1947 the Institute had a small operating deficit covered for the most part with money borrowed from its own special funds. The deficit was expected to rise with increases in salaries and enlargement of the student body.

An attempt was made to increase tuition fees. Before the war the annual fee had been £25. It was generally figured that a doubling of 1946 prices over those of 1939 would be equivalent in real costs, and the fee was therefore raised to £50. When it was raised again the following year to £60 the students refused to pay and went on strike. They remained away from their studies for a week, and finally agreed to the £60 on promise of rebates for needy students and a delayed schedule of payments throughout the year. However, they then instituted a payment strike and withheld the proportionate fees according to the staggered schedule. In the absence of Kaplansky, Prof. Tcherniavsky conducted the negotiations. After consultations he ruled that if there were no payments, there would be no services, and he closed down the Technion. The students quickly came around. In 1948 the tuition fee went up to £75.

Even before Kaplansky's trip to the U.S. the *Vaad Menahel* had voted him a substantial salary increase in consideration of the many activities and the broad development of the Technion under his administration. The Board expressed its regrets that it had not taken such a step earlier. In his acknowledgment Kaplansky said he appreciated the comments more than the money itself.

The casual events chronicled in this period gave little indication of the climax that was impending. May, 1946: a large delegation from the Technion participated in ceremonies marking laying of the cornerstone of the Weizmann Institute, and Kaplansky spoke in praise of the Institute. October 26, 1947: Henry A. Wallace spent the better part of a day on an in-depth tour of the Technion laboratories. October 30: Tova Shevelov, widow of the Technion's first watchman, and herself operator of the restaurant for the construction workers on the site before the First World War, died at the reputed age of 110. Prosaic activities filled the days, but tension in the air was obvious to all. The British Mandatory Government appeared to be both unable and unwilling to maintain stability, and armed clashes became more frequent, involving Jews, Arabs and the British forces. Investigating commissions had come and gone through the years, but no proposal for pacification of the country had been found practicable. There was no great surprise when Britain announced that it was divesting itself of mandatory responsibility, and the United Nations was called upon to make a decision. This it did on November 29, 1947, with adoption of the resolution calling for establishment of separate Jewish and Arab states in a partitioned Palestine.

The entire *Yishuv* waited with bated breath for the announcement from Lake Success, and then all joined in a night-long celebration of the imminent establishment of a Jewish State. The following day, without sleep, the 89 members of the 20th graduating class sat for the first of their final examinations and then, after ten days of gruelling tests, most of them went out to join their Haganah units.

U.S. Vice President Henry A. Wallace, left, is escorted by Dr. Kaplansky on his visit to the Technion. 1947.

231

When he addressed the diploma ceremonies for the 19th graduating class on January 26, 1948 (conditions permitting, granting of diplomas usually took place about a year after the examinations), Kaplansky soberly listed Technion's losses in the undeclared, but actual war, and warned there would be more to come. At the same time, he recorded further academic growth and progress. Engineering diplomas were granted to 61 graduates. The number of undergraduate students remained the same as in the previous year, 670, which was itself an achievement. New senior faculty members engaged included Dr. Markus Reiner, who had long been close to the Technion.

Dr. Kaplansky paid tribute to the American friends who had provided $400,000 thus far and announced that the money was being used for the construction of two chemistry laboratories, one for analytical and micro-chemistry, and the other for physical chemistry. These were to be the first of the laboratories in memory of Brigadier Kisch. The construction project included also an additional story for the Technical High School building and four new drafting halls built over the workshops. Judge Bernard Rosenblatt represented the Americans at the ceremonies.

During the same week the *Vaad Menahel* approved the recommendation of the Teachers' Council* that the Technion be authorized to confer Master and Doctor degrees. Appropriate regulations were drawn up and in 1952 the Institute's first degree of Doctor of Technical Sciences was granted to Dr. Eliezer Mishkin.

Dr. Kaplansky kept a low profile within the Technion on his personal political opinions, though these were well known. He had occasional meetings in his office, but his political activity never intruded on his Technion work. Yet he was often subject to public criticism for his views which were not always popular. Kaplansky was interested in all efforts to promote closer and more friendly relations between Jews and Arabs. He was also associated with the "V League for Soviet Russia", which sought friends and supporters for the Soviet Union. In the early years of the Second World War he played host at the Technion to two Soviet diplomats from the Embassy in Ankara. A reception for them which he tendered at his home was marked by clashes with some of his labor colleagues, who apparently saw things in a different light.

Nevertheless, he continued to be active in the affairs of the group, that later changed its name to "League for Friendly Relations with the U.S.S.R.", and served as its secretary. The League stated as its avowed purposes to further feelings of good will on the part of the Jewish population of Pales-

---

* This is literal translation of *Moetzet Hamorim*. The body was in effect a Senior Faculty Council, and later evolved into the Senate.

tine toward the U.S.S.R. and its achievements in liberating parts of Europe from the Nazis; to secure the support of the U.S.S.R. for the Zionist cause, and the cause of national and social liberation of the Jewish people in Palestine; to participate in the fight against any form of anti-Semitism and Fascism; to establish ties of brotherhood with Soviet Jewry.

In June, 1947, on behalf of the praesidium of the League, Kaplansky wrote to Andrei Gromyko, then head of the Soviet delegation to the U.N., expressing deep appreciation for Gromyko's "historic declaration" at the U.N. General Assembly, in which he had voiced Russian support for Jewish national rights in Palestine. Ironically, Kaplansky's statement also said that "Gromyko's declaration has once again proven that the U.S.S.R., as a workers' socialistic country, the stronghold of peace and progress in the world, is the true friend of every people, great or small, which is striving toward autonomy, freedom and peace."

Bombs were exploding and bullets crackling in almost every part of the country. In Haifa attacks were being mounted against Jews in the streets and it was decided to strike at the Arab military headquarters located in a stone house known as *Bet Najada*. A barrel of explosives (all they had was black gunpowder) was carefully prepared in the Technion workshop and gingerly loaded on a truck. The intention was to offload it as the truck passed the Arab building on the other side of Wadi Rushmiyah. The driver was nervous, however, and as he careened too rapidly down the steep, curving incline just before the Rushmiyah bridge, the barrel rolled off and came to rest in the middle of the road. It did not explode. The driver continued, passed *Bet Najada*, and kept on going to safety.

The problem now was what to do about the dangerous barrel. If the detonator were to be activated, it would go off with a mighty blast in the midst of a residential area. Yaakov Salomon, prominent barrister, and one of the leaders of the Haganah in Haifa, was the contact man with the British. Together with Max Gill, the Haganah man responsible for the barrel, Salomon at once informed the commander of a British platoon stationed at nearby *Bet Hataassiyah* what the barrel contained and told him they would like to approach the object and remove it. The commander refused to permit this, claiming he did not want them to endanger themselves; in reality, perhaps he did not want them to retain the bomb for future use. Instead, he assumed responsibility for its disposition. The British poured gasoline around the barrel, ignited it, stood far back and kept curiosity seekers at a distance. The barrel exploded with a mighty roar, shattering windows and causing damage to the Jewish homes on both sides of the road. Perhaps this is what the officer had intended.

The source of the bomb could hardly have been a secret. Another

Device produced for the Haganah by Simshon Turetzky and Menahem Haklai to create artificial fog at sea in the event of need for a smokescreen. 1947.

attempt to get at *Bet Najada* was through an automatic wagon devised by Yehuda Naot, a Technion staff member, who had for some years been engaged in producing ingenious and successful devices of various kinds for the Haganah. The wagon was to be lowered by pulley across Wadi Rushmiyah. It was never used, and for some time it remained in the Technion courtyard, an unknown and forgotten relic of an era. More than ever before the Technion was becoming the nerve center of the defense activity. There was much coming and going on the campus. Classes continued, but attendance fell off sharply again as students and teachers hurried to their units.

British troops were withdrawn from Palestine, section by section, as the Mandatory Government prepared to abandon the country to anarchy. Haifa was the last jumping off point.

On the fifth day of the Hebrew month of Iyar, May 14, 1948, as the last of the British left, the Technion, together with the entire *Yishuv*, welcomed the formal establishment of the State of Israel. A new era had opened in Jewish history.

* * *

Rear courtyard of the old building.

Statehood CHAPTER EIGHT

A Challenge

The full story of the struggle for independence and the emergence of the State of Israel must be told and retold elsewhere. Like everyone in the *Yishuv*, the Technion and its people were deeply involved in every aspect of the historic events of the years 1947 and 1948. Everything that happened had its impact on the Technion both immediately and in the long term as well. This period in the history of the institution must almost be read in parallel to a history of the early years of the State of Israel.

The birth of the State, with all its military and economic needs, confronted the Technion with the greatest challenge in its existence. The destiny of a nation depended to a very large extent on the skilled manpower, without which no modern country can long exist. The measure of the challenge and the demands made upon the Technion became more evident with the passage of each year, and made almost inevitable the development of a university which could meet the highest international standards.

The Institute's own celebration of the proclamation of statehood took the form of a convocation of all branches of the Technion family, held on May 20, 1948. As Dr. Kaplansky pointed out, the principal element, the student body, was to a large extent missing, and he again had the morbid task of reading out the list of names of students who had fallen in defense of the homeland. The list had grown since its last rendering. When the final figures were recorded, and it was not certain that these were complete, it was found that no less than 57 Technion students had given their lives in the national cause. Scores more were wounded and invalided.

A few days later the *Vaad Menahel*, meeting jointly with the Council of Professors and the Teachers' Council, formally placed the Technion with all its facilities and latent powers at the disposal of the new Government. The Technion also sent greetings and encouragement to its sister institution, the Hebrew University, beleaguered in Jerusalem, with hopes that it would be protected and defended from the dangers facing it.

The first Chief of Staff of the fledgling army of Israel, General Yaakov Dori, paid public tribute to the Technion for its role in the national defense. He singled out for special mention the organization of the Science Department in the Ministry of Defense by Prof. Ratner, and the mobilization of the scientific and technological skills so urgently required. The accomplishments of the professors, graduates and students of the Technion were most distinguished, he said.

In the United States the resources of the American Technion Society were being tapped. Moses D. Heyman, Chairman of the Engineering and Technology Committee of the Society, headed a campaign for arms procurement, all of it irregular, in which Technion people participated with the Haganah leaders who had set up the program. One of the giants in the effort was Charles Frost, who continued his valuable cooperation for many years thereafter.

The creation of the Israel Air Force was an epic in itself. The first aircraft technicians who enabled the early planes to take to the air were trained on the Technion grounds. A shack in the back courtyard was placed at their disposal and transformed into a mechanical workshop. At the outset the air force had only a few Piper Cubs (Primuses, they were called), and the engines were brought to the Technion from time to time for repairs or overhauling. Yehudah Hashimshony was in charge of the workshop.

The "aircraft ear" developed by Prof. Ollendorff for use during the World War was improved with the cooperation of Leo Schaudinischky and Baratz. It was set up in Nahariya to provide early warning of the approach of hostile Arab planes and was later sent south and erected on the roof of a house on the road to Iraq Sueidan.

The academic schedule at the Technion was greatly reduced, and eventually classes were discontinued because of the absence of students and many teachers and staff members. It was decided to continue to pay salaries to all full-time tenured teachers. Payment was discontinued to part-time teachers on the assumption that they had other sources of livelihood. Staff members who were in the armed forces were permitted to continue to draw their Technion salaries, but were expected to refund to the Institute their military salary and allowance. On the one hand Kaplansky encouraged every member of the Technion to fulfill his national defense obligations, and on the other hand, he pleaded with the military authorities to release students or teachers whose services were not necessary.

Because of the nature of the defense work going on at the Technion, the Hadar campus was put under special guard, and patrols circled it constantly. Conditions were tense in Haifa, where Jews and Arabs lived in close proximity to each other. Admission to the campus was by special permit or identification. At the time, there was only one Arab student, Rustum Bastuni, who was studying architecture. He asked for a permit to enter the grounds so that he could continue his studies. Yosef Karni, who represented the Technion on the guard committee, bluntly refused. He never told Bastuni the reason for the refusal. There was no cause to suspect the loyalty of Bastuni, who eventually graduated, and became a member of the Knesset. To the contrary, Karni feared that if Arab extremists were to learn that Bastuni had been given free and unquestioned access to the campus, they would have suspected him of being a collaborator and might have sought to kill him. Karni's decision, therefore, was made out of a desire to protect the student.

While the military and security problems facing the Technion were more obvious to the eye, there were also administrative and financial problems no less serious and in many ways more fundamental to the operations of the school. The emergence of an independent national government meant that all the quasi-governmental functions which had hitherto been carried on by the *Vaad Leumi*, as well as some of the responsibilities of the Jewish Agency and the Zionist Organization, now devolved upon the Government. This included, of course, the entire educational network in the country, and indeed the newly-formed Ministry of Education inherited the system which had been operated by the *Vaad Leumi*. The situation with regard to the institutions of higher learning was not as clear.

It was Kaplansky's opinion that the Government should set up a special Department of Science, Research and Planned Development. He saw as the purpose of this Department "to provide for the consolidation, improvement and expansion of the scientific institutions; to examine their schemes for future organization and development with a view to avoiding overlapping and duplication; to utilize their facilities and possibilities for the benefit of Government departments and thus to prevent the creation of parallel research and testing units by different Government agencies (symptoms of such tendencies leading to duplication and waste of money and effort had shown themselves already in the first few months of the formation of Government departments); to guide the scientific institutions toward a rational distribution of tasks and to coordinate scientific work carried out with public means from the point of view of the requirements of the national economy and the State."

At the same time, he emphasized the need to maintain the autonomy of the several institutions and their academic freedom. It will be recalled that at the time the only other institutions of higher learning, in addition to the Technion, were the Hebrew University and the Weizmann Institute.

The precise form to be taken by the proposed new Department remained open to discussion. There were some who thought it should be a full Government Ministry, headed by a member of the Cabinet. Others felt that a Department connected with the Prime Minister's Office or the Ministry of Finance would be sufficient. Others believed it should be associated with the Ministry of Education, though the scientists of the country for the most part opposed this suggestion on the grounds that the philosophy and attitude toward elementary and secondary education were quite different from those required for higher education and research.

Basing himself on "bitter experience", Kaplansky returned again and again to his warning that needless competition between the institutions should be avoided. In November, 1948, he wrote that even before establishment of the State each institution had sought to expand and broaden its scope as it saw fit. The result was the establishment of laboratories which duplicated each other, each with insufficient equipment. Instead they should have combined their meager resources for the maintenance of one central laboratory, on a rational basis.

"Our poor little country cannot afford the system prevalent in America where they have the means to maintain parallel setups in government, in private industry, in the armed forces and in the universities," he warned. As to cooperation between the institutions themselves, he recalled that under the Mandate there had been a Palestine Council for Scientific and Industrial Research, composed of representatives of Government, industry and the universities.

He suggested reviving and broadening such a body. By August, 1950, he could report to the *Vaad Menahel* that in response to Technion's proposals the Prime Minister's Office had set up a committee to look into the status of the institutions of science and higher learning, and important conclusions were anticipated. As a matter of fact almost 20 years were to elapse before any effective steps were taken.

In his greetings to the Hebrew University that year on the occasion of its 25th anniversary, Kaplansky made it clear that he did not regard the Jerusalem institution as a competitor of either the Technion or the Weizmann Institute. Each in its way was part of one university of Israel, he said, and he looked forward to the time when such a relationship would be formalized constitutionally. In an interview in London he went even further and declared that the idea of a University of Israel, embracing the three institutions, was very near realization. In the meantime, the three schools did indeed practice cooperation and coordination.

Technion's return to normalcy after the War of Independence was gradual. Graduation exercises of the 20th class took place on March 9, 1949. Most of the graduates and many of those in the audience were still in uniform. Four diplomas were conferred posthumously. Other recipients were absent because they were still on active duty. It was announced that 660 students had served in the armed forces, more than half of them as officers or non-coms. Because of their educational background, the Technion personnel played a significant role in almost every branch of the Israel Defense Forces, and it was only fitting that the diplomas should be distributed by the Chief of Staff, General Dori. Kaplansky utilized the occasion to ask him to accelerate release of students wherever possible, so that classes could resume. Indeed it became possible to commence the academic schedule on April 24, after suspension of operations for almost a year.

Statehood, huge waves of immigration and a rapidly expanding economy confronted the Technion with new challenges. Visionary goals which Kaplansky had suggested in the preceding years suddenly became not only possible but even necessary. He repeated his call for establishment of new departments, singling out agricultural engineering, economics and administration. He also pinpointed the need to train the teachers required for the technical high schools of the country.

Aeronautics and Goldstein

First on his list, however, was aeronautics. Israel's recent experience showed that "transport and defense by air are as vital to us as air for breathing," he said. The subject had been on his agenda for several years. Dr. Fondiller had broached the suggestion in 1943. It came up again two years later when Dr. Assaf Ciffrin, associated with the Aeronautics Research Division at N.Y.U., wrote Kaplansky proposing the establishment of a chair in aeronautics, the setting up of an appropriate library, a wind tunnel, an aeronautics workshop and the acquisition of trainer planes, with a pilot instructor. Fondiller, too, returned to the subject.

Gen. Yaakov Dori attends the diploma ceremony in his capacity as Chief of Staff of the Israel Defense Forces. 1949.

In 1946, at the request of Kaplansky, Prof. Ollendorff met with von Karman in Paris to solicit his interest and participation. The latter suggested that Technion establish contact with a Prof. Sydney Goldstein, who was Professor of Applied Mathematics at Manchester, Chairman of the Aeronautical Research Council of Great Britain and author of classical studies in fluid mechanics and aerodynamics. In the meantime Dr. Fidia J. Piattelli was invited to deliver the first lectures. Piattelli held a degree in aeronautical engineering from Rome and had been actively associated with aircraft industries in Italy and England. Interned in Italy during the war, he went to Palestine immediately thereafter with visions of establishing an aviation industry. He founded a company, Maof, to build gliders, but his efforts both in industry and at the Technion were premature.

In his formal report to the 22nd Zionist Congress at Basle in 1946, Kaplansky called for establishment of an independent Transport Engineering Department to comprise chairs and laboratories for automobile, marine and aeronautical engineering. "This new department will demand large investments in buildings and equipment," he wrote, "particularly the provision of extensive grounds for an aeronautical station with a wind tunnel, and an aeroplane and glider repair shop...." He had spelled out similar details for the American Technion Society earlier in the same year.

From California, where he had moved, Dr. Assaf Ciffrin bombarded the Technion with letters urging the introduction of aeronautical engineering, and went into some detail regarding the functions of the proposed subsonic and supersonic wind tunnels.

During his American visit in 1947 Kaplansky met with von Karman and others and it was agreed that the first steps should be the construction of a wind tunnel and the assembly of a collection of engines. Again Goldstein's name was mentioned.

The British scientist had been a lifelong Zionist. Years earlier his home had been a center of Zionist activity, and visitors to Manchester frequently lodged with him. Among them had been Dr. Kaplansky, when the latter had been a member of the Zionist Executive.

Goldstein's first serious bid to settle in Palestine was when he submitted an application to the Hebrew University for a lectureship in Applied Mathematics. He received no written response, but the University authorities sent word to him through the eminent philosopher, Prof. Samuel Alexander, that they were unable to offer him the post since "there was no such subject as Applied Mathematics."

Because of the Manchester connection his relations with Weizmann were good. In 1949 the latter, now President of the new State, formally invited him to visit Israel. The formal invitation was necessary to enable Goldstein to get his passport endorsed for a visit to that dangerous corner of the world. In March he went to Israel, and during the course of his call on Weizmann at Rehovot it was suggested that he visit the Technion. Both Weizmann and Ben Gurion passed the word on to Haifa.

There was an immediate meeting of minds on major principles and he was invited orally to join Technion's staff. On his return to Manchester Goldstein wrote out a formal application for an academic position with the Technion. Ben Gurion was enthusiastic and wrote to Kaplansky: "I highly esteem Prof. Goldstein's coming to Israel and joining our scientific powers. It would be well if he could begin his work at the Technion as quickly as he can free himself from his present duties."

In accord with proper procedure, Goldstein then submitted his resignation from the Aeronautical Research Council and from Manchester University, in each case giving a year's notice in advance of his withdrawal. The University officials begged him to remain. They could not understand how a brilliant scientist could abandon the western world for the wilds of the Middle East. Goldstein tried to explain something of his Zionist motivations.

His Manchester colleague, a devout Christian, sighed. "I suppose it's something like taking a missionary assignment," he said.

Von Karman had a similar experience which he frequently narrated. When it was formally announced that the Technion was to open a Department of Aeronautical Engineering, and von Karman's name was associated with it, one of his distinguished colleagues at Princeton turned to him in surprise.

"Surely," he said, "something like agricultural engineering would be of much more importance to Israel, along the lines of the famous *halutzim*. Why, in God's name, aeronautical engineering?"

Von Karman informed his friend that the idea had not come suddenly. It had been proposed in the days of the British Mandate, and he had been involved at the time. Assistance had been sought from the British High Commissioner, but the latter had waved the project aside with scorn.

"Everyone knows," he said, "that the Jews have no talent for aeronautical engineering."

Von Karman smiled grimly at his colleague. "Now," he said, "we're going to show them!"

After hearing Piattelli testify before its sub-committee on Transportation Engineering, the Teachers' Council formally decided on March 4, 1949, to set up a separate Department of Aeronautical Engineering. Goldstein was nominated to the post of Professor of Applied Mathematics, the *Vaad Menahel* approved, and Kaplansky wrote asking if Goldstein could come to take up his duties by October 1, 1950. In the meantime the Hebrew

Prof. Sydney Goldstein (1903–   ).

University awoke to the realization that something had slipped through their hands, and President Selig Brodetsky sought an arrangement with Goldstein. The Technion improved its offer, and appointed him also Dean of the new Department of Aeronautical Engineering. It was made clear that he would be too busy in Haifa to have time for Jerusalem.

Friendly letters with information were not sufficient; Goldstein was a stickler for form and he insisted on an official letter of appointment, with duties clearly spelled out. Such a letter finally reached him 13 months later, a month after both his resignations in Britain had gone into effect. He came to Haifa in November, 1950.

In the meantime two Americans had been added to the staff, the first of a large number to follow. They were Prof. Menahem Merlub-Sobel, in Chemical Engineering, and Prof. Judah L. Shereshefsky, who served as Visiting Professor in Physical Chemistry.

Changes With the Times

There were academic developments along other lines as well. It was decided that effective with the 1949–50 school year the Department of Technology should be divided into two units, Mechanical Engineering and Electrical Engineering. Together with the existing departments of Civil Engineering, Architecture and Chemical Engineering the Institute now had five major academic units.

Dr. Theodore von Karman (1881–1963).

Once again proposals were made for establishment of a Department of Industrial Management and Public Administration. Kaplansky had first raised the issue as early as 1939. The Teachers' Council had discussed it subsequently. By 1950 request had come from the students for studies in that field.

A decision to grant Master of Science degrees was adopted with some misgivings. There was fear that it would reduce the value of the professional engineering degree (Ing.) the Technion had been giving. The M.Sc. was approved at the outset for Mechanical and Civil Engineering only.

The need for an academic press had led the Technion into a publications partnership with the Association of Engineers and Architects but it did not function as expected, and the Technion withdrew. Kaplansky drew attention to Technion's need for an agency which would publish the necessary textbooks and the scientific works of the staff. A full-scale textbook in Hebrew, *Cement and Concrete*, by Rahel Friedland (Shalon) had been published by the Technion in 1939, but despite its success, other faculty members did not prepare additional texts.

The program of adult vocational training was expanded with an agreement between the Haifa Labor Council and the Technion for the conduct of evening classes for working people. The Council was to recommend up to 75% of the students, assuming they met with Technion's criteria. This was intended to be an eight-year program leading to a special diploma. A two-year trial period was agreed upon.

A reorganization of entrance requirements, long under study, was put into effect with the 1950 decision to abolish the comprehensive entrance examinations and to require examinations only in mathematics, physics and Hebrew.

Academic standards came into conflict with practical problems involving students whose studies had been interrupted by military service or the general disturbed conditions of the preceding several years. Should they be punished because they had not achieved satisfactory grades under such conditions? How could standards be maintained without penalizing those who had rendered patriotic services? The problem was discussed long and earnestly, and most cases were settled on an individual basis.

There was need for administrative changes as well. Kaplansky was crushed under the growing responsibilities of his position, and his health began to fail. His illness, first revealed in 1948–49, left him constantly fatigued. More and more he needed periods of rest and recuperation to gather his strength but he refused to slacken his pace. In June, 1949, the *Vaad Menahel* approved a three-month leave of absence during which time Prof. Shlomo Ettingen was named Acting Director and Prof. Merlub-

Sobel was put in charge of relations with the friends in the U.S. Some months later Kaplansky found it necessary to deny reports in the press that he was planning to resign, and that the Government was going to transform part of the Technion into a military academy. In 1950 he spent some time in a convalescent home. Ettingen and Merlub-Sobel were appointed to assist the Director on a permanent basis, the former to handle administrative matters, and the latter academic affairs. Kaplansky expressed a need for a full-time Deputy Director, but there was no candidate in sight. The *Vaad Menahel* itself was broadened to include industrialists and bankers as well. Additional professors were added to the membership.

Max Hecker, on the other hand, resigned from that body, pleading old age. He resented the tenor of a historical review of the Institute which had mentioned that in the early days there had been anarchy at the Technion. He expressed the hope that some day someone would write a history of the institution so that people could learn the truth about the founding of the Technion. A month later he was again a member of the *Vaad Menahel*, this time as representative of the Jewish Agency. Dr. Mordecai M. Levy became head of Public Relations and Ben Zion Tooval took over the post of Acting Academic Secretary.

Affiliation of the Nautical School with the Technion had not been successful. The original tri-partite agreement had called for the sharing of the financial burden by the Jewish Agency and the Palestine Maritime League, while the Technion was to handle the academic, professional and administrative operations through its Technical High School. In the latter years a new financial partner had joined, the American Fund for Palestine Institutions. The necessary funds had not been forthcoming in sufficient amount from any of the partners, and the School had become a financial burden on the Technion. Kaplansky proposed that the arrangement be terminated by mutual, friendly agreement. It was not until 1952 that the actual divorce took place.

There were nuisance incidents as well, not important in themselves, but they consumed much precious time, and tested patience to the limits. One such was the wooden hut in the Hadar front courtyard that eventually became known as *Shachefet* (Tuberculosis). The building had been put up by outside agencies in 1938 as an emergency center to house victims of Arab terrorism. Kaplansky was not happy about having a hospital on the grounds, and it was agreed that when the emergency was over the hut would be removed. However, it continued as a First Aid Station during the war, operated by the Magen David Adom. In 1949 the Anti-Tuberculosis League "occupied" the building and it was only after more than two years of police and court action that they were evicted.

Haifa Schoolchildren planted 500 trees through the Jewish National Fund, in honor of the Technion's 25th anniversary. Dr. Shlomo Kaplansky expresses his thanks to the representative of the children. 1950.

Prof. Shlomo Ettingen (1897–1963).

The Technion paid for the structure and put it to its own use.

There was time for celebration, as well. The Technion marked its 25th anniversary with a belated gala event on February 2, 1950, at which Dr. Weizmann was the principal guest, and spoke in praise of the Institute.

He did not know it, but this was to be Kaplansky's last major public appearance. He gave a broad review of the growth and development of the Institute, reporting that student enrollment had reached 800. He announced new appointments, chief among them that of Prof. Goldstein. For the first time the Technion was referred to as Israel's Institute of Technology. Kaplansky drew attention to the amazing developments in every field of science and technology, and warned that the nation must learn to use the tools of technology for its existence. A highlight of the event was the conferment on Kaplansky of an honorary doctorate.

The student representative at the celebration, Avraham Peled, in addition to the usual greetings, took the occasion to address the Government with an urgent appeal from the student body. To enable the Technion to fulfill its vital national mission, he called upon the Government to make possible increase of the teaching staff by mobilization of teachers and experts from Israel and abroad. Investment in improved laboratories would be quickly reflected in local industry, he said. He called for funds which would make it possible for the academic staff to publish their books and research studies in Hebrew. "Place the Technion at the head of the national industrial development program," he appealed, "for this is the source of our industrial strength."

Seeking New Friends

No less than in the past, financial problems still faced the Technion. The budget had grown through the years. Kaplansky noted that in the first decade of the Institute it had been around *IL 8,000 annually. This had grown to an average of IL 32,000 in the second decade and to about IL 140,000 in the ensuing five years. In 1950 it stood at IL 350,000, of which IL 105,000 was an anticipated deficit. The Jewish Agency was being asked for IL 50,000 and the Government for IL 55,000. Tuition fees were IL 75 a year, and the suggestion that they be raised to IL 100 was greeted with the expected protest of the students. Decision was deferred.

Even funds properly due to the Technion could not always be collected. During the years 1946–48 the Institute had been assessed taxes on the commercial property, *Bet Hakranot*, that it had built along the Herzl Street

* Hereafter IL refers to Israel pounds.

Main building "camouflaged" as an aquarium for the Student Ball. 1950.

side of its campus. Payment was made under protest, and the Technion brought suit against the District Commissioner for return of the funds, some IL800. The court found for the Technion, but the Mandatory Government appealed to the Supreme Court. In the meantime, the Mandate had come to an end. Because the court records still showed an appeal pending, collection could not be enforced. The money was lost.

Steps were being taken to protect the employees of the Institute, and a Provident Fund *(Kupat Tagmulim)* was established. This too cost money.

It was clear that help had to be sought elsewhere, and once again the efforts were directed to the Jewish communities overseas. A program to develop and expand the organizational base of support was indeed already under way. Outside of the United States there was activity in at least two countries. On the initiative of American engineers, a group of Friends of the Technion was organized in Mexico early in 1949. An engineer, Samuel Dultzin, played host to the group at his home, and he was elected the first President. Other officers included Leon Gerson, Vice President, and Isaac Grabinsky, secretary-treasurer. Dultzin was followed by David W. Mehl as President.

Great Britain had been a source of generous support in the 1930's, and an attempt was made to set up a Technion organization, independent of the Hebrew University group. Dr. E. Alexander-Katz, a faculty member, spent many months in London in 1947 and 1948 and set up a committee of about 45 members, a third of them non-Jews, drawn from science and technology. The list included such distinguished names as Field Marshal Jan Smuts, Lord Nathan of Churt, and Sir Robert Robinson, President of the Royal Society. A 44-page brochure was published, describing the work of the Technion. In the foreword, Redcliffe N. Salaman wrote: "Churchill, at the time of Britain's danger, no more threatening than that confronting Israel today, appealed to our brethren across the Atlantic, 'give us the tools and we will finish the job.' The leaders of Israel, with no less sincerity or urgency, call to their brethren the world over, 'give us the means to forge the tools and we will train the men to use them.'"

The results were meagre, less than £ 3,500, much of it payable in annual installments. Some equipment was purchased from the Depot of Army Salvage. The committee did not last long. Kaplansky visited London in 1950 seeking to reorganize it, but without success. The next effort was made in April, 1951, when the Technion Society of Great Britain was born in the House of Lords. Lord Silkin, former Minister of Town and Country Planning, played host to the gathering in a House of Lords committee room, and he was elected first Chairman. Eliahu Elath, Israel Ambassador to Britain, was present. Cabled messages of greeting were received from the U.S., Mexico and Canada.

Activities had begun in Canada as well. As early as 1944 Myron Samuel was listed as Acting Chairman of a Technion Society in Toronto. Nicholas M. Munk was treasurer. By the time of establishment of the State the activity had grown, and D. Lou Harris, President of the Canadian Technion Society, was already touring Canada and parts of the U.S. to spread the message of Technion. He held his post with distinction for over 25 years. President of the Toronto Chapter in 1947 was Asher Pritzker. In Montreal C. Davis Goodman, S. S. Colle and Abe Benjamin were early chairmen.

The first Technion Society in South America was set up in Argentina in 1950, with Jorge Chapiro as President. Within a few years there were functioning groups in Colombia and Uruguay. Harry Goldberg headed a

U.S. Supreme Court Justice William O. Douglas speaks at the Technion. Left to right: U. Shalon, Douglas, Prof. S. Ettingen. 1949.

Technion Society in South Africa. The Groupement Francais des Amis du Technion was formally established in 1951 with P. Dumanois, President; Armand Mayer, Vice President; Israel Leviant, Secretary; Robert Munnich, Treasurer; Albert Caquot, Hon. President.

The Technion was becoming well known at last and the campus was becoming something of a tourist attraction as well. Visitors during 1949 included U.S. Supreme Court Justice William O. Douglas, Lord Herbert Samuel, Lady Reading and other personalities.

More American Involvement

The main center of interest and activity was in the United States. The organization which Prof. Samuel had founded continued to flourish. Successive presidents of The American Society for the Advancement of the Hebrew Institute of Technology at Haifa, Palestine, Inc. were Lazarus White (1941–44), William Fondiller (1944–1947), Joseph W. Wunsch (1947–1950). In the latter year Col. J. R. Elyachar assumed the presidency.

Dr. William Fondiller (1885–1975).

More of the Americans were becoming personally involved. In the spring of 1949, in the presence of a number of leaders of the American group, a public ceremony was held marking laying of the cornerstone for a new electrical laboratory in the central court of the old campus. Harry Fischbach, who had pledged a major part of the funds required, was also present. Application had been made to the Government of Israel for a loan of IL 100,000 to make possible immediate construction. Plans were drawn up by Prof. Alexander Klein. Had the cash been available it is very likely that the building would have been put up on the old campus, and the eventual move to a new and larger campus might have been delayed for many years.

At the 1949 ceremony two honorary degrees were given, the first Honorary Doctorate, to William Fondiller, and an Honorary Engineer to Alexander Konoff.

In his remarks Kaplansky addressed himself especially to the Americans present. He acknowledged that many of the things that had happened might be called "miracles" but he referred to Weizmann's oft-quoted quip: "Anyone who does not believe in miracles in this country is not a realist."

His review of the industrial growth of the country was illuminating, and emphasized the need for the kind of manpower produced by the Technion. In 1933, he said, 19,600 persons had been employed in "industry", largely small workshops. In 1945 the figure had gone up to 57,000, and invested capital had increased four-fold. Consumption of electricity in industry had risen from 6.6 million KW in 1933 to 86.1 million KW in 1947.

Midshipmen of the S. S. Empire State, *training ship of the N.Y. State Maritime College, inspect the electrical laboratories of the Technion under the guidance of Prof. Ollendorff. 1950.*

He thanked the Americans for their support and referred to reports that they were becoming weary of fund-raising. Who had greater reason to be weary, he asked rhetorically, you over there, or we carrying on the endless work here?

In the period between 1945 and 1948 the American Society had raised about a million dollars at a cost of some $140,000. Because of the multiplicity of campaigns in America the Jewish Agency sought to enforce some degree of discipline against the separate institutions, whose efforts were said to be harming the fund-raising of the United Israel Appeal.

In November, 1949, Nahum Goldmann informed Jerusalem that a committee in New York was exploring the possibility of a joint fund-raising campaign by the three institutions of higher learning. Kaplansky was thunderstruck some weeks later to learn that at the importuning of Brodetsky on behalf of the Hebrew University and Meyer Weisgal on behalf of the Weizmann Institute, the Agency Executive had proposed that in return for suspending their independent fund-raising programs, the institutions should receive grants directly from the United Appeal as follows: Weizmann Institute, $1,500,000; Hebrew University, $1,000,000; Technion, $250,000.

Technion's friends in New York were asked to use all their influence to increase Technion's allocation, and cabled back a new offer: the three institutions to launch their own united drive, from which Technion would derive $350,000 annually, or 15% of the collections, whichever would be higher. Kaplansky preferred the $350,000, but if there was to be a percentage, he asked that they hold out for 20%. Since all the figures referred to fund-raising in the U.S. only, a new offer was made. Brodetsky and Weisgal offered the Technion 10% on a worldwide basis. Technion's American friends urged acceptance, but Kaplansky referred to the offer as utterly unacceptable and injurious to any future activity.

He had no choice, and early in 1950 announcement was made of the establishment of the American Committee for the Hebrew University, Weizmann Institute and Technion, known as UIT. Technion was to get 10% of the global collections, or 15% of the American campaign, whichever was higher, with a minimum guarantee of $350,000. But this meant also that all individual contributions to the Technion, as to the other institutions, would have to go into the pot.

UIT held its big dinner in New York on November 29, 1950, seeking $5,500,000. Speakers included Secretary of Defense George C. Marshall and Albert Einstein. The event was a social and publicity success but it was never clear how much money was actually raised.

The institutions charged each other with holding back on major gifts, and the agreement fell apart. Back in Haifa the *Vaad Menahel* objected to the proportionate division of funds and adopted a resolution to the effect that Technion would not consent to a worldwide agreement on the basis of the American percentages. Either Technion would be accepted as an equal partner with the others, or the Institute would remain outside.

Less than a year after the joint agreement had been set up, a simple announcement appeared in the press to the effect that the UIT had officially ended its existence on January 31, 1951. Later attempts were made to renew a UIT, with somewhat better percentages, and the new agreements lasted for some years, albeit on a much more limited basis.

The activity on the American scene was encouraged and helped by visitors from the Technion. Even before he took up his duties in Haifa, Prof. Sydney Goldstein went to the U.S. and was officially welcomed by the Society there. His plans for aeronautical engineering in Israel inspired and enthused his audiences.

Another whose efforts were harnessed was Prof. Ollendorff, who discovered that fund-raising was not as easy as many Israelis thought. In Chi-

Prof. Franz Ollendorff (1900–1981).

cago he was taken to an electrical equipment plant, where he hoped to get a good donation from the wealthy owner. He told the American of his plans for the new electrical engineering laboratory, whereupon the industrialist handed him the firm's catalog and told him to choose what he liked, adding, he was sure Ollendorff would find the prices most reasonable.

On another occasion Ollendorff attended a U.J.A. dinner in New York and was seated next to a man who had announced a very large contribution. The Israeli scientist engaged the man in conversation and told him about his research seeking to develop electronic instrumentation to aid the deaf. Later Ollendorff recalled wryly that he had worked hard and long to enable deaf people to hear, but this was the first time that as a result of his treatment a person with normal hearing suddenly became stone deaf!

More encouraging was Ollendorff's visit with Einstein, whose interest in the Technion was constant. Einstein even sought Ollendorff's help for a member of his family who was then a student at the Technion, but having difficulty with the language.

One bright light in 1949 was the allocation to the Technion of close to $400,000 from the loan given to the Government of Israel by the American Import-Export Bank for industrial development. The money enabled the acquisition of much needed new equipment.

More Land Needed

Despite the laying of the foundation stone for the Electrical Engineering Building on the Hadar campus, Kaplansky had not given up his dream of expanding Technion's facilities to other, larger areas. He still had his eye on the land in the Zebulun Valley, around the curve of Haifa Bay, which he felt would be ideal for the landing strips he thought necessary for aeronautical engineering.

In November, 1948, he met with the head of the Israel Air Force, which was in temporary occupancy of the area and buildings where he hoped to move his "greater Technion". It became clear that the military authorities were not prepared to leave the place. There was also a proposal from the Air Force that the Technion operate a school for technical aviation services. Such a school was eventually established and operated by the Air Force itself on that very site.

The need for room for expansion was becoming ever more acute, especially in view of a growing opposition to the construction plans for the old campus. Under pressures from both the Hadar Hacarmel Neighborhood Council and the Israel Engineers' and Architects' Association, the Municipal Engineer's Office refused to issue the building permit required. The objections were based on two issues. For one thing, it was maintained that the open land surrounding the old Technion building should be preserved as a green park, in view of the congestion in the area. In the second place, Paul Nathan's original concern to establish an unbroken parcel of land was now causing severe traffic problems in the growing metropolis. Various solutions were offered for the latter problem. One was to build two vehicular tunnels under the existing campus, one to run from Melchett to Ben Yehuda Streets, and the other from Pevzner, under Balfour, to Herzlia Street. Technion had no objection if the digging involved no cost nor damage of any kind to its facilities. Another proposal was simply to cut a road straight through the middle of the campus; the Institute of course objected to this most strongly. Still another suggestion was that any new Technion buildings be erected on high pillars, under which traffic could move. It should be noted that in the mid-1950's the city fathers did succeed in closing the main vehicular entrance to Technion from the Balfour Street side, opening the entrance from Schmaryahu Levin Street instead. This proved inconvenient to those driving into the campus but certainly was easier on town traffic.

In February, 1949, the Municipality granted a permit for construction of the new Electrical Engineering Building only, on condition that its foundations would not impede the digging of a tunnel at a later date.

During Kaplansky's absence abroad, the negotiations were carried on by Prof. Ettingen, as Acting Director. In June, 1950, he sent an urgent request to the head of the Science Department of the Ministry of Defense, Gen. Yaakov Dori, soliciting his endorsement for a Government loan to help Technion's construction program. Dori wrote a strong letter of support to Finance Minister Eliezer Kaplan. Ettingen also invited Dori to come to the Technion and to spend a little time there. Dori did indeed come, not long thereafter, and remained for 14 years — as President.

Technion's space problems became more acute. The Ministry of the Interior also entered the picture and reported that it would refuse to authorize any further construction on the old campus, and the request for a IL 100,000 loan earmarked for that purpose was accordingly rejected. The differences of opinion within the Technion were between those who favored seeking an additional campus for the overflow, and a minority who felt that the move should be complete, involving eventual abandonment of the original site for a much larger area elsewhere. Kaplansky sided with the majority view. He had cast his eyes on a new site, located at Tirat Hacarmel, adjoining the Haifa-Tel Aviv highway, about six miles south of the city. By August, 1950, he was able to announce that agreement had been reached in principle with the Jewish National Fund and with the Government to transfer to the Technion an area of 1000 dunams

Prof. Alexander Klein (1879–1962).

of land located near Military Camp 148 and Kfar Tira. Two-thirds of the land was flat country, about 65 feet above sea level. The remainder rose into the Carmel foothills, about 500 feet above sea level on its eastern boundary. The hill plateau was to be reserved for staff housing. Prof. Klein was asked to prepare a new Master Plan for the Tira campus. The many problems involved were to be revealed only later.

As the pace quickened and the pressures increased, Kaplansky's health continued to fail. In October there was a round of receptions and meetings with the President of the American Society, Col. Elyachar, and ambitious plans were drawn up. Many days were spent with Arthur Blok, who had come on a visit. The new Government committee on institutions of higher learning involved meetings in Jerusalem. The kaleidoscope of problems included the proposed opening of the new Department of Aeronautical Engineering, the new campus at Tira, the opening of another school year, the experiment with evening classes for adults, the complications of the new pension plan, the Technical High School, money matters, budgets, the T.B. squatters, more visitors.

Technion's Director was hospitalized briefly early in November and it was ascertained that he required an operation. On the 25th of the month he visited the newly-arrived Prof. Goldstein in his apartment. On the 26th he had a telephone talk with General Dori on Government policy toward the Technion. On the 28th he entered the hospital and the operation took place the following day. Dr. Shlomo Kaplansky died on December 7, 1950.

Pragmatist With Vision

The 19-year administration of the Technion by Kaplansky undoubtedly marked a turning point in the history of the institution. He brought to his post both leadership ability and a single-minded dedication to the mission which he correctly conceived for the Technion. His previous active role in the Zionist movement stood him in good stead, but he inherited the bundle of problems which had faced the Technion almost from its inception. He overcame these one by one.

Kaplansky was both a pragmatist, dealing with each day's objective problems, and a man of vision. He foresaw Technion's future and laid the groundwork for the great academic, physical and administrative development that was to take place.

He was not an easy man to work with. He insisted on personally supervising almost every aspect of the growing operation. Each letter that left the office had to be proofread by him. Arrangements for every public ceremony had to be checked by him. He took part in many of the oral examinations personally. No detail was too small to escape his attention, or to warrant referral to an associate. This was all well in the early years, but his one-man domination of the scene continued even after the institution had grown enormously.

His attitude to the staff was puzzling to some. A dedicated Socialist, he nevertheless cracked down on the staff, and was parsimonious in approving benefits of any kind. There were occasions when employees had to turn to the Haifa Labor Council to appeal against decisions of their "Socialist" Director. He sought to discourage the laboratory assistants from organizing and affiliating with the Histadrut. He preferred to deal with them on an individual basis. When they asked for a raise he told them they should consider their work at the Technion an honor, and could seek their livelihood from outside sources. He was unsympathetic to students who sought reductions in tuition fees and suspicious of those who advanced social/economic reasons for their requests.

He once confessed that perhaps he was not a Socialist in everything. Appearance meant a great deal to him and he always dressed impeccably. He had the manners and politeness of a European gentleman. He was frequently formal and dignified in conversation and often addressed people as *Adoni* (Sir), or in the third person. On the other hand those close to him remembered him as a warm, cultured human being.

He did not always subscribe to what would today be called women's rights. A man was entitled to higher pay because it was his duty to support a family, whereas that was not a woman's task. One colleague referred to him as an aristocratic Socialist.

He administered frugally, and was constantly budget conscious. He used to say: "The greatest disaster is that we have become accustomed to treating public funds like dirt." He worked long hours, and frequently was the last one in the building, late into the evening.

He directed with a firm, some have said autocratic hand. Despite the growth and progress made under his leadership, not all his colleagues agreed that he was a good administrator. Some believed that he did not have sufficient initiative in advancing the cause of the Technion, and as a result other institutions in the country continued to eclipse it.

In summation, he was a man of character and of the highest integrity, who appeared on the Technion scene at a crucial point in its history. He took over the leadership of an institution that was on the verge of disintegration, subject to internal squabbling, and in an economic sense bankrupt. He was able to arrest the process of deterioration and create harmony within the organization. As a result of his efforts the Technion, for the first time, attained financial stability, and as a consequence of his vision it developed academically to meet the demands which history was to make upon it.

A central corridor in the old Baerwald Building. Bust of Schmaryahu Levin is at further end.

Kaplansky was buried in Kibbutz Merhavia. His memory is preserved at the Technion by virtue of the physical existence of the institution which he rescued from near disaster, and in commemorative form by Kaplansky Square, the broad central courtyard of the Technion in Hadar Hacarmel, as well as by the Shlomo Kaplansky Chair in Agricultural Engineering. At the memorial meeting held on the first anniversary of his death a recording of the second movement of Beethoven's Second Symphony was played, in accordance with a request he had once expressed.

At its meeting on December 18, 1950, the *Vaad Menahel* grappled with the problems created by the death of the Director. A provisional directorship was named (in English, co-Presidents) composed of Prof. Ettingen and Prof. Merlub-Sobel, who had jointly carried responsibilities during Kaplansky's illness. At the same time the Council named three of its members, Uriel Shalon, Matityahu Hindes and Shabbetai Levi, as members of a committee to make recommendations for the new Director. Prof. Ettingen informed the Prime Minister, Mr. Ben Gurion, of the action taken, and promised that the name of the candidate proposed would be submitted to him for his approval. As a matter of fact it was the Premier who suggested the candidate.

The search committee lost no time, and on February 13, 1951, recommended to the *Vaad Menahel* that General Yaakov Dori be named Director, with Prof. Sydney Goldstein as his deputy. The Council approved, but asked for Government approval as well.

The new Dori–Goldstein administration was formally installed on February 27, 1951.

* * *

Signing the scroll for the Foundation Stone of the Electrical Engineering Building at the old campus. The building was eventually constructed at Technion City instead. Left to right: Eric Moller, Dov Givon, B.Z. Tooval, Prof. E. Schwerin, Dr. Mordecai Levy, Dr. William Fondiller and Dr. Shlomo Kaplansky. 1949.

From Adolescence To Maturity

CHAPTER NINE

Alternative Sites

Yaakov Dostrofsky was one of the Haifa children who had watched with curiosity as the Technion cornerstone had been laid in 1912. He attended the local Reali School, joined the British army in the First World War, and served in the 40th Battalion of the Royal Fusiliers, popularly known as the Jewish Legion, where he reached the rank of sergeant. He remained in the army after the War to gain further military experience, but was dismissed in 1921 for his activities in defense of the Jews of Tel Aviv against Arab rioters. He went to Belgium and graduated as an engineer from the University of Ghent. On his return he was quickly re-involved in the Haganah and in 1929 became Haganah commander in Haifa. By 1931 he was called upon to devote himself full-time to military matters. Upon the outbreak of the War, he was named Chief of the General Staff of the underground Haganah. In 1948, with establishment of the State, he emerged from the anonymity of that position as Yaakov Dori, Chief of Staff of the Israel Defense Forces. His responsibilities in that post took a heavy toll of his health, and in 1950 he retired from military service to assume new duties as head of the Scientific Department of the Prime Minister's Office.

As already noted, Dori was Premier Ben Gurion's choice to head the Technion, and both the search committee and the *Vaad Menahel* agreed with alacrity. The only other name that had been seriously considered was that of Prof. Goldstein. Dori agreed to accept on condition that Goldstein serve as his Vice President. Dori was 51; Goldstein, 47.

The new President took over the reins of duty at a modest ceremony attended, among others, by the Minister of Education, David Remez. In his remarks Dori said: "More than ever before we need an army of great spiritual strength, technical skills and ability, an army of scientists and experts in technology and agriculture; it is the task of the Technion, in fraternal cooperation with the Hebrew University and the Weizmann Institute, to educate, instruct and train this army of the Israeli people which will fashion the image of the nation and entrench and safeguard our ancient homeland in its rebirth."

Ill health continued to trouble the new President. He was absent for various periods during his first year and it was not until the following year that he took over fully. The major problems facing the administration were three: the need for physical expansion of the campus; the need for reorganization both academic and administrative; and the need for funds to enable the institution to meet these challenges. All three problems were attacked simultaneously.

Dori had not yet taken over when the Planning Department of the Government informed the Technion that the proposed site at Tira was not considered suitable. The area was too small, hemmed in by roads, and

Inauguration of the new administration of the Technion. Left to right: Judge Bernard Rosenblatt, Dr. M. Merlub-Sobel, M. Hindes, Prof. H. Neumann, Prof. Sydney Goldstein, David Remez, Minister of Education, Gen. Yaakov Dori, Shabbetai Levi. February 27, 1951.

without opportunity for future expansion. The Government was prepared to offer another site.

Professor Ettingen, Acting Director, replied on January 22, 1951, that the Technion would not yield its demand for Tira, and would proceed with construction plans. He charged that the authorities were more concerned with protecting some inconsequential housing project than with the necessary development of Israel's technological institute. He served notice that the Technion was not prepared even to look at another site.

The discussion continued in a rapid exchange of letters. The Government spokesman sought to reassure the Institute that it was precisely because the Technion was held in such high esteem, and its importance to the nation so valued, that the authorities were concerned and would recommend another site more suitable for future expansion.

Prof. Ettingen was not convinced, and responded that the Technion was familiar with other sites that had been mentioned, but none was satisfactory. The search had been going on for some time, and the Technion had now decided not to delay any longer. It was proceeding with site surveying and preparation of construction plans for Tira.

Realities dictated otherwise. An existing highway, which it was impossible to change, fixed one boundary irrevocably. While the discussions were going on, authorities of the Jewish Agency blithely went ahead and constructed hundreds of *Maabara* huts as temporary homes for the thousands of immigrants then streaming into the country. *Maabarat Tira* became a huge transit camp.

There were also other claims on the same property. A quarry developer and a fish processing plant maintained that they had liens on part of the area. A real estate operator presented plans for construction of badly needed permanent housing. The Defense Department, which had planned to utilize one section for military purposes, would also not yield.

There were further complications. Part of the property, it appeared, was owned by a religious denomination, and expropriation would do harm to Israel's image abroad. In any event, who was to pay all the land owners in the event of expropriation? The Technion was advised to whittle down its plans to the limited area which was not under dispute, or look elsewhere. While they talked, new facts continued to be created on the site itself. Contrary to standing instructions, many of the immigrants began putting up stone houses, without regard to legal niceties of land title.

At the end of 1951, despite the problems and obstacles, Prof. Alexander Klein was still going ahead with full-scale master planning of Technion's new campus at Tira. Col. Elyachar, President of the American Technion Society, turned to the Jewish National Fund for support. Gen. Dori assigned Yosef Karni, then a staff member of the Building Materials Laboratory, to help accelerate the construction plans and Karni went to Tira with Elisha Shklarsky to survey the site for the aeronautical engineering building, which was to be the first to go up. The survey was completed but Karni expressed his doubts as to the suitability of the area, citing the various reasons which had already been advanced. There were others as well: the dangers of humidity and corrosion due to proximity to the sea.

The new Board of Governors, meeting in Haifa in April, 1952, with the participation for the first time of overseas members, tackled the problem. Most of the Israeli members urged authorization to begin construction immediately. The Americans, troubled by the lack of clear title to much of the land and the various conflicting claims, were more cautious. The mood of the Board was to begin building, and continue the negotiations with the Government and the Jewish National Fund in the meantime. Abraham Tulin, speaking for the overseas members, took a different tack. "I am against starting the building until we know that we are going to have the site. If this site should give too much trouble, we shall look for another one." And he added: "We are going to build on land we do not have with money we do not have."

Replied one Israeli impatiently: "We cannot be guided by American standards of procedure. The means affect the end. The Board wishes to start building as soon as possible." Formal resolution was adopted to begin construction as quickly as expedient.

The adoption of a resolution did not settle the controversy, and the differences of opinion reached the press. In the very month of the Board meeting Yekutiel Baharav wrote in *Haaretz* that the Tira site was unbefitting a great institution such as the Technion. The campus should have a grand and splendid setting, somewhere up on top of the Carmel, overlooking the city and the Bay and the industries which the Institute served. He pinpointed a 10,000-dunam area stretching between Khreiba, at the highest point of the Carmel, to Naveh Shaanan. The location had been earmarked for a forest and a national park, but part of it could be made available for the Technion, he wrote. Indeed, the whole isolated area would be opened up to growth and development as a result of the location of the Technion there, just as Tel Aviv had grown around the Herzlia Gymnasia. The area had once been spoken of as the site of Herzl's Tomb, and the building of a great university on the mountain which had meant so much to Herzl would be a memorial to the founder of the Zionist movement. Don't push the Technion down into Tira, Baharav concluded.

Baharav was in a position to do something about the matter. As a member of the official Regional Planning Committee he opposed the application for Tira and recommended that an inter-ministerial committee be set up to designate a better location for the Technion. Early in May the Ministry of Finance appointed a committee to study the possibility of finding an alternative site for the Technion on Mount Carmel, or to recommend which of the Tira plots should be expropriated for the purpose.

The committee met twice that month and came to the following conclusions:

A.   A Mount Carmel site was unfeasible because the Government-owned Naveh Shaanan area was too hilly, and in any event was already earmarked for a public forest which had in part already been planted. Title to the Khreiba site was held by many individual owners, and expropriation would present the same problems as at Tira.

B.   Other alternative sites which were considered, such as Ein Hod, Shfar-Am or Yaarot Hacarmel, were too far from Haifa. Furthermore, there had not yet been any basic decision regarding the development of those areas.

C.   The only remaining alternative was Tira, and steps should be taken to lease to the Technion what was then available, and to expropriate

The new Technion campus before beginning of construction. 1952

and then lease to the Technion additional acreage as acquired. The size of the campus would have to be reduced considerably and Technion would have to curtail its plans accordingly.

Prime Minister Ben Gurion, then vacationing in Tiberias, met Baharav in the hotel lobby and commented that he had read his letter in *Haaretz*. Baharav seized the opportunity to press his case, and on July 23, 1952, Ben Gurion was taken on a visit to the Carmel slopes. Of the two sites he saw, the Prime Minister preferred Naveh Shaanan. The Committee on Forestry Preserves objected, but Ben Gurion overrode their objections.

The site chosen consisted of 550 dunams and Karni asked for permission to try to get more. With the help of Aryeh Sharon, head of the Government Planning Office, Arthur Glikson, a lecturer in Architecture, and Avraham Akavia, he managed to extract further concessions, and the site was increased from 550 to 850 and eventually to 1150 dunams. The land was for the most part Government property, taken over from the British

Inspection of the new Mount Carmel site for the Technion. Left to right: Prof. Y. Ratner, David Ben Gurion, Carl Alpert. 1953.

Mandatory Government. One corner had been earmarked for construction of a workers' rest home, but this, too, was released.

Karni had to convince the Land Registry Office that no damage would be done to the young forest planted there by the British. The Technion would also be bound by municipal regulations which forbade uprooting of trees, and it pledged to plant another for every tree that had to be removed because of construction. The promise was more than kept, and at a later stage the Technion put in 40,000 additional saplings. The master plan also called for a public thoroughfare straight across the area, but Karni warned that the Technion could not be responsible for forest fires or other damage to the woods if the public were given free access. The proposed road was struck out of the plan.

On a subsequent visit to the site in February, 1953, Ben Gurion was shown an adjacent area of some 200 dunams and was advised that the Government must be sure to keep the place clear because of possible danger from various experiments to be conducted on the Technion campus. Thereupon Ben Gurion gave instructions to add the 200 dunams to the Technion area, for a total of 1350 dunams, equivalent to more than 300 acres.

Naveh Shaanan had first attracted attention shortly after the First World War when Patrick Geddes, the British town planner, visited the country.

He toured Haifa with Arthur Ruppin, was struck by the many climatic and topographical advantages of the east ridge of the Carmel, and recommended that an area of about 1000 dunams be purchased. The asking price was £1 per dunam. Veteran residents of Haifa were opposed. They maintained the area was too remote; there was no connecting road, and no water. It would cost a fortune to build a road and pipe line. Furthermore, no Jew would risk his life to build a home in such a desolate spot. Nevertheless, with Ruppin's help some 50 families, poor people most of them, banded together and bought plots alongside the Government preserve, paying in deferred installments. They named the area Naveh Shaanan (Serene Habitation).

In October Dori informally reported that the site had been approved. He was hopeful and enthusiastic but was moved by mixed feelings of joy and fear since if anything should go wrong they would have to start all over again. On November 18, 1952, at exercises marking opening of the new school year, Dori made public announcement of the impending expansion move.

All the plans prepared for the Tira location were scrapped, and Prof. Klein was asked to draw up a new Master Plan for what soon became known as Technion City. Klein was in the United States at the time but his young assistant, Yohanan Elon (later to become a full Professor and Dean of the Faculty of Architecture and Town Planning) quickly adapted the Tira program to the new site with its special topographical problems, following Klein's principles. Immediately there were differences of opinion as to which building should be the first to be constructed. Voices were raised on behalf of Civil Engineering, since there was a boom in the building industry. Others favored Electrical Engineering. It was Prof. Goldstein who insisted on Aeronautical Engineering, for reasons associated with national defense, and in April, 1953, the cornerstone was laid for the new building in conformity with plans drawn up by Prof. Ratner. Work was accelerated on one wing and by November of that year the first students were already in occupancy. As the structure grew, it provided room also for the Faculty of Architecture and Town Planning and for some years the two departments were in joint occupancy.

New Constitution

The need for organizational and administrative reforms in the Technion, parallel to the physical growth and development, was obvious. It was pointed out that the By-Laws under which Technion was operating were basically those adopted in 1928. The *Vaad Menahel* accordingly asked Prof. Goldstein to bring these documents up to date. Drawing on his knowledge of the organizational structure of British and American universities, with such changes as were necessary to adapt the regulations to conditions in Israel, Prof. Goldstein proceeded to draft an entirely new constitution and appended statutes.

One of the first steps taken was the decision to set up a new administrative body to be responsible for governing the Technion. Prior to establishment of the State, when the Technion had been under the jurisdiction of the Jewish Agency and then the *Vaad Leumi*, the members of the Board had been appointed jointly by these two bodies. The number averaged about 15. The new constitution provided for the setting up of a representative Board which would renew or broaden its membership as necessary and would be self-perpetuating, with safeguards for adequate representation from Technion institutions, public agencies and the public at large. The new body was called Board of Governors in English, and *Kuratorium* in Hebrew, thus retaining the original Latin term used by the Hilfsverein. The *Vaad Menahel* became in effect the smaller executive committee, or Council, functioning between meetings of the Board. Composition of the first Board, as approved by the *Vaad Menahel* on August 13, 1951, and at subsequent meetings, was as follows:

From Israel, by virtue of their official posts at the Technion: Gen. Dori, Prof. Goldstein and Prof. Ratner; representatives of various bodies: Berl Locker, Yitzhak Wilentchuk, Yaakob Ben Sira, Alexander Hassin, Dov Givon, Prof. Aharon Tcherniavsky, Shimshon Turetzky; ad personam: Eliyahu Berligne, Matityahu Hindes, Gustav Levy, Eric Moller, Jacob Reiser, Uriel Shalon, Shabbetai Levi, Dr. Arthur Biram, Dr. Aharon Barth, Yaakov Geri, Hillel Dan, Israel Abramski, Judge Moshe Landau, Avraham Klir, Avraham Rutenberg, Zalman Shazar, Aryeh Shenkar, Shimon Bejarano, Dr. Pinhas Rosen, Yaakov Peled, Herman Hollander and Heinz Gruenebaum. Five Ministers of the Government were also invited to serve.

From Abroad: U.S. — Col. Jehiel R. Elyachar, Charles Frost, Alexander Konoff, Abraham Tulin, Joseph W. Wunsch, Elias Fife, Harry Fischbach; Alternates, B. Sumner Gruzen, David Rose, Maurice Spertus, Prof. Abel Wolman, Ben Cooper; Great Britain — Lord Silkin.

Twenty-one members were in attendance at the first meeting of the reconstituted Board which was held in Haifa, April 20–22, 1952. Overseas members present were Elyachar, Fife, Konoff, Rose and Tulin, with additional visitors.

The Constitution and Statutes which were formally adopted by the Board at its meeting in April, 1953, spelled out the objectives of the Institute:

"1 To disseminate knowledge through teaching and to advance it through pure and applied research in pure and applied Science,

Students ready for class in the first building at Technion City. 1953.

Engineering, Architecture, Technology and related activities including the Humanities, Social Science and Education.
"2. To impart to its students a broad general education. The Institute shall pursue these objects without discrimination against any person on the ground of race, religion, nationality or sex.
"3. To serve the State of Israel and its economy by counsel and research, and by other appropriate means, and to serve the people of Israel by the provision of courses of instruction and lectures, the publication of books and similar activities in the areas specified above."

The adoption of the Constitution was but one step in the organizational metamorphosis of the Technion, which had already gone through several "ownerships" and identities. In January, 1940, it had been formally registered under the then prevalent Ottoman Law of Societies under the name of *HaTechnion HaIvri b'Haifa* (Haifa Hebrew Technion), but was generally known as the Hebrew Technical Institute. In 1945, with the consent of the Education Department of the Mandatory Government, the English version was changed to Hebrew Technical College. In April, 1953, the name was officially changed to *HaTechnion, Machon Technologi LeIsrael* (Technion — Israel Institute of Technology). The final stage in establishment of its corporate identity came in August, 1962, when it was recognized as an Institute of Higher Education, pursuant to the Law of the Council for Higher Education of 1958.

Academic Changes

Re-evaluation of academic forms and standards took place parallel to the physical and organizational development of the Institute. Prof. Goldstein maintained that the existing academic regulations were as inadequate for the requirements of a modern university as the management by-laws, and he initiated discussions leading to a complete revamping of what were to become the Academic By-Laws. A first draft was placed before the Teachers' Council in May, 1951, and a prolonged and sometimes bitter discussion followed in ensuing months. There was near unanimity on the need to elevate academic standards, but the differences of opinion were sharp as to how that was to be brought about. These differences were based, in part, on the contrasting backgrounds of the parties to the discussion, who represented three points of view. One group was composed of the veteran staff members, educated in Eastern Europe, and schooled in a particular philosophy of academic organization. A second group consisted of the teachers who had come to Technion during the Nazi period, and represented the thinking of the German school of science and education. Prof. Goldstein served as the spokesman for the British-American concepts of university education, and clashes were inevitable.

Prof. Niels Bohr, right, received at the Technion by Prof. Sydney Goldstein. 1953.

The resolution adopted by the Board of Governors in 1953, after receiving the Goldstein proposals, read as follows: "The Board approves the course of action indicated and stressed by the administration for raising standards at the Institute, namely the intellectual standard, the improvement of teaching methods and guarantee of education toward human and social responsibility, all in addition to professional responsibility. Emphasis must be placed on promotion of creativity and research, and an atmosphere of trust, respect and mutual cordiality must be created among all members of the Technion family, teaching staff and students alike."

Some years later Gen. Dori recalled the original Goldstein plans and itemized their essence in these words:

"1. There should be a substantial change of emphasis in the education in the Degree courses, not with emphasis on the transfer of information, but emphasis on the transfer of intellectual attitudes and systems, and on professional attitudes and systems; on independent thought, analytical and synthetic thinking, on-the-job (clinical) training, training to deal with design criteria, training for responsibility for human life, economic assets, social and aesthetic values.

"2. Graduate schools should be opened for Research Management and Planning, Development and Design, with corresponding changes in diploma regulations.

"3. Alternative courses to the Degree courses should be opened. Before the level of such courses is finally decided, more information should be obtained on what is needed."

Misunderstanding of these principles was based in many instances on diverse outlooks and on terminological differences. Since academic titles were drawn from various European sources, Goldstein proposed a new series of grades based on the American system: Professor, Associate Professor, Senior Lecturer, Lecturer, Instructor and Assistant. The latter was considered the lowest rank, one just beginning his academic career. But in certain European universities "Assistant" had the suggestion of a senior grade, and some Technion faculty members were very proud of their rank as "Assistant". They felt that the new definition would adversely affect their status, and they called a strike in protest. Goldstein's manner of handling the strike was certainly unorthodox by Israel standards. He called in the whole committee which was authorized to negotiate on behalf of the Assistants. He ushered them into a room, locked the door and put the key in his pocket. He gave orders that neither food nor drink was to be brought in and informed the committee they were now engaged in non-stop negotiations which would end only when agreement had been reached. The talk went on for about six hours without a break. In the end it was settled. Academic beginners became known as Assistants. The old "Assistants" became Lecturers or Senior Lecturers, as the circumstances warranted. Some of them eventually rose to the rank of full Professor.

No less difficult was the problem of placing all members of the academic staff in the appropriate new category befitting their education, experience and duties.

Inauguration of post-graduate studies was a major step forward. In 1951 the first three theses for a doctor's degree were submitted by junior staff members on the subjects of "Radio-activity of the Tiberias Hot Springs", "Water Storage in the Negev", and "Theory of Induction Machine Derived Directly from Maxwell's Field Equation". A year later Technion's first Doctor of Technical Sciences degree, as previously noted, was conferred on Eliezer Mishkin, who had presented the latter thesis.

At the opening of its new era the Technion had five Faculties* and Departments: Civil Engineering, Architecture, Mechanical Engineering, Electrical Engineering and Chemical Engineering, the latter previously known as Industrial Chemistry. The new academic development of the Institute was in breadth as well as in depth. Until 1951 the basic sciences

---

* At the Technion "Faculty" is generally (but not consistently) used in the European sense to indicate a large academic unit. "Department" is a smaller academic unit. We shall use these terms accordingly. Members of the academic staff will be referred to as faculty in the American sense.

Prof. Walter C. Lowdermilk (1888–1974).

had been taught only as service courses to the engineering disciplines, and it was difficult to obtain good faculty members under such circumstances. In that year it was decided to set up a separate Faculty of Science comprising four divisions: Mathematics, Chemistry, Physics and Mechanics. Degrees were to be offered in these subjects, research would be fostered, and the sciences would be taught as ends in themselves, not merely to meet the basic requirements of the other departments. In explaining the innovation at the graduation exercises in January, 1952, Gen. Dori emphasized that the sciences are "the foundation stones of technology" and declared that they must "take first place in the whole system of instruction and research."

The new Department of Aeronautical Engineering also began to take shape. Upon his arrival in Haifa Goldstein found that there had been preliminary discussions regarding the program of the Department. In June, 1946, a meeting had laid down a basic outline for the teaching of aeronautics. In March, 1948, Dr. F. I. Piattelli gave Kaplansky an eight-page memorandum on his proposals for the new Department and referred to the need for new facilities, including a landing field capable of handling at least small planes and gliders; an aerodrome, a suitable wind tunnel, an aircraft workshop and a hangar.

Following the death of Kaplansky the interim administration, headed by Professors Ettingen and Merlub-Sobel, was burdened with the day-to-day problems of the school and was hardly in a position to help the professor newly arrived from Manchester, who had come to head a non-existent department and teach a non-existent course. In 1951 the department comprised Prof. Goldstein and two research assistants. He consulted with Prof. Ernst David Bergman, head of *Hemed*, the Science Division of the Ministry of Defense, who agreed to release two bright young men on his staff. In preparation for their roles at the Technion Abraham Kogan was sent to Princeton and Meir Hanin to Cornell. In 1952 Goldstein was joined by two young lecturers from the U.S., Dr. Hirsch Cohen and H. Jerome Shafer. A course was offered in "Theory of Flow", taught to students from Mechanical and Electrical Engineering, but it was not until 1955 that the Department accepted its own first undergraduate students.

In October, 1953, the new Department of Agricultural Engineering was opened. Proposals for inauguration of such studies at the Technion had been made by Dr. Kaplansky as early as 1938 and repeated in greater detail in 1941. In his memorandum in the latter year Kaplansky wrote that the new Department would accept ten students per year. Major fields of study were to be irrigation and water problems. It was again made clear that the Department was not to teach agronomy or overlap in any way the program of the Faculty of Agriculture of the Hebrew University. The intention was to complement that program by providing instruction in the application of engineering to agriculture. In the first stages this was done through three divisions: Farm Mechanization, Farm and Rural Architecture, Water and Soil Conservation.

The Department was organized by Herman Finkel, who had come to Israel from the U.S. in 1949 and had served for three years as Chief Engineer of the Soil Conservation Service of the Israel Ministry of Agriculture. He developed the program and curriculum with the help of a U.N. expert from the Food and Agriculture Organization (FAO), and teaching began with first, second and third year students simultaneously, so that the first agricultural engineers were graduated within two years. In 1955 Prof. Walter C. Lowdermilk was sent to Israel by the U.S. State Department as a Distinguished Scientist, and he was appointed head of the Department, serving for two years. In 1959 the Department was named in his honor, in recognition of his services to Israel over a period of many years. Prof. Abraham de Leeuw succeeded Lowdermilk as head of the Department. The demand for graduates was so great that by 1960 there were 81 undergraduate students, and 19 studying for higher degrees. In 1970 the number had risen to 180 and 49, respectively. The special program for students from developing countries is described on other pages.

Broadening of the academic program took other forms as well. In September, 1951, the Research and Development Foundation was formally established and took under its wing all the test work and the sponsored research which the Technion had been doing for some years. A five-year grant of $10,000 per year for research given by the American Fohs Foundation in 1948 and continued for many years thereafter, as well as a 1953 research grant from the Ford Foundation, made possible a marked increase in the volume of research activity. The Foundation also undertook to coordinate all the work carried out at the Technion for local industry and Government departments. Through the years the R. & D. Foundation developed into Technion's most valuable agency for promoting close relations between industry and the academic world. Its commercial turnover ran into many millions of Israel pounds. Successive directors were Yehosua Tsimhi (1952–53); Dr. Joseph Ben-Uri (1953–55); G. Majevski (1956–59); Prof. David Ginsburg (1959–60); Prof. Yosef Karni (1960–67); Dr. David Armon (Acting 1963–64); Prof. Joseph Hagin (1967–70); Dr. Herbert Bernstein (1970–78).

The expanded program was manifested also in the addition of new faculty members. A natural source of academic personnel was of course from the ranks of Technion's own graduates, and through the years a number of the Institute's alumni did indeed join the staff. Among the earliest graduates who decided to make Technion their careers, and eventually rose to senior positions, were Prof. David Yitzhaki, a member of the first graduating class, Prof. Rahel Shalon, Prof. Joseph Edelman, Prof. Mordecai Mitelman, Prof. Elisha Shklarsky, Mordecai Geisler, Prof. Yosef Karni, Prof. Aviah Hashimshony, Prof. Aharon Zaslavsky and Prof. Avshalom Thau.

To prevent academic inbreeding, additional academic personnel were brought in from abroad. In 1952 the Dori–Goldstein administration arranged to bring to the Technion from the United States Prof. Nathan Rosen in Physics; Dr. William Resnick and Dr. Jacob M. Geist, Chemical Engineering; Dr. Otto Schnepp, Physical Chemistry; and Dr. Arrigo Finzi, originally from Italy, Mathematics. In all some 40 appointments were made to teaching positions. A clear picture of what this meant may be obtained from the following table:

Faculty Members

	Full Time	Part Time	Total
1941–42	22	24	46
1950–51	49	68	117
1952–53	108	98	206

A special Technion housing project, consisting of a number of wooden prefabricated buildings, was set up in Ahuza to accommodate the new faculty members. The structures were added to and expanded later and the Technion *Shikun* remains in use to this day. Provision of housing for faculty members was a perpetual problem.

Goldstein was proud of the fact that he had brought Prof. Jeremias Grossman "back from limbo" where he had been since his clash with Kaplansky some years before. Grossman was elected Deputy Vice President in 1953 and served in that post with distinction until his death in 1964.

Money

The spectacular growth of the Technion had its implications for the budget as well. Development of a new campus, whether at Tira or on Mount Carmel, was going to cost a great deal of money. Even rough estimates were that $20,000,000 would be required. Where was the money to come from?

The President of the American Technion Society, Col. Elyachar, had seized the opportunity of meeting with Ben Gurion when the Prime Minister was vacationing in Tiberias and had told him of Technion's ambitious plans and of the willingness of the Americans to help raise the funds required. He offered the Prime Minister a "deal". Would the Government allocate $10,000,000 toward construction of the new campus if the American Society would raise a similar amount? Ben Gurion agreed on the spot, and Elyachar hastened to inform both his American colleagues and the Technion authorities in Haifa. At its meeting in New York on November 17–18, 1951, the American Technion Society endorsed its President's spontaneous pledge in Israel and agreed to raise $10,000,000. Elyachar solicited and obtained from Ben Gurion a warm letter of endorsement which was effectively used in the ensuing campaigns. Wrote Ben Gurion:

"An Institute of Technology of high level is one of the most vital needs of the State of Israel at this time. We are a small, poor people; our country is largely desolate; we are surrounded by hostile nations who threaten our existence. We are compelled to create, at an unprecedented rate, sources of livelihood for masses of our brethren who are returning to their homeland. We must build up and develop the whole country speedily, and especially the desolate regions. For these great and difficult tasks we must call on the aid of the latest achievements of both pure and applied science, and the most advanced improvements of technology. Small and poor as we are, we must submit to the very strictest austerity in our daily lives; but for those very reasons we can have no austerity in science and technology . . . . I wish to express to you and your colleagues on the Board of Directors of the American Technion Society my deep appreciation of the great task you have undertaken to raise a special fund of ten million dollars for the building of the Technion on its new premises . . . ."

The message was greeted with enthusiasm in the U.S. but there were still many questions. The Board meeting held in Haifa in April, 1952, noted that despite the warm endorsement, there was no (written) commitment from the Prime Minister to match the $10,000,000 being raised, and it was suggested that the matter ought to be clarified.

When first demands were made by the Technion upon the Minister of Finance, Eliezer Kaplan, he rushed to the Prime Minister. Was it true that such a matching promise had been made? How could he do it? Ben Gurion's reply was diplomatic. Elyachar had pressed hard, and since it seemed hardly likely that the Americans would indeed raise ten million dollars, the Premier felt perfectly safe in agreeing to provide matching funds. The American Technion Society did fulfill its pledge, and the Government eventually kept its promise.

A pattern was set for large contributions. In December, 1950, it had been announced that Alexander Konoff was giving $100,000 for *Bosmat*, the Institute's Technical High School. Not long thereafter Harry Fischbach announced $100,000 for the new Electrical Engineering Building. Both men were to make much larger gifts in the years ahead.

Obviously the annual operating budget was affected as well. Whereas the actual operating costs for the year 1951–52 were IL678,000, the Board adopted a budget of IL1,059,000 for the following year, an increase of 57%. Anticipated income to cover this was uncertain, and it was decided to raise tuition fees from IL100 to IL150 per year. Additional income would have to be obtained from fund-raising over and above what had been pledged for capital expansion. Dori was absent from the Board sessions because of illness, and all eyes turned to Goldstein. He drew the line. He was an educator and an administrator, he said, not a fund-raiser, and he threatened to resign as Vice President. The incipient crisis was quickly smoothed over, but the forebodings for the future were ominous.

Financial problems continued to mount. The 1953–54 budget was set at IL1,889,447. Government grants toward the operating budget were small and slow in coming. Overseas aid was concentrated largely on the physical development of the new campus. The Israel Technion Society stepped up its activity, but the results were meagre: in 1950–51, IL7,000; 1951–52, IL 15,000; 1952–53, IL60,000. Dr. Biram, now once again deeply involved in Technion affairs, was responsible for most of this income. In 1954 he was elected Chairman of the Board of Governors.

Gen. Dori was becoming pessimistic over the possibilities of receiving operating funds from abroad, since the trend appeared to be toward contributions to central agencies, rather than to individual institutions. For this reason the 1953 Board meeting called for joint fund-raising activities of the three institutions of higher learning, this despite the previous unhappy experiences in such joint effort. The major sources of support, Dori felt, should be the Government and the Jewish Agency.

The attempt by the Hebrew University earlier in the year to raise tuition fees had been met by protests and student strikes which came to an end only after intercession of the Knesset. Nevertheless, the Technion saw no alternative but to implement the Board decision, raising the fees to IL200.

The undergraduate student body now numbered 1200, and the increased financial burden led to the expected reaction. Student discontent had been mounting since the new administration had taken over. At a Teachers' Council meeting in April, 1951, Prof. Tcherniavsky had served as student spokesman and had raised a series of questions which he felt deserved more than passing attention, among them: financial problems, relations between teachers and students, need for sport and music activities, student health problems, extension of the general, non-technical education, easing of the heavy academic burden, and improvement of the general atmosphere on the campus. A committee had been set up to study the situation, but little progress had been made.

In February, 1953, first-year students went out on strike in protest against new examination regulations which required re-examination in the Fall in all courses failed in the Spring as well as in some courses that had already been passed. Some felt that Goldstein was moving too rapidly in his introduction of academic reforms, and that the changes had not been adequately explained.

The mood of revolt among the students came to a head in December, 1953, in a dispute whose issues were both academic and financial. The Government announced that it was drastically reducing its grant to the Technion, and the deficit, which already stood at IL250,000, would rise accordingly. The new tuition fees were put into effect at once, but the students refused to sign the required obligation to pay the IL200 annual fee. Dori came to the conclusion that it was impossible to continue to operate without adequate funds. Classes were suspended for three weeks and a long public debate ensued. It was not immediately clear whether this was a strike by students, refusing to pay tuition fees, or a lockout intended to pressure the Government.

The students picketed the main building carrying posters reading: "We want to study!" They linked the tuition issue to other matters and drew up a proclamation of student demands, including request for adequate housing. From a tactical point of view their mistake was in linking the controversy to academic matters as well. Unlike the students of 1924–25, they complained that the standards were now too high and they asked that the demands made upon them be eased. Dori stood firm. He declared that the academic standards of the Institute would not be lowered under any circumstances, and while the students might not appreciate it at the time, he told them, long after graduation they would be grateful for his firmness.

A compromise proposed by the Minister of Education and accepted by both sides, fixed the tuition fee at IL200, with a pledge by the Technion that more scholarships and loan funds would be made available to students unable to pay the increase.

An Abortive Report

The growing costs of the universities and the mounting student unrest rang a warning bell in Jerusalem, and a study committee, headed by Jacob Arnon, was named to analyze the financial problems of the Hebrew University and the Technion and to make recommendations to a Ministerial

In the fuel testing laboratory: determination of sulphur content in oil struck at Heletz. 1956.

Committee. The key recommendations are of interest now, years later, despite the fact that many of them were never put into effect. The Arnon Committee report was but the first in a series of studies to be made in the years that followed. Highlights of its findings, submitted in March, 1954, were:

A.   The committee believes that the development of the Hebrew University and the Technion should be geared to the real needs of the country and in accordance with budgetary possibilities.

B.   In view of the financial difficulties and the shortage of academic personnel, the committee views with concern the tendency to create new institutions of higher learning, which will only aggravate the situation of the Hebrew University and the Technion.

C.   The committee approves the idea that the two institutions merge as one national institution, with joint administration. Such union could result in budgetary savings.

D.   The committee recommends:

1.   To transfer funds intended for long-range investment from the operating budget to a special development budget.

2. To set up a joint committee of the two institutions to seek more efficiency in operations. It would be desirable for the Weizmann Institute to be included also.

3. a. To organize the salary scale of the academics clearly along the same lines as the Civil Service, and to eliminate all the extra payments which are actually salary by another name.

   b. A committee composed of the chairman of the Knesset Finance Committee, the chairman of the Knesset Education Committee, and the head of the Government Civil Service Commission should be set up to carry out 3-a.

4. To reorganize the wage scale of the administrative personnel along Civil Service lines.

5. Such salary and wage reorganization should be done without delay because wages are a large part of the budget, and without such clarification budgets cannot be set properly.

6. In view of the financial difficulties, to reconsider if it was really necessary to appoint all the new teachers now proposed by the institutions.

7. To make sharp increases in tuition fees, which at the existing rate were much lower even than High School tuition. Part of the tuition fee income could be set aside for loans to needy students.

As noted, little was done about the recommendations. The two institutions themselves had only limited comments or reservations.

It was not often that the specific problems of the two major institutions were taken up jointly. Though the Technion had long since emerged from its relegation to inferior status, it was at a continued disadvantage because of its location in Haifa, remote from the lobbies and chambers of the capital where decisions were made and budgets approved. In the friendly rivalry between the two, Technion was no longer the ignored step-sister, and lost no opportunity to advance its cause. There was pride in Haifa that the first building put up by the University in 1954 on its new Givat Ram campus was built according to the design of Abraham Yasky, who had graduated from the Technion only two years earlier.

Technion's developing sense of self-awareness led also to the first attempt to write the history of the school. Avraham Alsberg was asked to prepare a manuscript on the early period of the institution but his paper was most critically read by Hecker and many changes made. Some difference of opinion then ensued as to who could claim authorship of the material, but Hecker declared he was willing to forfeit all credit, so long as the story was properly told. Dr. Eliyahu Zakon, a teacher of mathematics, was asked to be the impartial editor and to prepare the manuscript for publication. It was published in 1953 as a 38-page brochure in Hebrew under the title *Toldot Hatechnion B'reshito 1908–1925* (History of the Technion from its Beginnings, 1908–1925), but without credit to any author.

Pride led also to greater efforts in the direction of public relations, sometimes with an occasional lapse, as in the exaggerated publicity given to reports of the discovery of radioactivity in the Tiberias Hot Springs, based on the Technion doctorate thesis of David B. Rosenblatt.

An expansion of the public relations program called for increased release of Technion news to the press, and improved reception of visitors to the campus. Mrs. Goldstein headed the program and gathered a small staff composed of Mrs. Rivka Watson, who also laid the foundations for Technion's organized scholarship system, Dora Barkai and Gedalia Anoushi, a recent immigrant from Egypt. The need was felt for a professional director to take the program in hand and link it with a worldwide fundraising effort. In 1952 Professor and Mrs. Goldstein interviewed in New York Carl Alpert, who had been Editor of *The New Palestine*, Director of the Education Department of the Zionist Organization of America, and a journalist. He was engaged and arrived in Haifa in August, 1952, as Director of Public Relations, later taking on additional duties as Executive Vice Chairman of the Board of Governors. Soon thereafter Otto Stiefel was added to the Public Relations staff.

The 1953 meeting of the Board of Governors seemed aware of its responsibility as the supreme administrative body of a major institution of higher learning. It approved the final revised draft of the Constitution and Statutes, approved the establishment of Technion City on the new Mount Carmel site, approved the budgets, formally named Prof. Goldstein Vice President for Academic Affairs, for Student Affairs and for Research, and appointed additional members and alternate members: Bern Dibner, Joseph Wertheimer, Nathan Schooler, Ben Halperin and Benjamin Wilk, from the U.S.; D. Lou Harris, Canada; David Mehl, Mexico; S. J. Birn, Great Britain; Harry Goldberg, South Africa; Gershon Gurevitch, David Hacohen, Max Hecker, Yaakov Klebanoff, Pinhas Sapir and J. Raczkowsky, from Israel. Special tribute was paid to Prof. Ratner, who had served as Acting President during absences of Gen. Dori. Ratner had returned to the Technion after serving as Israel's first military attaché to Moscow in 1948–49, and was for many years Dean of the Faculty of Architecture and Town Planning, and Acting President on many occasions.

The Board also grappled with the retirement problem. For lack of an adequate pension plan and regulations governing retirement, many veteran members of the staff continued with their academic program. It was noted that at least two were still teaching at the age of 75, and one at the age of 82.

Reunion on campus of Max Hecker (left) and Joseph W. Wunsch. 1954.

The Technion's educational activity was expanding in several directions.

An attempt was made to set up an evening school, at which students would be able to begin their studies for a Bachelor's degree. This was to be available for the equivalent of the first two years only, after which the students would be expected to complete their studies in the regular day program. The courses were extended and continued for some years, but later dropped.

Negotiations were begun with the Municipality of Tel Aviv for the opening of a branch of the Technion in that city. The idea was to inaugurate first-year studies for 50 or 60 students in Civil and Mechanical Engineering. The project never materialized.

Better progress was made in other directions. By the end of 1954 some 50 students were already registered for graduate studies, and the Graduate School was in effect a reality. The Department of Chemical Engineering was broadened to include studies in food technology. And in the same year the Technion launched a Department of Extension Studies, to provide afternoon and evening courses on high academic level, but not necessarily of university standard. The purpose was to give working people an opportunity to improve their skills and advance themselves professionally.

There was acknowledgment from another direction that the Technion had "arrived" in academic circles when in 1954 the prestigious Israel Prize in the field of Exact Sciences was awarded to Prof. Franz Ollendorff for his book, *Calculations of Magnetic Fields*. Other staff members were to be recipients in ensuing years.

The influence of the Technion in the country was felt primarily through the services of its graduates who provided the skilled manpower required by expanding industry. Additional services were provided by the Research and Development Foundation, but in small ways, as well, the Institute made its presence known and felt. Considerable attention was given in 1953 to suggestions made by Alex Taub, of Washington, a noted designer of automobile engines, who visited the Technion and recommended that the country ought to develop methods of producing alcohol locally, and thus free itself to a large extent from full dependence on imported petroleum. Taub demonstrated the feasibility of the use of alcohol as a source of power, and also indicated easily grown agricultural crops which could serve as a source of supply. His proposals were interesting, but the energy crisis had not yet become acute. Indeed, the very suggestion was ridiculed in some quarters and the whole country laughed at the satire written by a columnist in *Haaretz*:

"The longed-for day arrived. The planning was very efficient and everyone began to travel on alcohol. Government Ministers travel on cherry brandy, high officials on Carmel Hock, and the general public on *spritz*. The oil from the Negev drillings is reserved for cigarette lighters only. I go into the street and see a Dan bus moving along with elegant ease. The driver is very satisfied and from time to time, through a special tube, takes a taste to see if he has enough fuel. The train runs without rails, and the engine belches mightily at each station. All the problems of industry are solved. Vodka is put into the engines, and a sandwich with a pickle is added. The workers lie around and sleep after receiving fuel for their own consumption. The international expert, too, lies there and snores at intervals."

Taub also drew up a proposal for the teaching of engine design at the Technion, leading to the establishment of a local industry for the manufacture of engines in Israel. He felt that within a five-year period Israel could produce from 15,000 to 20,000 engines a year. There was no serious reaction.

A year later there were proposals for a desalination plant, harnessing the rays of the sun. This, too, was premature.

The lighter touch was not absent. For Mount Carmel residents who were disturbed by the howling of jackals who came up from the wadis nightly, Dr. Nathan Robinson rigged up an electrical scarecrow. It was a magnified recording of barking dogs, and when activated was most effective in sending the wild animals in precipitate retreat.

A Rebel Generation

Student unrest caused by both academic changes and financial difficulties, continued throughout the decade. Mrs. Rosa Goldstein, wife of the Vice President, was the first to try her hand in the role of Student Counsellor. She was followed by a young former military chaplain, Y. Zaira, in 1954, and then by Ben Zion Tooval who was formally known as Commissioner *(N'tziv)* to the students, a post which later bore the title of Dean of Students. It was not easy to establish lines of communication with a generation which had just come through the bloody struggle for the establishment of the State of Israel. The students were of a rebel generation, and highly individualistic. Their lack of discipline infuriated the President of the Technion, who had been a lifelong military man, and their lack of cultural standards upset the Vice President who had been nurtured on British standards of university education.

Gen. Dori gave frank expression to the confrontation in his report to the Board in April, 1954, just a few months after the three-week strike:

"Though the efforts of the Technion are concentrated primarily in the

M.I.T. President Karl Compton on his visit to the Technion with Mrs. Compton; at left, Mrs. Sydney Goldstein. 1954.

fields of science and the practical professions, this is not merely a technical school. Even in the general course of education we do not seek only to provide our students with a profession; our primary task, as in every institution of higher learning, is to introduce our student body into the temple of learning from whence come the wellsprings of man's culture. More important than the mere acquisition of knowledge is the promotion of true understanding and a critical mind, as well as the desire for learning and the striving to create .... To our great regret we cannot report a satisfactory state of affairs in this area. We have not yet succeeded in arousing within the student body a spirit of cooperation in our efforts to raise the standards of the Institute. Our efforts in this direction are nullified by the great majority of the students, whose lack of understanding is evidenced by the absence of any consciousness of higher culture or self-fulfillment on their part .... We resist with might and main the trend now current in the student body to level off the standards on the plane of the average student ...."

The basic issue which he pinpointed was one already familiar from the early history of the Technion: "We must decide between a Technion on the level of a mediocre technical school, or a Technion with the standard and stature of an Institute of Technology on the level of a University . . . ."

In November of that year the students again went out on strike, still protesting the new and as yet unfamiliar changes in examination regulations. They did not understand the implications of at least one of the reforms: if grades achieved in the week-to-week class exercises and in the occasional mid-term quizzes were sufficiently high these could compensate for poor results in the final examination, and still allow the student to pass. It was not until after several years of observation that the students realized the changes were actually in their own best interests. In the meantime, they reacted to the Technion administration with what Dori called "trade union tactics". The President was patient for seven days and then he set a deadline. If the students were not back at their desks by the date set, he would declare the semester ended because of the time already lost, and all students would have to repeat the complete semester. The strike crumbled, and the students went back to their studies, but their feelings toward the President were not improved. In their various strikes and protests he was frequently caricatured as a military administrator. In new forms they were repeating the cruel line which had been followed by the students of 1925 and 1926. It was the classical struggle between different generations.

Help from Abroad

The academic differences were matters of principle. Most of the other student grievances — tuition fees and lack of dormitories — were matters that could be solved if funds were available. Hence a renewed stress was laid on overseas fund-raising. In 1950 Dr. Mordecai Levy was sent to France, and a year later the Groupement Francais des Amis du Technion was established. The group in London had already undergone several metamorphoses and in 1951 took shape as the Technion Society of Great Britain. Since this was a cultural organization only, a British Committee for Technical Development in Israel was established alongside it in 1953. By 1956 the two merged to form the British Technion Committee, and four years later it was reorganized as the British Technion Society. For a long time the emphasis was on scientific and academic contacts between Israel and Britain, with a reluctance to engage in fund-raising. Once this reluctance was overcome, the British launched a series of highly successful projects which enhanced the campus of Technion City and provided great help to the operations of the Institute.

Michel Polak (1864–1954).

278

Israel's President Yitzhak Ben Zvi, left, receives two veteran members of the British Technion Society, S.J. Birn and Arthur Blok. 1956.

The honor roll of British Technion leadership was impressive: Technion Society of Great Britain, 1951, The Rt. Hon. Lord Silkin, President, Arthur Blok, Chairman; British Committee for Technical Development in Israel, 1953, Sir Louis Sterling, President, I. J. Lindner, Chairman; British Technion Committee, 1956, President, Lord Silkin, Chairman, Sir Michael Sobell, Hon. President, Victor Mishcon; British Technion Society, 1960, Hon. Life President, Sir Isaac Wolfson, Hon. President, Victor Mishcon, President, Lord Silkin, Chairman, Edward E. Rosen. In 1962 Sir Leon Bagrit succeeded Lord Silkin as President. The first Executive Director was Edgar Stern in 1953, who was succeeded by his assistant, Joseph Cohen, in 1957.

The first major gift to the Technion from Great Britain, £100,000, came from The Humanitarian Trust, which had been established by Michel Polak, founder of the Nesher Cement plant in Palestine. The Trust continued Polak's long-standing support of the Technion for many years, and the Building Research Station on the campus was named for him. Further

Ground-clearing on the site of the Senate House and the Churchill Auditorium. 1956.

large gifts were later received from Harry and Abe Sherman, of Cardiff.

Toward the latter part of 1954 the British decided to raise the funds for construction of the campus central auditorium and they sought permission from Winston Churchill to name the building after him. Sir Louis Sterling procured Churchill's interest and consent. Later, the Senate Building on the campus was named in memory of Sir Louis.

Churchill had previously refused all efforts to involve him in endorsement of various organizations and agencies, but his message of consent in this case was unreserved. "I feel truly honored that some new buildings of the Israel Institute of Technology are to be named after me, and that my name will be associated with an undertaking devoted to the advancement of knowledge and human well-being," he wrote. "Israel has no lack of skillful professional men, scientists and artists but those with all their gifts cannot alone solve Israel's present economic problems. She needs also technicians and craftsmen to build new towns and factories and to bring what is desert under cultivation. I am sure that the Israel Institute of Technology has a great contribution to make to Israel's future prosperity and that Israel's prosperity cannot but be of great benefit to other countries, as well."

At conferment of an Honorary Doctorate on Dr. Albert Einstein and Dr. James Franck at Princeton, N.J. Front row, left to right: Abraham Tulin, Dr. Franck, Alexander Konoff, Dr. Einstein, Col. J.R. Elyachar, Dr. Ben Zion Dinur, Dr. Philip Sporn and Dr. William Fondiller. October 3, 1954.

A year later his son, Randolph Churchill, laid the cornerstone of the Churchill Auditorium. He wielded the silver trowel with skill, and to the amusement of the audience declared that as a youth he had gained his first experience as a bricklayer under his father's guidance. Many years later, when he revisited the Technion, he was asked if he remembered the occasion. Remember? He still had the trowel, with its commemorative inscription, he said, and made frequent use of it — as a cheese slicer.

South African Jewry was represented on the campus by facilities contributed by Morris Mauerberger, Woolf Senior and Isidor Cohen. For many years the Technion group in South Africa was guided by its Honorary Secretary, Gus Osrin.

A Far Eastern element was introduced onto the campus with a generous contribution from Singapore for the Reuben Manasseh Meyer Auditorium in the Aeronautics Building.

Across the seas the American Technion Society was gearing itself for its major campaign. Judah Wattenberg, who had been Executive Director for ten years, was succeeded in turn by Israel (Jacobson) Gaynor, Philip Chasen, William Schwartz and Edward Vajda. Lazarus White, first President of the A.T.S., 1941–44, was followed by William Fondiller, 1944–47; Joseph W. Wunsch, 1947–50; Col. J. R. Elyachar, 1950–55; David Rose, 1955–59. In addition to these, the records of the American Society contain the names of scores of generous and dedicated individuals whose efforts contributed to the development of that organization as one of the prestigious Jewish organizations on the American scene. During these years men like Elias Fife, Charles Adelson and Harold E. Beckman guided the financial affairs of the Society either as Chairman of the Finance Committee or as Treasurer. Legal counsel was provided then, and for many years thereafter as well, by Murray Rubien.

On the initiative of Elyachar, the organization acquired a six-story building at 1000 Fifth Avenue into which it moved in 1952.

The reputation attained by the American Technion Society was further indicated on October 3, 1953, when Dr. Albert Einstein received a delegation from the Society at a special convocation at Princeton University, where Technion's honorary degree of Doctor of Science in Technology was conferred on him and on his colleague, Dr. James Franck.

The support from the United States reached a new peak in 1956–57 with the establishment of the Swope Fund, an historic achievement which must be ranked with the munificent assistance received from Jacob Schiff in his day.

The Swope Story

Gerard Swope, long head of the General Electric Corporation in the United States and for almost half a century prominent in American industrial affairs, was not generally known to be Jewish. There is nothing to indicate that he ever hid or denied his origins, but neither did he go out of his way to associate himself with Jewish causes. For that reason, his reaction to British tactics in Palestine was a newsworthy event. In 1947 the British Government decided to appoint him an honorary commander of the Most

Gerard Swope (1872–1957).

Excellent Order of the British Empire (OBE), presumably on the basis of his contributions to the Allied war effort. He expressed his gratification at the recognition. The formal presentation was to take place in Washington in June, 1948.

In the meantime, the British were engaged in turning away from the shores of Palestine shiploads of Jewish survivors of the concentration camps. The United Nations decision for the establishment of a Jewish State had not been received with good grace in London, and Swope felt that the British were being obstructionist. He used even stronger language. In a letter to the British Ambassador in June, 1948, Swope announced his rejection of the decoration because of the British Government's "vacillating, reprehensible and non-constructive attitude" toward the creation of the new state.

It can be presumed that the American industrialist continued to take a close interest in all that was transpiring in the Holy Land. Nevertheless, when he and his wife took a trip to Europe in the early 1950's he extended the voyage to include a week in Israel almost reluctantly, and at her insistence, because she wanted to visit the Christian holy spots. The visit drew no attention from anyone.

Early attempts to interest him in the Technion had been made by the American Technion Society. At one stage the Society prevailed on Dr. Einstein and on Dr. Harold C. Urey to write Swope, soliciting his membership and one day a $100 membership was received from him. His name was kept on the mailing list.

His interest became channelized in an historic way as a result of his meeting with Abraham Tulin. The latter, a distinguished attorney, collaborator with Brandeis in Zionist matters, and active in Technion affairs ever since Professor Samuel had arrived in the U.S. in 1940, felt that as a graduate of M.I.T. Gerard Swope would understand the role played by the Technion in the development of the young state. At one of their frequent meetings Tulin told Swope about the Institute, and about higher education in Israel in general. As a result, in 1954 Swope made a contribution of $100,000 to be divided between the Technion and the Hebrew University for a student loan fund.

The authorities in Haifa lost no time in using the funds and soon thereafter the first ten fortunate students were informed of the timely help they were receiving. Their letters of thanks, promptly forwarded to Mr. Swope, made a deep impression upon him. In the following year he gave the American Technion Society $100,000 and an equal amount each year until his death in 1957.

However, provision had also been made for the Technion in his will and that of his wife, Mary Hill Swope. Mrs. Swope died on October 28, 1955.

She bequeathed a large amount of money in trust to her husband, the principal to be used as he should direct by his own will upon his death. And on March 23, 1956, after legal consultation with Tulin, Mr. Swope executed a new will in which he left substantially the whole of his own estate, as well as the principal of his wife's trust fund, to the American Technion Society, for the benefit of the Technion in Israel.

Undoubtedly the warm friendship, understanding and trust that had developed between these two brilliant men, one the scientist and business genius, the other the humanist and organizer in the legal field, fused into making the Swope Fund possible. Both men, from their different approaches, understood clearly the absolute need of the newly created State for modern science and technology if the State was to defend itself and prosper under modern conditions. Thus it was made possible for Gerard Swope to create and for Abraham Tulin to perform the magnificent legal craftsmanship to maximize and establish firmly the Swope Fund of the American Technion Society.

In June, 1956, the Technion Board, with the approval of the Senate, voted to confer on Gerard Swope its honorary degree Doctor of Science in Technology. A year later he came to Israel, together with his daughter, Miss Henrietta Swope, and received his degree, in the presence of Prime Minister David Ben Gurion. Five months later, on November 20, 1957, at the age of 84, he passed away.

The total of the two estates amounted to something in excess of eight million dollars, yet there is no building on the Technion campus which bears his name. Swope had general ideas as to the use to which the money should be put, and he directed that within the discretion of the American Technion Society it be used "for the educational and scientific purposes" for which the Society was organized. He knew what he did not want it for. On more than one occasion he had made it clear that the money was not to be used for "bricks and mortar". Technion would always find donors who would contribute for construction of buildings, he said. It would be more difficult to procure funds for aid to students or to pay salaries to faculty members, or to foster research, and he therefore stipulated that the income from his funds should be utilized for these purposes.

It should be noted that three Chairs at the Technion bear the name of Gerard Swope and one bears the name of Mary Hill Swope. A major part of the Student Loan Fund at Technion had its origin in Swope funds. Enormous help was provided to faculty members in need of housing in Haifa. So many of these apartments have been built on Einstein Street that the street is sometimes jocularly referred to as Swope Boulevard.

In view of the fact that there have been many "claimants" to the credit for having been instrumental in obtaining this princely gift for Israel, it should be noted that Gerard Swope himself placed on record his statement that Abraham Tulin alone was responsible. A major wooded section of the upper stretches of the Technion campus is known as the Abraham Tulin Forest, and a Chair bears his name.

News that the Technion was to benefit from such a large amount of money confronted both the institution and the American Technion Society with a problem. There were some who held that once word got around it would effectively kill all further fund-raising. Who could be induced to give money for an institution which was already so richly endowed?

The news did get around, quickly, but the effects were quite different from what had been feared. People who had never heard of Technion before became curious. What was this institution which merited the generous support of one of America's industrial giants? Why was it so important?

New avenues of support were opened up, and the fund-raising operations of the American Technion Society flourished.

Abraham Tulin (1882– 1973).

The interest and activity of the American Society extended beyond fund-raising as well. Technion's friends sought to involve themselves in every aspect of the growth and development of the Institute. In 1952 they launched a program to provide post-graduate professional training in the U.S. and in the years that followed many Technion graduates were enabled to acquire practical experience in their fields as temporary employees of American firms. The American State Department cooperated by providing the necessary work permits for such employment of aliens.

For some three years a group of friends in New York, operating as the Technion Food Club, systematically sent Food Package certificates as gifts to members of Technion's staff, and during this time many hundreds of families living on marginal salaries were helped through periods of food austerity.

The American Society reached broad scientific and industrial circles in the United States through its Technion Year Book. From modest beginnings in 1941, the Year Book became a prestigious journal reflecting not only Technion activity, but also developments in the world of science and technology at large, as recorded by distinguished participants in those developments. The Year Book had various editors, but much of the direction was provided by Dr. Bern Dibner who served as Editorial Chairman from 1954 until the journal was discontinued in 1968.

Still, the emphasis remained on fund-raising, and the annual transmittals to Haifa rose from year to year. Up to 1961 the Technion recorded

The Technion family turns out for memorial exercises on the death of Albert Einstein. 1955.

more than $6,000,000 received from the A.T.S. for capital construction alone. Buildings began to sprout in various corners of the new campus, and Technion City began to assume a most impressive appearance. The basic need was for operating funds.

Goldstein's Departure

Technion had become a big business in more ways than one, and the pressures began to take a toll of the personnel involved. The first to be affected was Prof. Sydney Goldstein.

Goldstein had come to Technion with the primary aim of establishing a Department of Aeronautical Engineering. Almost immediately after his arrival Kaplansky died and the new professor was drawn into administrative affairs of the Institute. He accepted the post of Vice President with the understanding that he would be spending his time on academic matters. Hence, when Dori went abroad he refused to serve as Acting President, and asked that he be relieved of the Vice Presidency. The Council urged him to continue in office, and spelled out the specific functions of his Vice

Presidency. It will be recalled that Dori had agreed to accept the Presidency only on condition that Goldstein serve alongside him. Dori realized that he was not an educator, and needed Goldstein to deal with academic matters. However, the realities of the situation, and the irregular state of Dori's health, resulted in drawing Goldstein frequently into the vortex. Furthermore, his attempts to bring about a revision in academic standards, and to institutionalize academic procedures brought him into frequent conflict with some of his colleagues. He and Dori also came from two different worlds, and their outlooks differed. Goldstein had been nurtured on the British tradition of the human relations ethic. He accepted promises at face value. A gentleman's word was sacred. If the Government made a promise, he expected it to be kept. It was a PROMISE. When there was failure to keep promises, he took it as a personal affront against himself. Principles were important to him, and he felt let down — by the Government, which failed to provide funds that had been promised; by his academic colleagues who resisted his reforms, and sometimes resorted to personal animadversions; by the students, who went on strike instead of sacrificing themselves for ideals; by Gen. Dori who, unable to implement some of the academic reforms, kept piling administrative tasks on him.

Dori was part of the fabric of Israel life, and a product of the philosophy of improvisation out of which grew the Haganah and the State and its early institutions. He was not shocked when promises were not kept because he understood the pressures under which the promises had been made, and the conditions and circumstances which made fulfillment impossible.

In July, 1954, Goldstein informed Dori that he could no longer continue, and the President announced that Goldstein had at his own request been relieved of his duties as Vice President "so that he may be free to pursue his scientific work." At the same time it was reported that Goldstein was taking a year off to be a visiting lecturer at Harvard. Dori explained further: "His contributions to the academic, scientific and administrative activity at the Technion during his tenure as Vice President have been outstanding, but we must face the fact that his full devotion to his scientific work holds forth even greater promise for the Technion and for Israel as a whole." He had hopes that the break was temporary, and that the Vice President on whom he had leaned so heavily would be back again soon. Hence no formal decision was taken as to Goldstein's departure.

Goldstein, ever the formalist, could not reconcile himself to this attitude and asked that his resignation be accepted. For obvious reasons Dori was unwilling, and it was not until 1960, when Justice Moshe Landau reported to the Board of Governors that without formal acceptance of his resignation Goldstein would not consider the question of returning, that the Board belatedly accepted the resignation which had been tendered in July, 1953, and was considered to have gone into effect retroactively as from the beginning of the summer of 1954.

As will be seen, Goldstein rendered additional valuable services to the Technion some years later, but he never returned as a staff member. In a review of his first decade at the helm of the Technion, Gen. Dori summarized the contributions of the scientist from Manchester: "Dr. Goldstein was the architect who designed the academic framework of the Technion. It was not only that he penned the academic constitution and by-laws, set up the academic and administrative operational bodies, defined the functions and authority of such bodies and created the tools and broad outlines for the operation of the Technion; he also gave content and spirit to these outlines, drawn from his own experience and from the fine traditions of the institutions of higher learning in that country where he lived and worked as distinguished professor, scientist and academician . . . . Goldstein was my teacher and mentor . . . . He introduced me into the world of scholarship and learning, into which I had previously only peered from the outside . . . ."

Dori required top-level administrative help, and for some years Professors Ratner and Grossman served as Deputy Vice Presidents. He introduced a new administrative position to which he gave the ancient Aramaic title of *Amarcal*. This was occupied from 1954 by Bezalel Yaffe, from 1957 by Jacob Quat and from 1959 to 1966 by Basil Herman. For many years the position of Director of the President's office was held by Ilana Lenji, an immigrant from Hungary, who had first been engaged as secretary to Prof. Goldstein after his wife had met her in the Ulpan where both were studying Hebrew. The search for a full Vice President continued, and in the Spring of 1957 Prof. Sebastian B. Littauer, of Columbia University, arrived to take up his post as Vice President for Academic Affairs. He remained for about a year. Prof. David Ginsburg, of the Chemistry Department, was in 1959 named Vice President in Charge of Research, and when he went on Sabbatical leave the following year the Board voted to name Prof. Kurt Sitte to the same position. It was a post which Sitte was destined not to fill.

During this period other administrative changes were taking place as well. The venerable Dr. Biram stepped down as Chairman of the Board in 1956 and Justice Moshe Landau, of the Israel Supreme Court, was elected in his place, serving through 1962, and again later in 1965–66 and 1969–71.

The course on which the Technion had been set had a dynamism of its own, and the Institute grew, both qualitatively and quantitatively. During Dori's first decade in office new academic disciplines were introduced or expanded. The courses in Production Engineering introduced by Prof. Jules Cahen in the Faculty of Mechanical Engineering developed into a separate Department of Industrial and Management Engineering and the first undergraduates received their B.Sc. in 1961. The studies in Metallurgy were broadened and steps taken to open a Department of Mining Engineering. The 1956 meeting of the Board called for immediate inauguration of studies in Nuclear Engineering.

New Sources

The urgent need for top-level academic personnel to fill teaching and research posts at the Technion was met in large part by visiting staff members sent by various overseas organizations. These included Technical Assistance agencies of the U.N. such as the FAO (Food and Agriculture Organization), WHO (World Health Organization), UNESCO (Educational, Scientific and Cultural Organization), ILO (International Labor Organization), as well as U.S. Government agencies such as the Point Four Program, the U.S. Operations Mission, and the U.S. International Cooperation Administration. Some of the experts sent to Technion on these assignments eventually decided to remain and became permanent members of the Technion staff, among them, Prof. J. G. Zeitlen, who had come in 1952 as a U.N. expert to advise the Department of Soil Mechanics, and Prof. Alberto M. Wachs, from Argentina, an expert in Sanitary Engineering, sent by the WHO in 1956. Prof. Jay Tabb had independently joined the faculty in 1954 as an expert in Labor-Management Relations and continued until his death more than 20 years later.

The assistance provided by U.S. Government agencies took many other practical forms as well. One of the largest sources of help in the 1950's was the U.S. Operations Mission which helped establish the Israel Institute of Metals at the Technion, provided aid for the Plastic Pilot Plant in Chemical Engineering and set up the Israel Institute of Industrial Design which was at first associated with the Technion, all this in addition to providing visiting professors and lecturers in many fields.

The funds made available from the American Special Cultural Program (popularly known as the Katzen Fund because it was administered by Bernard Katzen), made possible construction of the building to house the Department of General Studies, establishment of a Chair in Industrial Relations and the setting up of a U.S. university and teacher training scholarship program over a four-year period. Research programs of considerable value were sponsored by the U.S. Air Force, the U.S. Office of Naval Research and other Government Departments. Within the framework of the Fulbright Program many visiting professors were sent to the

Technion and travel grants provided for Technion personnel to study at American universities. This broad and continuing program of assistance had an enormous influence on the development of the Technion, but it also served to put into contrast the difficult problems of day-to-day financing.

Two events in the Fall of 1956 had an effect on the Technion. One was the revolt in Hungary, quelled by Russian troops between October 23 and November 4. In the aftermath of the uprising there was an influx of both students and faculty members from that country, repeating a now familiar cycle in history.

The second event was also on the international scene. Between October 29 and November 7 students and teachers were suddenly called into uniform to take part in the surprise Suez Campaign. Classes continued without interruption and laboratories operated, albeit with reduced staff. This military action, too, was followed by a rise in immigration of students from Egypt.

Growth and Deficits

The second half of the 1950's was marked by steady growth, progress and development. In reporting on activity Dori could often tell the Board that the previous 12 months had been a "good year", but the progress was being achieved under great strain. Deficits increased from year to year, and it was not always possible to meet the monthly payroll. Every new source of income was explored.

In 1955 the Council had approved adding two more stories to *Bet Hakranot* but the new income received turned out to be minimal. The operating budget of the Technion was by now soaring far ahead of available resources. Tuition fees, last set at IL 260, had lagged behind all other increases, and at the Board meeting in May, 1958, it was proposed to raise the fee to IL 400.

Expenditure for construction had also exceeded income, in part because pledged contributions were slow in coming in, and in part because Government payments due were likewise being deferred. Dori expressed regret to the 1958 Board that the outlay of IL 14,000,000 on capital development was some IL 1,700,000 above the resources at his command. Beyond that he could not go, he said, because both means and credits had been exhausted.

There was criticism of Dori, not for his deficits, but for what was called his conservatism. Other institutions of higher learning in the country (and their number had increased) were mounting deficit on deficit, secure in the confidence that a university would never be declared bankrupt, and

Ceremonial opening of the Fischbach Electrical Engineering Building. Left to right: Harry Fischbach, David Ben Gurion, Mrs. Fischbach, Gen. Yaakov Dori. 1957.

that the Government would in the crucial stages always step in to help out. Other campuses were being built at a far more rapid rate than Technion City, and what Meyer Weisgal of the Weizmann Institute referred to as the "edifice complex" was rampant.

Technion's physical growth on the new campus had been slow but steady during Gen. Dori's first decade in office. The Samuel Fryer Aeronautical Engineering Building and the Charles and Bertha Bender Aeronautical Laboratory had been the first structures erected in 1953–54. The Seniel Ostrow Building was added to the aeronautics complex later. In 1956 the Michael and Anna Wix Student Hostel, and the Philadelphia Hostel, the latter a project of the Women's Division in the U.S., were occupied. A year later construction was completed of the Harry Fischbach

It is a source of great regret to me that I cannot be present with you at the Opening of the Technion Institute. I have followed with keen interest the progress of this far-sighted and enlightened endeavour.

I have been a Zionist for many years, and I view with pleasure and admiration the maturing of the State of Israel. So to increase the technical aptitude of your people is indeed commendable. It is perhaps the most urgent requirement of any free country who wishes to preserve its standing, dignity and independence.

I pray that your efforts may be crowned with success, to the detriment of none and to the lasting benefit of all the peoples of the Middle East.

David Ben Gurion affixes the mezuzah on the doorpost of the Churchill Auditorium. 1958.

Electrical Engineering Building and the Michel Polak Building Research Station.

The big year was 1958. Dedication of the Winston Churchill Auditorium was a festive event, with participation of the Prime Minister of Israel and members of the Churchill family. Britain's wartime leader sent a special message which was read by his daughter, Sara: ". . . I have been a Zionist for many years, and I view with pleasure and admiration the maturing of the State of Israel. So to increase the technical aptitude of your people is indeed commendable. It is perhaps the most urgent requirement of any free country who wishes to preserve its standing, dignity and independence."

The dedication of the Albert Einstein Institute of Physics in the same year recalled when the father of relativity had visited the old Technion campus in 1923 with his wife and had planted two palm trees there. On this occasion Einstein's son, Hans, planted two cypress trees at the entrance to the new building named for his father. The Lidow Building for Research in Physics was added to the Einstein complex. Other buildings occupied in that year were the David T. Siegel Hydraulics Laboratory, and the Eva Frost and Henrietta and Stuard Hirschman Student Hostels.

In 1959 the Alexander and Anna Konoff Junior Technical College Building, the Helen and Morris Mauerberger Soil Engineering Building and the Rose Mazer Hostel were added. The following year saw dedication of the Benjamin Cooper School of Industrial Engineering and Management, and following that the Jacob Ziskind and Sylvia and Barnett Shine Student Hostels.

This vast construction program was under the direction of Yitzhak Ben Gera, himself a Technion graduate, who served as head of the Building and Maintenance Department of the Technion from the inception of Technion City in 1953 until his death in 1967.

In retrospect it can be said that a move from the old campus in Hadar Hacarmel to Technion City, which began in 1953, and had not been completed by the time this book went to press, was certainly indicative of a conservative policy, especially considering that the "reckless" institutions did indeed get the help necessary to bail them out of their difficulties. Later historians must reach conclusions whether it might not have been better to heed Dori's critics who said, in effect, "Damn the deficits — full speed ahead!"

When money and credit ran out construction could be stopped, but salaries and wages had to be met each month, and Technion's representatives beat a path to Jerusalem with wearisome monotony, seeking increase in allocations, or at least more prompt payments of what was due. On many occasions Nessyahu, or his aide, Naomi Lev, sat for hours in the waiting rooms of the Ministry of Finance, until receipt of the precious payment which made it possible for them to hurry back to Haifa and pay salaries ten days beyond the first of the month on which they were due. The sums involved were much larger, but the tensions were no less than those in the critical days of the early 1930's.

The situation was further aggravated by the complaint of the academic staff that their salaries were lagging behind other sectors of the economy. In June, 1955, the staffs of the Hebrew University and the Weizmann Institute went out on strike to enforce their demand that the salary scale in the institutions of higher learning be separated from any linkage with the usual Civil Service grades. The Technion staff and other professional groups joined the protest. As a result the Government agreed in principle to salary rises but postponed actual implementation, and the result was a new wave of strikes.

Matters reached a head in February, 1956, in a "revolt of the intellectuals". A general strike of all professionals and academic personnel paralyzed not only activity at the universities, but also all other professional practice in the country. Gen. Dori met with representatives of the Technion staff, and agreed that he would pay them according to the increases approved by the Government as yet only in principle. He asked them to return to work. The agreement was at once disavowed by the Government, which said that unilateral raises of this nature were unacceptable. The strike lasted for eight days, and finally a compromise was reached whereby the increases were granted, but were made payable on a staggered basis over two years.

It was the Technion's contention, with all due respect to the other institutions, each of which had its own special functions to fulfill, that allocation of Government funds should take into consideration Technion's unique role. A division of the global budget for higher education on a per capita basis, by number of students, did a gross injustice to the Technion which provided the skilled manpower needed for the country's industrial development, since the actual cost per student at the Technion was far greater than that required for education in the humanities or social sciences. Successive Prime Ministers and Ministers of Finance and Education paid tribute to the Technion, but the practical results were small. In January, 1956, the Israel Post Office issued a special postage stamp honoring the Technion on the occasion of the thirtieth anniversary of the beginning of classes. (The stamp was two years late.) The compliment was appreciated, but the philatelic issue did not help the Technion in its financial difficulties. As one member of the staff pointed out, the Institute did not even get complimentary samples of its own stamp.

The Technion demanded that the Government appoint an official com-

Israel postage stamp issued in honor of the Technion. 1956.

293

mittee whose function it would be "to study the Technion's situation and produce recommendations for establishing the Institute on a sound financial basis, with full consideration of the tasks which the Technion must fulfill in the service of the State."

In the fall of 1958 Dori suggested shock treatment: not to open the school year until the financial situation had been clarified.

Since most of the funds raised overseas were for construction purposes, income for operations fell far behind the rising budget. The following excerpts from the budgets for the years indicated give a picture of the problem: All figures are in dollars, according to the rate of exchange in the respective periods:

Year	Annual Operating Budget	Government Participation	Contributions from World Jewry	Income from Tuition Fees
1951–52	$1,046,387	362,745	314,558	136,753
1954–55	1,859,577	788,452	252,292	238,435
1956–57	2,978,933	1,140,000	459,956	317,166
1959–60	4,769,213	2,141,666	794,171	494,079

The relative weight of the various sources in meeting the budget is seen from the following table which reflects the cumulative total for the 13-year period from 1951 to 1964:

Annual Budget	Government Participation	World Contributions	Tuition Fees	Other	Deficit
100%	43.7%	14.8%	9.7%	24.7%	7.1%

Dori was extremely sensitive to criticism of his economic policies. When a Board member in 1956 observed quite innocently, and certainly with no malice, that the new student dormitory buildings "looked" luxurious, the President bristled. He demanded that an investigating committee be set up at once to "look into desirable building standards as applied to the student dormitories." He wanted no whitewash; the committee should report objectively what it found. The committee presented its report to the Board the following year. It went into great detail, and found no basis whatever for the charge of luxury. To the contrary, some members even found that the construction, based on utility-per-student per square meter, was more economical than in other institutions.

Prof. Ratner confirmed rumors about the use of marble in the Aeronautical Engineering Building which he had designed. The building committee had severely limited the quantity of marble available to the architect, and he came to the conclusion there was not enough to line the entrance lobby. Hence the only rooms in the building which were walled in marble were — the toilets and washrooms. Ratner maintained this would reduce long-range maintenance costs there, and would also reduce to a minimum the graffiti which was customary in such places.

The watchful self-criticism at the Technion led the following year to the appointment of a committee headed by Jacob Reiser, with a far wider scope. The Reiser Committee was asked to make an in-depth study of all of the Institute's operations, in view of the financial crisis, and to ascertain where economies could be effected. The committee did a thorough job, and its recommendations were far-reaching. It called for a reduction in academic staff, conversion of fellowship grants to repayable loans, and reduction in the enrollment in the Graduate School. Other retrenchments were advised, including a study of all administrative operations, with a view to cutting budget.

The report, submitted in 1959, raised a storm. On the one hand were those who maintained that implementation would lower academic standards and set the Technion on the path of scholastic decline. Others insisted the intention was only to raise efficiency. Decision was to implement a number of specific economies, and though the major elements affecting academic matters were not put into effect, the result was a drastic tightening up of administrative operations in the various academic units.

A significant change in the academic organization of the Technion recommended by the Reiser Committee called for abolition of Divisions, which had multiplied in number, and for concentration of administrative authority in the hands of Faculty Deans or Heads of Departments. The result was both greater efficiency in administration and economy in operations.

Student enrollment also reflected the dynamic growth of the Technion. In the academic year 1951–52 there were 966 undergraduates. In 1954–55 this had risen to 1,571; 1956–57, 1,837; and in 1959-60 the number of undergraduates had reached 1,973, with an additional 440 students registered in the Graduate School.

The number of candidates for admission to the Technion almost always exceeded the number of places available, sometimes quite considerably. Accordingly, the Technion authorities were frequently subjected to pressures which were couched in diplomatic words of warm recommendation. One prominent member of the Knesset urged the President to accept as a student one who had failed the entrance examinations in mathemat-

Graduation exercises at Technion City. 1959.

ics. The boy's principal qualification: he was of Yemenite extraction, and should be encouraged. One of Israel's most distinguished artists asked as a special favor that the daughter of an artistic colleague not only be accepted, but also excused from payment of tuition fees. Instead, her father would contribute a painting to the Technion. Human and emo-

The new Kisch Memorial Laboratory in Electrical Engineering. 1959.

tional factors were often cited as compensation for lack of necessary academic qualifications. On the whole the Technion was able to withstand this exercise of *protectzia*, though occasionally the strict regulations were stretched somewhat as mercy tempered rigid bureaucracy. There were also occasions when after the Technion had refused to stretch the regulations, and had rejected a candidate who had failed to measure up to the standards, the boy went abroad, enrolled at a prestigious American university, and graduated with honor grades.

One of the milestones of the decade was the formal establishment of the Graduate School as a separate academic unit. Graduate studies had been offered since 1948, and doctorate degrees had been awarded but it was not until August, 1956, that the Senate approved a set of regulations formalizing the academic framework for award of the degrees of Master of Science, Doctor of Science and Doctor of Science in Technology in virtually all of the Faculties and Departments of the Technion. Enrollment in

A 225-foot model of the 16-mile-long Yarkon River built at the Hydraulics Laboratories, Technion City. 1958.

Gen. Moshe Dayan, as Chief of Staff, on a visit to the Technion, where he was received by Prof. Y. Ratner. 1958.

these programs rose rapidly and the student body of the Graduate School provided a valuable source of Instructors and Assistants for the undergraduate program. Such opportunities for academic service to the Technion likewise helped the graduate students augment their income. The Graduate School also gave great impetus to the advancement of scientific research at the Technion.

Prof. Nathan Rosen was named first Dean in 1956. He served in that post for three years and was succeeded by Prof. Rahel Shalon (1959–62). Successive Deans were Prof. Rosen again (1963–66), Prof. Shragga Irmay (1966–68), Prof. E. Amitai Halevi (1969–71), Prof. Asher Peres (1972–74), Prof. Frank H. Herbstein (1975–77), Prof. Alexander Solan (1978–80), Prof. Zvi Rigbi (1981–   ). Secretary of the Graduate School for 14 years, from 1958, was Mrs. Erika Schlosser. In 1972 she was appointed in charge of the Office for Academic Staff and served until her retirement in 1976.

Parallel to the emergence of the Graduate School was the development of research procedures at the Technion. In March, 1960, the Senate formally adopted the regulations which have since then, and subject only to minor amendments, served as guidelines for research administration. The regulations were intended to encourage pursuit of both basic and applied research, to provide for proper coordination, and to assure the necessary financial support. It was established that each full-time member of the academic staff possessed the basic right to engage in research of his choice, and to receive support to the extent that the Institute's resources permitted. The regulations set forth detailed procedures for submission of research proposals and for determination of the proper academic level. It was at this time that the Board of Governors created the new post of Vice President for Research, to administer this aspect of the Technion's activity.

Student Life

Student life at the Technion has never been easy. Even with the construction of the first hostel buildings there was still no campus life, and the students commuted each day from their homes, or in the case of out-of-town students, from their rented quarters. The hope was that when the majority of the student body was housed on campus a school spirit would develop, as well as campus-based extra-curricular activities. In 1955 the students themselves initiated a project to involve them in tree-planting on the new campus, and on *Tu B'Shevat* some hundreds of students did swarm over the hillside setting saplings in place. In this they were renewing a tradition dating back to 1926, when students had planted trees on the old campus. For lack of a sports field athletic activities were minimal.

First-year students of civil engineering get their first lessons in elements of home construction. 1957.

A few years later the students contributed a total of 5,500 work days in ground preparation for a municipal sports field to be located in a wadi just below the campus. The field was completed, but it was not made available for student use. Various volunteer movements commanded student support, and many of the future engineers registered for help to new immigrant settlements. In 1958 a student orchestra was formed and this, together with a student chorus, ultimately developed into one of the more successful activities.

Professionally and academically the students did well. When the professor of architecture commented in 1957 that two of his brightest students, Eitan Kaufman and Amnon Gelfman, did not seem to have their minds on their studies all year, they replied that they were "working" on something. Later in the year the secret came out. The two boys were announced as winners of a IL 1,300 consolation prize in the contest for design of the new Knesset building in Jerusalem. They did not get first prize, but then they were only third-year students. And their professor also won a IL 1,300 prize in the same contest.

Pressures on the students were both academic and financial. The very demanding scholastic requirements, product of the earlier years, had resulted in a program of studies which called for 42 contact hours per week. Horrified at this excessive amount of time spent within the classrooms and laboratories, Prof. Goldstein had scaled the schedule down to a maximum of 36 hours per week. Among the changes inaugurated by Prof. Littauer was a further reduction in number of contact hours required down to about 32. Student cartoons at the time sometimes pictured the embryonic engineers as slaves engaged in hard labor. One year, at about the time of final examinations, an exasperated student placed a poster at the entrance to the campus: "Technion City — the Town Without Pity!"

Problems such as these, added to the financial burdens, and the emergence of an aggressive student leadership, combined to bring matters once again to an open confrontation with the administration. Dori did not have to defer opening of the school year; the students seized upon the imminent increase in tuition fees and called another strike — the first of two that year — and school did not open as scheduled in the Fall of 1958. The struggle was conducted with militancy, and was directed both against the Technion and against the Government, which had forced the Technion into financial difficulties. As the days went by the students engaged in personal vilification of Dori. Some of his associates urged that he break the strike by threatening that first-year students who did not appear would forfeit their precious acceptance into the Technion, and last-year students who remained away would not be qualified for their diplomas. Yet despite the charges that he was displaying military rigidity in his attitude to the students, Dori refused. He looked upon a strike as a legitimate form of social expression, though he felt the students were misusing it. The problem of tuition fees was submitted to a Government committee headed by Supreme Court Justice Shimon Agranat, and both sides agreed to accept the findings of the committee. The Agranat Committee recommended acceptance of the IL400 annual tuition fee for all universities, called for a widespread program of scholarship assistance to needy students, and suggested that in the future tuition fees be linked to the Cost of Living Index. The latter formula was indeed adopted and remained in effect for many years.

Despite importunings from Haifa to their colleagues in Jerusalem to join them in protest, the students of the Hebrew University remained on the sidelines. However, when the findings of the Agranat Committee were applied also to the University, the students there went out on strike and asked the Technion student body to extend moral support. But by that time Technion's students had already gone through their second strike of the year, and had had their fill.

The second strike had broken out in April, 1959. The immediate issue

Mrs. Eleanor Roosevelt on her visit to Technion City, with General and Mrs. Yaakov Dori. 1959.

was academic standards. The students took exception to new procedures and regulations which, they claimed, constituted a burden. Their specific grievances were directed against the following changes: 1. Greater emphasis on the basic sciences of mathematics and physics; 2. Introduction of a system of more frequent quizzes during the year, so that the final grade would not be dependent exclusively on the final examination; 3. Reduction of the number of contact hours, with greater emphasis on self-study; 4. Elimination of the "second chance" for students whose final grade was so low as to make it obvious they could not satisfactorily continue; 5. Raising of the passing grade.

The picture of the hard-working student, who was being pushed to the wall by extraordinary demands which he could not meet, was calculated to win public support, but the Student Association strategy committee chose to link their campaign with a vigorous defense of two students who had been apprehended in flagrant cheating during examinations. The students pointed to suspension of the two as indicative of the "Prussian rule" at the Technion and demanded that they be restored to good standing immediately. Public and press turned against the students. Industrial leaders declared they would think twice before hiring Technion graduates, who might have obtained their degrees by such methods. After almost a month away from school, the chastened students flocked back to their classes, seeking to make up for lost time.

It had been a difficult and bitter year for Dori, but he could report that the administration was determined to continue with the new academic system designed to raise standards; nevertheless, he was prepared to make such changes, adjustments or improvements as might be dictated in the light of experience.

The repetition with increasing frequency of student unrest over grievances, real or exaggerated, had lit a red light in the administrative offices of the Technion. Was it possible that almost exclusive emphasis on science and technology was resulting in the production of automatons? Gen. Dori thought out loud with respect to further implications of a narrow education: "We must be certain that the skilled men and women who graduate from our classrooms and laboratories are aware of the extent of the powers which they have been taught to exercise, and are responsive to the need for utilizing such powers in the interests of the nation as a whole and mankind in general .... Too little attention has been paid to the functions of the Technion as an institution for training citizen leaders. Not necessarily the ability to make an eloquent speech, or to write beautiful rhetoric or to gain a following by demagoguery, make for the wise leader. How much more important it is that a man shall have integrity; that he shall have been taught to revere truth — that truth which science and technology regard as fundamental."

Toward the end of 1958 announcement was made of the establishment of a new Department of General Studies, to offer courses in languages and the social sciences and other subjects outside the fields of technology and science. It was a beginning, but several more years were to elapse before the program received proper impetus.

The national need for secondary school teachers trained in the exact sciences and technological subjects led in 1959 to the setting up of special courses in teacher training, given within the framework of the Department of General Studies. Later this was to become an independent department.

If the student body had grown, the teaching staff had more than kept pace. The total number of teachers, including part-time, had risen from 132 in 1951–52, to 312 in 1954–55, to 427 in 1956–57, and to 503 in 1959–60. There were now three distinct elements in the faculty, each with a different background and outlook. The original teachers from Eastern Europe, who had been responsible for the launching of the Technion's academic program more than 30 years earlier, had for the most part passed on. The veterans now were those who had come from the German school of education, with its own attitudes and philosophy of pedagogy. The second group was of younger teachers whose education and experience had been gained in English-speaking countries since the Second World War. The third group were Israelis, graduates of the Technion, who were beginning to rise in the ranks of the academic staff as local products, most of them having acquired some postgraduate training abroad as well.

The differences of outlook were sometimes confusing to the students. The story was told of one class which was expected to stand when the distinguished Herr Professor entered the room, and to stand again when he left. They did so. The following hour they were lectured by a young American professor who sat informally on the corner of his desk, dangling his legs as he spoke, and when the bell rang raced the students to the exit, shouting: "Last one out is a donkey!"

Each teacher had his own system, and it was impossible to create a standard pattern. The hope was that in practice the students were receiving a distillation of the very best from all. Some teachers were singularly successful in presenting their scientific lectures in a manner easily understood. When Prof. Markus Reiner was descending from a bus in Haifa the doors closed suddenly on his coat and he was thrown to the ground. The bus gathered speed, dragging him along the road, and at a crucial moment the coat tore in half, leaving the remaining half still held securely by the closed door of the bus. Reiner noted that he owed his life to the fact that the coat had slit tails, which provided the first weak point for the tear. He saved the torn half as visual evidence for his lecture to students on stress concentration. He also asked the bus company to replace the coat.

Professorial participation in international conferences, honors received, interesting discoveries made were all reported in the press, and reflected to the honor and glory of the Technion. Early in 1956 the scientific world took note of the successful construction by Prof. Reiner of a centripetal pump, product of a research project sponsored by the U.S. Air Force. The theoretical possibilities of such a pump had been noted some years earlier but it had not been considered possible to build a working model. The American Air Force interest was based on the fact that the effect increases as the density of the air decreases, thus making it most useful in the upper layers of the atmosphere. It was felt that the phenomenon would have to be taken into consideration in all future design of guided missiles. Prof. Reiner continued with his work in this field, but no one picked it up after his death in 1976.

In 1959, Prof. Rahel Shalon, the first of Technion's graduates to rise to

Prof. Markus Reiner (1886–1976).

the rank of full Professor, and since 1952 head of the Technion's Building Research Station, was elected President of RILEM, the International Union of Testing and Research Laboratories for Materials and Structures. She thus became the first Israeli to assume presidency of an international scientific organization. The annual meeting of RILEM's permanent commission and an international symposium were held at the Technion in 1960, the first time any activity of that organization had been held outside of Europe.

Publicity of quite another kind was created in July, 1960, when the campus, and indeed the entire Israel scientific community, were rocked by news of the arrest of a senior member of the academic staff, who was ultimately convicted on spy charges.

Prof. Kurt Sitte had joined the staff of the Physics Department in October, 1954. A citizen of Czechoslovakia, he had been Professor of Theoretical Physics at the University of Prague. Because of his anti-Nazi record he was arrested by the Germans when they invaded Czechoslovakia and spent the entire period of the war in concentration camps. He was released by the Allies from Buchenwald only after the defeat of Germany. After the war he testified against the Nazis at the Nuremberg trials. Thereafter he continued his work in physics, first in England and then at Syracuse University and in Latin America. His joining the Technion staff was seen as something of a coup, strengthening the Physics Department.

In the years that followed, Sitte made himself felt in the academic and scientific community. Not long after coming to Haifa he was sent to Pisa as Technion's representative to a gathering of more than 400 nuclear scientists from all parts of the world. He was invited by the Argentine Atomic Energy Commission to organize and supervise the research activities of the high energy group of scientists there. Early in 1956 he was named head of the Division of Physics at the Technion.

His special interests were in cosmic rays. He was in charge of setting up an ionosphere recording station, which was part of a world network. In the framework of the International Geophysical Year he directed Technion's team of scientists studying the rays from outer space.

In the Spring of 1960, speaking at the second annual Conference on Aviation and Astronautics at Technion City, he told of the contributions of the Division to efforts to find some effective protection against cosmic radiation in manned space flights.

Sitte was accepted in the community not only because he was a scientist of international repute, but also for his personal charm. Respected by his colleagues and students, he was in demand socially because of his continental grace.

On June 9, 1960, the Board of Governors voted to appoint Prof. Sitte Vice

President for Research for the year beginning July 1. Five days later he was taken into custody.

The report of his arrest was a thunderbolt. It appears that he had for some time been under surveillance by the security authorities, and when they arrested him they had a full and detailed record of his alleged misdeeds. He was officially charged with communicating information calculated to be useful to an enemy, for a purpose prejudicial to the safety or interests of the State, and with delivering secret information without being authorized to do so, with intent to impair the security of the State. Evidence was produced of dozens of clandestine meetings with personnel from Eastern Europe.

Despite the wide press coverage, none of the details were ever divulged, because of the nature of the charges. The defense, conducted by Yaacov Salomon, based itself on several points. For one, while Sitte did not deny that he had had contacts with those said to be foreign agents, he maintained that the exchange of scientific information between scientists on the international scene was a vital part of scientific development and progress. Furthermore, the information he had passed on was not really secret, since much of it was available elsewhere. He denied there had been any harm to the security of the State, or any intent to cause such harm. The contacts had been continued, so it was said, because of Sitte's concern for his family behind the Iron Curtain, a threat which had apparently been made clear to him.

His colleagues flocked to his defense. Though the two-month trial in the Haifa District Court was held in camera, a representative of the Technion academic staff was permitted to be present, so that he could be assured this was no fabricated, star-chamber proceeding. On February 1, 1961, Sitte was found guilty.

An appeal to the Supreme Court against the conviction was of no avail, and the five-year prison sentence stood. One justice commented he felt that in the circumstances the sentence had been too lenient.

In April, 1963, after he had served for two years, the sentence was commuted. He was released and expelled from Israel. He went to Freiburg, and there continued his work on cosmic rays until his retirement.

In his report to the Board of Governors in June, 1961, Gen. Dori noted that ten years had elapsed since he had assumed the Presidency of the Institute, and he took the occasion to render a rather more detailed account of his stewardship. He singled out four vital developments, initiated by Prof. Goldstein, which in his opinion had had a fundamental influence on the academic life of the Technion: the establishment of the Faculty of Science, the founding of the School for Graduate Studies, the creation of the Research and Development Foundation, and the imple-

Battery of cylinders containing compressed air for operation of the wind tunnel in the Department of Aeronautical Engineering. 1957.

mentation of a broad program of reforms in the basic education of the engineer and scientist.

The accumulated deficit for operational activities had come to IL 2,623,000, but the overall report was one of growth and progress. Student enrollment had reached 1,900, and new buildings had been added to the campus.

Dori's health had continued to bother him. He had been absent for additional periods, and in September, 1961, the Council approved an extended leave of absence. Prof. David Ginsburg was elected Vice President for the academic year 1961–62 and was designated Acting President during Gen. Dori's absence.

*   *   *

Dynamic Expansion CHAPTER TEN

Ginsburg's Year

In preparing his report for presentation to the Board of Governors meeting in June, 1962, Acting President Prof. Ginsburg could look back upon a period beset by many problems, both academic and material, but marked also by serious grappling with those problems. As a member of the academic staff he found himself in a difficult position when he was presented with demands by the teachers for a 25% increase in salaries. Broad wage increases given to many labor blocs in the country which were able to exert pressures, had resulted in erosion of the comparative income of the academics, and dissatisfaction was rampant. Protest meetings were held and proclamations published. It was an issue on which the staff of all the other institutions were in agreement, and attempts were made by the leaders of the teachers' associations in the several universities to coordinate their activity. When the engineers of the country called a strike in February, 1962, the academic staffs nationwide suspended work for one day as an expression of their sympathy. A month later, in anticipation of an increased Government grant, the administration of each university reached agreement with its academic staff for an appreciable increase in salaries.

There was less to show in the effort to obtain the services of additional top-level personnel for a number of the disciplines at the Technion, and Prof. Ginsburg called for the allocation of more funds, particularly to enable inviting distinguished visiting professors, some of whom might be induced to become permanent staff members. Budgetary problems remained, and he called for increase of Government participation in the cost of operating the Technion to the extent of some 60%.

With regard to development and construction, Prof. Ginsburg minced no words. He referred to other institutions which found it possible to embark on ambitious construction projects by "financial manipulations which eventually led such institutions into impossible financial straits." But when the Technion sought a consolidation loan for its own modest debts, it did not meet with the cooperation expected. He charged that the Israel Treasury came to the help of those institutions which were in a worse position precisely because they followed a reckless financial policy. The net effect was that the Technion was penalized because of its economies and its conservative policies.

He proposed that the American Technion Society use the capital gains on its Swope Fund to finance construction, and the Fund as a whole as collateral for a consolidated loan for the Technion. Abraham Tulin expressed vehement objection on the grounds that, by terms of the bequest, the money could not be used for "bricks and mortar", and that the Society must use its discretion to enlarge, not diminish or endanger the Fund. The Acting President thereupon withdrew his recommendation.

Prof. Ginsburg gave considerable thought to the question of the nature of education, which had also disturbed Gen. Dori. He called once again for the introduction of humanistic studies and for the engaging of suitable teachers in the social sciences and humanities. He went so far as to recommend introduction of the humanities even at the expense of the scientific and engineering programs. The goal should be to educate for good citizenship, too, he said. At his instigation the Senate approved in principle the setting up of extra-curricular groups to raise the cultural level of the students, and soon thereafter he procured from the Senate a formal decision to inaugurate obligatory weekly courses in Humanistic Studies, to begin in the fall of 1962.

He addressed himself to other aspects of campus life as well. He called for establishment of a Faculty Club which could, among other functions, serve as the much-desired meeting ground for staff members from the science and engineering disciplines. In this connection, he opened his home for a monthly "punch and cookies" get-together for Technion faculty and staff. This open house, an innovation on the Israel academic scene, introduced a new fraternal spirit into campus life, but it was not continued after Ginsburg stepped down.

He advised early construction of a Student Union Building, recommended decentralization of the Technion library, and announced impending acquisition of a large computer. He pursued an idea advanced by Gen. Dori calling for a radical reorganization of the entire academic structure of the Technion, which then consisted of nine Faculties and Departments, into a School of Engineering, a School of Sciences and a School of Architecture. However, his exploration of the proposal with a considerable number of faculty members revealed that about 55% of the staff were against such a change, with some of the more distinguished members of the faculty in the minority. He advised Dori not to pursue the matter.

In conclusion he announced that upon the return of Gen. Dori to the Presidency he would go back to teaching and scientific research. He would continue to contribute to the Technion to the best of his ability in those fields, but not in the area of administration, since he knew this would monopolize all his time, to the exclusion of his academic interests.

Prof. Ginsburg's year was marked by dynamism and fresh initiatives, but out of loyalty to the President he was at all times careful not to venture into long-range planning.

In mid-year there was indeed danger that Gen. Dori might not return. In letters from Kenya he implied that unless he were given the assistance of a Vice President for Academic Affairs he would be unable to go on. An

Circle of hora dancers. A mural in one of the student dormitories, executed by Paul K. Hoenich.

unofficial delegation, comprising Professors Rahel Shalon, Nathan Rosen and William Resnick, flew to Nairobi to have a talk with him. He was persuaded not to leave, and not long after his return the Vice Presidential post which he desired was created.

The 1962 Board meeting took final action on a long-standing dispute with the Government on the matching it had promised Col. Elyachar if the ATS were to raise $10,000,000. The existence of the pledge was confirmed by the Treasury, and the Technion produced documentation showing that the Government was in arrears. There were differences of opinion as to the exact amount involved because of the shifting rates of exchange, but the Technion computed that by 1965 the Government would owe the Institute IL 15,000,000 to match the Society's $10,000,000. Ginsburg steeled himself for a confrontation in Jerusalem with Finance Minister Levi Eshkol, and armed himself with a Steinberg cartoon from *The New Yorker* showing an important executive harranguing a suppliant with a flood of verbiage which, when distilled, amounted to the single word "No!". Eshkol burst out laughing, and the tension was released. In the name of the Government he made a compromise offer of IL 12,000,000, to be paid over a five-year period. In the negotiations Nessyahu's experience proved invaluable. The Board voted to accept this offer "under protest" and with further proviso that the balance outstanding be linked to the Cost of Building Index so as to compensate the Technion for delays.

Yosef Almogi, right, hears an explanation on Technion's research in cosmic rays from Uzia Galil, Associate Member of the Physics Department. 1961.

Alexander Goldberg, Managing Director of Chemicals and Fertilizers, Ltd., was elected Chairman of the Board, succeeding Justice Landau.

At the graduation ceremonies in July, 1962, 556 degrees were awarded to graduates: 404 B.Sc., 58 Ing., 79 M.Sc., 15 D.Sc. Feature of the program was the conferment on David Ben Gurion of the degree of Doctor of Architecture, Honoris Causa, in recognition of his historic contribution as architect of the State of Israel.

In the Fall of 1962 Gen. Dori returned to his post after a year's absence that had included convalescence and a prolonged diplomatic mission for the Government in Kenya. He at once resumed his full duties. One of his first acts was to appoint Prof. Ginsburg Dean of Students. He accepted Ginsburg's suggestion that this post should be filled by a professor, and was grateful that Ginsburg volunteered for it. The precedent was adhered to in the years that followed.

The decision by the American Technion Society to advance a million dollar loan for new construction in anticipation of contributions, and the final agreement with the Government for its payment schedule of the IL 12,000,000 made possible intelligent planning of development and eased the President's administrative problems in his first few months back in office.

Growth

There were other encouraging reports of growth and progress in the months that followed. The Departments of Physics, Mathematics and Aeronautical Engineering, having grown appreciably, were accorded the status of full Faculties. Chemistry was separated from Chemical Engineering and likewise became a Faculty. The Division of Food Engineering and Biotechnology, first headed by Prof. Joseph Braverman, became a Department. Enrollment continued to mount. The 804 freshmen accepted in 1963 constituted the largest number admitted in any one year since the opening of the Technion, and total undergraduate enrollment reached 2500.

Technion entered the computer age. Gen. Dori announced that due to the efforts of Prof. Ginsburg an Eliott 803 Model had been procured for staff training, and a more sophisticated Model 503 was to follow. Financial help toward these acquisitions came from the Edmond James de Rothschild Memorial Group. Plans were already in hand to obtain an even larger computer. The Technion later adopted IBM equipment.

Tactile Controller, developed in the Faculty of Aeronautical Engineering, to regulate the flight of drones— unmanned aircraft—by remote control. 1963.

The affiliated Technical High School had also continued to grow. By 1962 it had an enrollment of 911, including girls, who had been accepted as students since 1959. Dr. Gershon Ahroni continued as Principal. In view of the Technion's own financial problems, its participation in the budget of the High School was small, and the deficit of the School increased. Efforts to obtain increased allocations from the Ministry of Education and from the Municipality of Haifa met with little success. At the same time steps were being taken to expand technician training on a high level, and a new framework, the Junior Technical College, was set up.

The Rothschild Memorial Group made a further gift to enable the Technical High School to provide tutoring for pupils from what was called the "Second Israel". These were the Jews variously known as Sephardim, or Oriental Jews, or those coming from Arabic-speaking countries. Due to their lack of cultural opportunities and their reduced economic status they were unable to overcome built-in handicaps to advancement. The problem was a major one in Israel, and led to occasional social clashes between the Ashkenazim, Jews of European descent, and the Sephardim, largely from the countries of North Africa and the Middle East. Provision of suitable educational opportunities for the young people of the latter communities was obviously one of the best ways of overcoming the handicaps, and the pilot plant program inaugurated at the Technical High School did indeed provide evidence of latent talents and abilities. Given the opportunity, these young people could reach satisfactory achievement levels.

New look in the workshops of the Technical High School. 1959.

An even broader program, which was eventually extended to all the universities of the country, was initiated at the Technion in 1962 at the suggestion of Prof. Haim Hanani, and with the full cooperation and assistance of Prime Minister Ben Gurion. Soldiers in the defense forces, coming from the disadvantaged communities, would have found it extremely difficult to pass Technion's rigid entrance examinations in mathematics and physics, especially in competition with candidates from economically and socially favored homes. The new scheme enabled selected Sephardi soldiers, possessed of certain basic qualifications, and in the final year of their military service, to register for a special pre-academic preparatory program at the Technion. Technically they were still in the army and subject to military discipline, but their training was academic and was provided on the Technion campus. The purpose was to raise their level in the basic sciences, to expand their intellectual horizons and to stimulate their capacity for study and thought, thus overcoming sociological and economic handicaps. The course took nine months, and those who completed it were able to sit for the entrance examinations of the Technion, with chance of success no less than their more favored colleagues.

The results more than vindicated the highest hopes for the experiment, and in the annual competitive entrance exams in the years that followed, graduates of these courses frequently performed better than the Ashkenazi graduates of the best high schools in Haifa, Tel Aviv or Jerusalem. The program was expanded and contributed to gradual increase in the percentage of Sephardi students enrolled at the universities of the country. It would never have been possible without the full cooperation of the military authorities. At the outset financial help was also required, and this was generously provided from the U.S. by Col. Elyachar, who had for years been agitating for greater opportunities for the Sephardi segment of the population, and by Mr. Marco Mitrani.

This kind of educational program, sponsored and made possible by the military establishment, is certainly not matched in any other country in the world.

Abba Eban and Col. J.R. Elyachar, at a Board meeting in Haifa. 1962.

Nature of the Student

The student body continued to be a heterogeneous group, comprising native-born Sabras as well as students who had come to the country as children from the four corners of the world after the establishment of the State. Almost all had first served in the military forces, and the common experiences there helped create the conditions in which young people with widely varied backgrounds could live and study together. The military connection was maintained even after the initial period of compulsory service, since almost all of the students were in the active reserves, and were called up for periodic service annually. There were also some students in active service, but assigned to complete their studies at the Technion. These were almost always in mufti, and it was impossible to tell who was who, or what military rank they held.

It was therefore not uncommon for a lecturer, who may have been a corporal or a sergeant, to hold forth before a class which included comissioned officers and possibly even his own commanding officer. The story has been told of one young man, holder of high military rank, who was frequently late for class. One day the professor lost patience with this student after the latter had wandered into class 20 minutes late and disturbed the lecture. Stopping in mid-sentence, the professor fixed his gaze on the tardy student and declared:

"Young man, I know that you are in the army. What do you think they would say to you there if you came late like this?"

Unblinking the student replied: "They would say to me, 'Good morning, Sir!' "

That the students had a sense of humor the teachers learned on more than one occasion, and practical jokes were not uncommon. The teachers were not to be outdone in quick-witted reaction. Many still recall with a chuckle the time a group of students found a runaway donkey in the Technion courtyard and brought it into the classroom, to the great glee of their colleagues. Anthony Peranio, then Senior Lecturer in Hydraulics, and himself possessed of a good sense of humor, was not daunted when he discovered the extra "student" in his class. From the podium he surveyed the "guest", then looked over the rest of the class, murmured "not such a big difference at that," and commenced his lecture. When a short time later the students became noisy, the lecturer turned to them and said:

"Gentlemen, at least one of you is listening, so please do not disturb him."

The character of the Technion student as well as the nature of the education he was receiving came in for wide public attention in mid-1963 as the result of a chance newspaper article. Prof. Hanani, then Vice President for Academic Affairs, had been asked how he thought Technion graduates would react in the following hypothetical case: Their employer, private or government, asks them to plan and construct a metal pipe suitable for the flow of blood from Eilat to Haifa. Would they look upon this as a challenging technical task, with interesting problems of corrosion, gravitation, viscosity of the liquid, etc., or would they stop to ask pointed questions about the source of the blood and the reasons for the pipeline? Hanani replied without hesitation that in his opinion 90% would ask no questions and would proceed diligently to solve the engineering problems involved. He made it clear he did not expect any employer to impose such a monstrous task on his engineers, but implied that this caricature did accurately reflect an existing state of mind with regard to human values. Hanani went further. He said that in his view most of the students at the Technion were political and humanistic illiterates. He had grave fears for a situation in which cobblers would be dealing with the atom.

A storm broke, and the pipeline for blood became a cause célèbre, discussed everywhere in Israel, in the press, in the cafés and among the students. Prof. Hanani was called on the carpet by Dori and asked to explain his utterance before a committee of the Senate. An official statement was issued making it clear that the professor had spoken in a personal capacity, and was not expressing any official Technion view. Further, the committee statement said, Hanani had no grounds whatever for voicing such severe criticism of Technion graduates. The committee disassociated itself from his views. It did not question the right of any teacher to voice his opinions, but since what Hanani had said was only a

Observation tower at Givat Hamoreh, where studies were carried out to determine economic feasibility of large-scale generation of power from prevailing winds. 1960.

313

personal thought, and was not based on any kind of proof or evidence drawn from thorough research, the result was the passing of judgment unjustifiably. "The opinion of the committee with respect to the ethical /human image of the students and graduates of the Technion differs completely from that of Hanani, although the committee, together with all of Technion's authorities, is convinced that much must still be done for the humanistic education of the students."

There was some talk that Hanani might resign; he denied that he was even considering it. *Haaretz*, in whose columns the original interview had appeared, charged that Technion was trying to silence a non-conformist, and this threatened freedom of thought.

The Technion Student Association also issued a formal statement commenting on the professor's "extreme" views. "If Prof. Hanani finds fault with the education of the student, he may seek the reasons in the system of studies at the Technion," the statement said. "The key to change the situation is in the hands of the Technion administration."

Three months later the controversy was still bubbling. Interviewed by the student paper, *Kol Hastudent*, Hanani declared the question had been a provocative one, and perhaps he should not have answered it the way he did. The point he wanted to make was that the liberal arts education of the young people was being neglected.

For several months more the subject continued to agitate public and press, as the discussion branched out into consideration of the nature of the general education system in Israel. If Hanani had intended to shock the public into awareness of the problem, he had certainly succeeded.

Undoubtedly the controversy provided added incentive to the plans for more widespread courses in the humanities at the Technion. Gen. Dori found occasion to comment that the Technion would be judged not only by the intellectual-professional standards of its graduates, "but also by their ethical image, their attitudes and behavior as scholars, and their understanding of the obligations imposed upon them in service to human society and the state of which they are citizens." He further noted that while the idea of humanities courses had been under discussion for several years, it had been Prof. Ginsburg who had "with bold decisiveness" implemented such a program.

Within the first few years of inauguration of the program, the courses offered included such subjects as Archeology, Literary Heritage of the Western World, History of Art, Personalities in the Field of Music, Theater, the Writer and his Environment, Bible, History of the Jewish People in its Homeland, Hebrew Drama, Medieval Renaissance of Classical Art, English Literature, History of Political Ideas, Hebrew Literature, Jewish Ideology, Sociology of Science, and others. By far the most popular

Student Choral Group, under direction of Dalia Atlas, sings for the Board of Governors meeting. 1967.

course was Music Appreciation, conducted by Frank Pelleg, and the Churchill Auditorium had to be used to accommodate the large throngs that his lecture/performances attracted.

The influence of the humanities was underlined beginning in 1963 in a series of annual lectures inspired and endowed by Dr. Joseph W. Wunsch, a former president of the American Technion Society. Dr. Wunsch sought to bridge "the gap which has deepened between man's mastery of science and technology on the one hand, and man's cultural and spiritual involvement in the arts and humanities." Dr. Wunsch maintained that the conflict was irrational and unnatural, and he hoped that his lecture series would help promote the close cooperation of the two cultures, so necessary to the fulfillment and survival of civilized man. The first lecture was given by the Nobel Prize Laureate, Prof. Isidor I. Rabi, and through the years that followed a distinguished list of scholars and scientists added their views to the growing library of literature on the subject.

Kind Words and Practical Help

"Extension of Remarks in the U.S. Congressional Record" is a fairly common matter, but when Rep. Emanuel Celler of New York had his

2300-word paean of praise for Technion published in the Record of the Proceedings of the 88th Congress on September 18, 1963, it read well in the thousands of reprints which the Technion Societies hastened to distribute to their friends. Among other warm statements, Rep. Celler declared: "Israel and Technion are indivisible. I do not say that one could not exist without the other, but I would say without fear of challenge that neither the State of Israel nor the Technion would be the same. ..." He went on to support American foreign aid to Israel and Technion as being in the best interests of the U.S. itself.

These were encouraging words to Technion's President, still grappling with his ever-mounting deficit. He had to apologize to his Board for not keeping a promise he had made in 1961: not to spend money he did not have. The growth of a living organism could not be artificially frozen, he explained, and the deficit had now mounted to IL5 million. There had been generous help from abroad—the million dollar loan of the American Society, followed by a two million dollar loan from American AID* funds in 1964—but the basic assistance had to come from the Government of Israel which, Dori said, must look upon the Technion not as an institution which it helps, but as one it maintains as a State tool for the development of the country.

On Dori's initiative, it had been decided to request the Government to cover 70% of the operating budget, and there was expectation that Ben Gurion would be understanding and helpful. But he had in the meantime resigned as Prime Minister, and the request was left suspended.

Valiant attempts were being made to procure financial support from domestic industry. Prof. Rahel Shalon, who had in 1954 become head of the Building Research Station, had approached Golda Meir, then Minister of Labor, whose responsibilities also included the Housing Division. She asked for help in obtaining assistance from the construction industry, which benefited from the research carried out at the Station. Mrs. Meir reproached her for seeking a larger portion of the national pie. To this Prof. Shalon replied that her intention was to make the national pie larger, so that all could have larger shares. She made little headway, and continued her efforts with the following Minister of Labor, Mordecai Namir. The effort was picked up and carried forward by Dr. Arthur Biram, Chairman of the Israel Technion Society. He made the point that in a period of boom in the construction industry, the contractors who were doing so handsomely ought to set aside a small percentage of their profits for the

---

* AID, the Agency for International Development, which since 1961 had been the U.S. Government agency administering foreign assistance.

benefit of the institution which provided their skilled manpower, as well as the testing and laboratory facilities for the industry. His suggestion that the Government levy a tax on the industry for this purpose was unfeasible, but another idea emerged: a voluntary tax, accepted by the industry out of a sense of obligation.

He took the proposal to David Tanne, Director of the Housing Division of the Ministry of Labor, who gave it his warm support. Finance Minister Levi Eshkol and Minister of Commerce and Industry Pinhas Sapir had initial doubts, but gave the idea their endorsement. Aharon Goldstein, head of the Contractors' Association, gave his full cooperation.

By January, 1959, the project took specific form. After further consultations with all concerned, it was suggested that the best measure of volume in construction was the quantity of cement used, and all the major parties involved in building agreed to a voluntary tax of 300 prutot per ton of cement. The four major groups were: Government Departments engaged in construction; the Jewish Agency; the Association of Contractors; and six large independent builders (such as Solel Boneh) which were outside the Association.

The collection was made by the central cement marketing agencies, and the funds were given to Technion for the expansion of services and operations of the Building Research Station, especially in all that pertained to the use of cement. The Cement Levy (Tav Melet), as it became known, was increased from time to time. The income to Technion was IL 61,000 in 1962, IL 560,000 in 1970 and IL 3,450,000 in 1978. As the sums grew, some of the companies questioned the legality of the procedure, and were told that the payment was voluntary. Solel Boneh had already withdrawn, and made an independent annual contribution to the Technion, but the smaller private contractors felt they were bearing an unfair share of the burden. In February, 1978, the system was replaced by an official Government Fund for the Cement Industry, based on a mandatory tax, per ton of cement purchased. The lion's share was earmarked for specifically ordered research projects at the Technion.

Whatever help the Building Research Station received from industry was more than justified by the work of the Station. Despite the local tradition of building with high ceilings, presumably for climatic reasons, research at the Station convinced contractors and customers alike that the ceiling height could be lowered appreciably without any ill effects, but with enormous savings. Another research project of the Station established that much more cement than necessary was being used in the concrete mixes, not only resulting in waste, but actually reducing the strength of the concrete. There were other similarly useful findings.

The help received from the builders served as example to another

Model study of foundation settlement effects on a concrete block wall, in the Building Research Station. 1958.

industry, and in 1962, on the initiative of Eliahu Sacharov, the country's plywood manufacturers accepted a levy in aid to the Technion, based on units of their production.

Aid to Africa and Asia

Preoccupation with meeting the industrial, technological and scientific needs of the burgeoning country did not prevent the Technion, precisely during this decade, from playing a unique role in coming to the aid of many of the newly emerging and developing states of Africa and Asia. Great pains were always taken never to refer to them as undeveloped or even as under-developed states.

The pioneer in the program was 21-year-old Beda Jonathan Amuli, of Tanganyika, a graduate of the Royal Technical College in Nairobi, Kenya. He enrolled in a full-time degree course in architecture at the Technion in

1960, paving the way for many who were to follow. After receiving his degree, Amuli would tell curious campus visitors that he was not sure how good an architect he would be when he returned home, but he was confident he would be the only Hebrew-speaking one in his country. He mastered the language, in addition to his knowledge of English and his native tongue, Swahili.

In 1961 Dori reported that the Israel Foreign Ministry had accepted Technion's proposal to offer studies in Agricultural Engineering for students from some of the new states of Africa and Asia. It was agreed that the program, to be conducted according to the Institute's full academic standard, would be in the English language and would lead to the B.Sc. degree. Nearly 30 students from 12 Afro-Asian countries as well as from Cyprus and the West Indies were enrolled, and on completion of the first intensive program in 1966, 24 degrees were granted. A banner across the platform at the graduation exercises that year read: "To Help Banish Hunger from the World".

New freshman classes were accepted each year. The students lived on the campus, and for the most part shared dormitory facilities with their Israeli fellow-students, who did everything possible to help ease the process of integration. Their problems were many because of the completely different way of life. Typical was the reaction of one. At a festive dinner given in their honor, each of the participants from abroad was introduced and was asked to say a few words. Each voiced similar sentiments along the lines of :"My name is so and so, and I hope to live up to your expectations," or "I hope I will enhance the prestige of your country. . . etc." When it was the turn of the last to introduce himself he stood up, looked earnestly around the table, and said: "Ladies and gentlemen, my name is Tundeh Osunsanya and I come from West Nigeria. I hope and pray that I'll get used to Israeli cooking!"

There were social problems as well. One Israeli girl told a friend that she really felt bad when she turned down an invitation for a date. "How can I explain to him that it's not because he's black," she said. "It's because he's not Jewish!"

Aryeh Freeman served as personal counsellor to these students, and when the program was dropped he continued as director of a summer science camp at Technion for high school students from abroad, and as adviser to foreign students generally.

The program was continued with great success for a number of years until 1967, when budgets were reduced in the International Cooperation Division (Mashav) of the Israel Foreign Ministry, and acceptance of new students was halted. The three remaining classes on the campus continued their studies until graduation, and the program was discontinued.

The first African student at the Technion, Beda Jonathan Amuli, who came from Tanganyika to study architecture; with his American room-mate, Daniel Litwin. 1960.

Mrs. Golda Meir addresses graduation ceremonies for students from developing countries. 1966.

Students from the developing countries of Africa, Asia and Latin America continued to come to the Technion, either on an individual basis, or in groups for special courses. Many enrolled in the Graduate School and earned higher degrees. Among these mention can be made of Zawde Berhane, of Ethiopia, who was awarded both Master and Doctor degrees in Civil Engineering, under the supervision of Prof. Shalon. When he and his family returned to Ethiopia in 1971 they took back with them a Sabra daughter, born in Haifa, and appropriately named Israela.

A unique event with international implications took place on the Technion campus in the summer of 1962 when Felix Houphouet-Boigny, President of the Republic of the Ivory Coast, was awarded Technion's Honorary Degree of Doctor of Science in recognition of his pioneering work in African independence and revival. The ceremony was attended also by Israel's Foreign Minister, Mrs. Golda Meir.

Technion's aid to the new countries took another form as well, and dozens of experts were sent abroad on special aid missions, some lasting a year or more. The honor roll of services rendered is long, but mention must be made of some of the missions as typical of the assistance provided. Yohanan Elon, Senior Lecturer in Town Planning, helped plan the industrial center in Port Harcourt and was a Consultant to the Ministry of Development of Northern Nigeria; Dr. David Armon, of the Faculty of Civil Engineering, headed the pre-engineering department of the University of Liberia; Ephraim Spira, Senior Lecturer in Civil Engineering, served as Dean of the newly established Imperial College of Technology in Ethiopia and was succeeded by Technion's Prof. Yehuda Peter. At the same college, Prof. Eri Jabotinsky was Dean of Science, and Dan Gonen taught Electrical Engineering. Gdalyah Wiseman, Senior Lecturer in Soil Mechanics, helped organize the teaching program in that subject at the Kumasi School of Technology in Ghana and was succeeded by Senior Lecturer Gabriel Kassif. Dr. Leslie Stoch later went to the same school to prepare the groundwork for the Department of Surveying. Senior Lecturer Yitzhak Alpan taught Soil Mechanics at University College in Nairobi; Yaacov Mevorach conducted teaching and laboratory work in Hydraulics at the University of Lagos, Nigeria; Dr. Mordecai M. Levy was adviser to the Federal Government of Nigeria on high-level manpower training.

The Prime Minister of East Nigeria, Dr. Michael I. Okpara, hears an explanation of Technion's program in hydraulic engineering from Prof. A. DeLeeuw. 1961.

Countries in other parts of the world also benefited from the technological and academic assistance of Technion personnel. Assoc. Prof. Jay Tabb, of Industrial and Management Engineering, was Industrial Relations Adviser to the Government of Ceylon; Prof. Jules Cahen drew up the first plan for industrialization of Singapore; Assoc. Prof. Haim Finkel went on a number of special missions to various countries in South America and elsewhere; Prof. Ratner headed a three-man mission to set up a program of technical and vocational education in Burma. Dov Nir, Senior Lecturer in Agricultural Engineering, helped set up the program at Chapingo Agricultural College in Mexico; Zeev H. Raphael aided Ceylon's authorities in establishing a science teaching equipment production unit and did similar work for the Government of Korea; Prof. Rahel Shalon and Itzhak Soroka advised the Government of Turkey in a program leading to establishment of a Building Materials Laboratory in Ankara; Benjamin Zur served as consultant on the use of atomic energy for agriculture in Brazil; David Carmeli was adviser on water planning in Turkey and Peru; Dr. Dan Zaslavsky helped plan irrigation, drainage and land development in Turkey; Assoc. Prof. Joseph Mouchly was adviser for Industrial Planning in Peru; Assoc. Prof. Ariel Taub was Professor of Metallurgy at the Middle East Technical University in Ankara; Prof. Aaron B. Horwitz was consultant for city and regional planning in various Latin American countries. Many of these missions were undertaken under the auspices of various U.N. agencies. All of these services were rendered by Technion staff members

Students from African countries, on a course in Food Preservation and Nutrition at the Technion. 1964.

Conferment of Honorary Doctorate on the President of the Ivory Coast, Felix Houphouet-Boigny. With the guest of honor, left to right, Prof. David Ginsburg, Foreign Minister Golda Meir and Prof. Shragga Irmay. 1962.

and do not include similar activities carried out by hundreds of Technion's alumni.

Occasionally the scientific assistance was involuntary, as in the case of Prof. Reiner who discovered that his textbook on Rheology had been pirated, translated into Russian, and published by the Soviet State Publishing Department, without the author's consent. Agreement was finally reached on payment of royalties, on condition that they be spent only within the Soviet Union, and Prof. Reiner made use of that income on his visits there.

The last two years of Gen. Dori's administration were characterized by dynamic expansion on both the academic and physical fronts, as well as by the accompanying financial problems. In the academic year 1964–65 a new freshman class of 880 was chosen from among 1700 applicants. M. Sc. degrees were awarded to 115 and D. Sc. degrees to 18, in addition to 422 B. Sc.'s.

New programs leading to undergraduate degrees were offered in Mineral Engineering, Teacher Training and Applied Physics. Agricultural Engineering was promoted to a full Faculty.

Large grants were received each year for sponsored research in specific

Elyachar Central Library.

fields from U.S. Government agencies such as the Department of Agriculture, the Bureau of Standards and the Air Force, among others.

A grant of over a million dollars from the U.N. Special Fund made possible establishment of a Center of Applied Research on the campus, to be operated with joint participation of the Government, the Technion and other agencies. A subsequent generous contribution from Anatol and Ganna Josepho, of California, provided the building to house the Center.

An ambitious program to propel the Technion into the atomic age was encouraged by Gen. Dori. Dr. Shimon Yiftach, of the faculty, drew up a memorandum entitled "Nuclear Science and Engineering Center at Technion City", in which he anticipated a need for nuclear energy in Israel in the years 1965–1975, and urged that the Technion begin at once with the technological and scientific preparations for the national requirements. The program received little support. An American energy expert, like Philip Sporn, considered the program too grandiose, and premature. In a continuing correspondence with Dori he urged that the Yiftach program be "rephased downward". He criticized what he felt was a tendency to neglect other fields of technology at the Technion and to "go chasing after

323

the beautiful rainbow idea of atomic energy in the hope, or what would be even more unfortunate, belief that Israel's future over the next 20 to 25 years is vitally tied in to its development of an extensive atomic program in its one technological institution, the Technion."

General Dori pressed for Government support, but when the authorities in Jerusalem turned to Sporn for his expert advice, the fate of the program was sealed.

By 1965 the Research and Development Foundation was operating a complex of laboratories, testing stations and other units dealing with Building Research, Building Materials, Chemical Processes, Farm Equipment, Geodesy, Electronics, Hydraulics, Soil Mechanics, Food, Chemical, Mechanical and Electrical Testing, Industrial Psychology, Rural Building Research, Fertilizer Development, Cartography, Scientific Translation and others.

New buildings occupied included the Elias and Bertha Fife Materials Testing Laboratory, the Rebecca Rose, the Beatrice Fischbach and the Louis and Rebecca Susman Student Dormitories, the Woolf Senior Sport Center, the Anne Borowitz Civil Engineering Building, three wings of the Chemistry Department bearing the names of Canada, Karl T. Compton and Samuel and Belle Bernstein, the Elyachar Central Library, the Senate Building and the Arturo Gruenebaum Metallurgy Building.

Master Plan Restudied

The extended construction program had again called into question the adequacy of the original campus Master Plan drawn up by Prof. Alexander Klein almost a dozen years earlier. In 1954 Prof. William Holford, of the Department of Town Planning of University College, London, had been invited to collaborate with Klein on aspects of the Tira plan which had been adapted to Technion City. Holford had found it difficult to get Klein to compromise, and found him "tenacious of certain planning principles which he had evolved over a distinguished career of more than 50 years." The choice was either to begin all over again with a new designer, or try to understand Klein's principles and try to persuade him to make changes where practical requirements so necessitated. The latter course had been chosen, but within a few years the basic principles were challenged by physical realities.

The Klein Plan provided for a ring road circling much of the campus, well within the perimeter, and serving as the major highway for heavy traffic. The various Faculty buildings were to be located on small side roads, most of them dead end, leading off the ring road, thus reducing to a minimum traffic movement in and around structures in which teaching

Technion City. 1964.

and research were being conducted. The magnificent campus, with its breathtaking view, also contained built-in difficulties because of its topography. The difference in elevation from the upper and southern end of the campus, down to the northern slope was so great that it discouraged easy pedestrian movement in the two directions. Adherence to the Klein Plan led to a dispersion of the new buildings over a large area, but liberties were taken with the plan to meet the specific needs of various Faculties.

Architect David Pinshow was invited to study and recommend revisions in the Master Plan and submitted his outline to the Board in 1960. There was much interest in his ideas, some enthusiasm, but considerable reserve. It was in effect sent back for further study.

A further revised plan was submitted by Architects Pinshow and Dov Carmi, but a Technion committee in 1964 concluded the proposals were impractical. The task was then imposed on Architect Shlomo Gilead, himself a Technion graduate. His proposal retained the basic principles of the Klein Plan but emphasized certain features: concentration and consolidation of built-up areas, preservation of the natural woodland, improvement of parking facilities, emphasis on a common architectural denominator, and creation of a campus center with shops, post office, bank, etc. The Gilead Plan was adopted and subjected to minor corrections thereafter, as necessary, though not all its features were immediately implemented.

An unusual though temporary change in the function of Technion facilities took place in 1964 at the old campus. The founding fathers would have blinked in disbelief. The main courtyard, with Baerwald's impressive building as a backdrop, had been transformed by a Hollywood staff into a bustling Middle Eastern marketplace, complete with exotic shops, jostling crowds and occasional animals. For a few days Technion was the locale for shooting scenes for the film, "Judith", starring Sophia Loren. Her meeting with Gen. Dori provided Technion's President with one of the lighter and more pleasant moments of those difficult years.

Surprisingly, there was little public criticism of this use of Technion's

Fischbach Electrical Engineering Building.

facilities, but another proposal raised political problems. In 1965 two of Israel's labor parties, Mapai and Mapam, proposed holding a ceremony marking the setting up of an electoral Alignment (*Maarach*) between them, and felt that the most appropriate site for this demonstration of labor unity would be the Technion building where 45 years earlier the Histadrut had come into existence. The Council debated the matter at length and then concluded, by a close vote, that this would be utilization of Technion's premises for political activity, a matter which they felt to be repugnant to Technion's detachment from political affairs.

The growth of the Technion made changes necessary in administrative structure, and the Vice Presidency was extended eventually to four posts. In 1961, Prof. Shragga Irmay was named Acting Vice President; in 1962, Prof. William Resnick was chosen Senior Vice President and Vice President in Charge of Research, and Prof. Haim Hanani, Vice President for Academic Affairs. In 1963, Professors Resnick and Hanani were re-elected, and Prof. Rahel Shalon was elected Vice President for Research. In 1964, Professors Hanani and Shalon were re-elected, the former with Senior rank, and Prof. David Ginsburg was elected Vice President for Development.

Other Universities

Questions were beginning to be asked within the Technion with regard to setting a ceiling on the growth of the Institute. There were some who felt that optimal size had already been reached, and any further increase in the number of engineers to be trained should be at other institutions in the country. There were also some who believed that the Technion should broaden its academic base and consider becoming a full-scale university. The latter point of view was stimulated with the reports that the Municipality of Haifa was planning to develop its educational center, Bet Erdstein, into an institution of higher learning. The Hebrew University of Jerusalem had been invited to extend its academic patronage, and steps had already been taken to establish what was known in the first stages as a University Institute to teach arts and the humanities and other subjects not within the scope of the Technion. Dori's first reactions were positive. He welcomed the project and said he would be delighted to cooperate with the new institution as proposed. Others at the 1963 Board meeting were not as pleased. Some expressed doubts as to the wisdom of having a second institution of higher learning in Haifa, and suggested instead that the Technion should expand and embrace the broader program of studies, thus becoming a comprehensive university.

A year later Dori reported that there had been little progress on the plans for a School of Education and Social Sciences in Haifa, but if the intention was to establish a second university he now had reservations. The matter smoldered.

Technion's 1964 Board meeting considered the implications, bearing in mind also that it meant creation of still another institution which would seek Government allocations. There was resentment that plans for the opening of another university in Haifa were apparently being made without any consultation whatever with the Technion.

In the meantime formal approach was made to the Technion by Beersheba University, asking for cooperation in opening a program of undergraduate studies in engineering for residents of the southern part of the country. The request was received sympathetically. The Technion decided to assist in setting up the curricula, but felt that decision with regard to development of a full-scale program in Beersheba should await the findings of the National Council for Higher Education, which was engaged in a study of the country's needs for technological education on all levels.

The subject of the Technion's relations with its Haifa neighbor, as well as with other institutions seeking to teach engineering and technology, was to be raised again in the years ahead.

One of the vexing problems of this period was the new building to house student laboratories in Mechanical Engineering. Technion's authorities had contracted with Prof. Alfred Neumann of the Faculty of Architecture to design the building and he, with the participation of a junior associate, Zvi Hecker, had submitted a set of novel plans. A press release described the building as "a radical and exciting break with conventional planning. The two-story building has no walls in the conventional sense. The exterior consists of a continuous row of triangular precast elements, which also compose the roof. On one wall of each triangle long, narrow jalousies have been inserted. The saw-tooth contours and the overhang protect the building from direct sun, wind and rain. Glare is eliminated since the face of the building is shaded by the design."

The design was undoubtedly novel, but the Technion's building committee, after study of the cost of useable space in the structure, came to the conclusion that the building would be too expensive, and asked the architect to present a different design, with a more economical layout. Funds for construction were available from the estates of Dan, Sadie and Joseph Danciger in the United States, but work was held up pending settlement of differences of opinion with the architect. The good offices were sought of the Israel Engineers' and Architects' Association in promoting an agreed settlement. A committee of experts set up by the Association ruled that Prof. Neumann's plans were mainly valid, but that

Prof. Harold C. Urey receives congratulations from Prof. William Resnick, on receipt of Honorary Doctorate from the Technion. 1962.

agreement ought to be reached on some changes. This only resulted in continuation of the deadlock.

The conflict between an architect who insisted on his artistic integrity and on fulfillment of a legal contract on the one side, and an institution which was paying the bill and wanted a structure which it could afford, on the other, was the talk of the community. The Danciger Building became something of a cause célèbre. The deadlock became even more rigid when the 1965 Board session expressed grave doubts as to the functionalism of the proposed building. "It is the sense of the Board that satisfactory studies of the plan be submitted to the Council, and the Council engage a new architect if it deems this necessary...."

The matter dragged on and into litigation. There is no doubt that the aggravation of the Danciger Building was not least among the many problems bothering Gen. Dori. Within the Faculty of Architecture and Town Planning there were already simmerings of an internal dispute regarding methods of teaching, but the full brunt of this controversy was to be felt by the next administration.

Dori Retires

The financial strains on the administration were enormous. "It is with considerable distaste that I must again serve as a peddler hawking his wares," declared Dori in reference to the protracted negotiations with the Government and the importuning for larger grants. Funds were indeed provided, but never enough. The academic staff returned to their demands for salary increases, and they were joined by administrative employees as well, all seeking automatic rises in grade similar to the automatic raises given to engineers in public employ and other civil servants. A delegation from the teaching staff appeared before the Technion Council to present its case, and the Council confirmed statements made earlier by Dori to the effect that university personnel could not be compared to other professions in the country, and should be given special treatment. In the meantime cash was running low, and the Institute again found it difficult from month to month to meet its payroll. Short warning strikes took place.

In March, 1965, Dori was given leave of absence for health reasons and Prof. Hanani became Acting President.

In May the Chairman of the Board, Alexander Goldberg, announced to the Council that he had received a letter from Dori requesting him to present to the forthcoming meeting of the Board his resignation from the Presidency. The post was one of national importance, and its occupant had to have health and vigor to fill it properly, Dori wrote. The deteriora-

Alexander Goldberg (1906–   ).

tion in the state of his health made it impossible for him to function properly, and he regretted that he could not be present for the Board sessions.

A nominations committee invited Prof. Ginsburg to stand for the Presidency, but he refused. From Geneva, where he was convalescing, Dori wrote again, expressing his regret and disappointment that Ginsburg had not been moved by consideration of Technion's destiny, but since the refusal was firm, Dori recommended to the Board that it choose Mr. Goldberg, whom he had seriously considered earlier. He had refrained from involving Goldberg because of the responsibilities which he carried as Managing Director of Israel's giant chemical industry complex. Because of his "personal character, his qualifications and his deep understanding of the needs of Israel, especially in the field of technology," Dori wrote, Goldberg was highly qualified to head the Technion in its necessary progress and development.

The nominations committee offered the appointment to Goldberg, but it still had to meet with Senate approval before being submitted to the Board. This was the first open election of a President, both Dori and Kaplansky having in effect been imposed from above. The Senate took its deliberations very seriously, and Goldberg was asked to be available in a nearby room in the event that any additional information would be required. After lengthy deliberations, Prof. Rahel Shalon emerged to inform the candidate that the Senate was prepared to approve him, but for a two-year period only. Goldberg balked. Since he would be giving up his position in the chemical industry, he felt that the term of office should be for a minimum of four years. The Senate finally agreed. The constitution of the Technion sets no limitation on the President's term except as may be determined by the Board of Governors from time to time, and the Goldberg case set a four-year precedent which has been followed ever since.

Dori's farewell message to the Board took the form of a 40-page addendum to the usual annual report, which he entitled "On Problems of Higher Education in Technology and the Sciences in the State of Israel."

Because of Israel's unique situation, he said, the level must be "of the very best possible in the world" and that could be achieved only if the highest of priorities were given to the needs. He sketched a broad review of the resources that had hitherto been available, and came to the conclusion that the major burden must inevitably fall on the national treasury. Guided by national interests only, the State must realize that there must be a careful process of selectivity, and an increase in the number of the scientific and technological elite in the country. He dealt with the problem of duplication and overlapping among existing universities. He found flaws in the situation whereby the universities of the country were placed under the supervision of the Ministry of Education and Culture since the institutions of higher learning dealt not only with teaching, but also to a great extent with matters that affected the national economy, defense and society as a whole.

He proposed the setting up of a Ministry of Higher Education and Science, thereby repeating a similar proposal that had been made 15 years earlier by Dr. Kaplansky.

In addition to the broad general principles, Dori singled out several specific elements which he felt merited special attention. He noted that the academic staff served as the central pillar for the fulfillment of Technion's mission. Academicians at all the universities, he said, should be treated as a special unit in the nation's economy.

With regard to the Technion's curriculum, he was of the opinion that serious consideration should be given to the addition of a fifth year, leading to the Bachelor degree, both because of the expanded requirements in the training of engineers, and because of the need for broadening the humanistic horizons of the students.

The Board conducted its proceedings in the knowledge that there was to be a change in the leadership of the Institute. The resolutions adopted dealt with the problems of the day, including academic staff salaries. It was decided to increase the intake of new students by 100. The approved operating budget totalled IL 28,000,000, leaving an expected deficit of IL 7,660,000 if the Government would not provide the anticipated extra grant.

Prime Minister Levi Eshkol was present. He heard the many speakers who addressed themselves to the Technion's financial problems, in the hopes that he would provide a sympathetic ear. He was not encouraging. "You should not expect that the Prime Minister would enter into competition with his own Minister of Finance," he said. As for deficits, "there is nothing to be frightened about. If not for deficits Israel would never have been built. Indeed, the country was built more on the basis of deficits than with money in the cash box. Our achievements burst all barriers, the way a growing child outgrows his clothing and the parents have no choice but to have new clothes made for him no matter what their poverty. It may be said that Israel seems to thrive on its deficits."

Minister of Labor Yigal Allon, who also addressed the Board, referred to the need, in his opinion, for a second Technion in the country, preferably in Beersheba.

The closing session took place on June 3, 1965. The Board formally accepted Dori's resignation and placed on record "its sense of deep obligation and gratitude to Gen. Dori for 15 years of unique and dedicated

Winston Churchill Auditorium. 1961.

service to the Technion. During these years his qualities of leadership and human understanding served to inspire all, both from Israel and other lands, who were privileged to be associated with him in Technion activity. When he assumed the Presidency of the Technion in 1951 he had already achieved fame as a military leader. He has added luster to his name by his outstanding success in raising the Technion to its present level."

Alexander Goldberg was duly elected President, but the beginnings were not auspicious. He was chosen in an atmosphere of labor unrest. Administrative and clerical employees who had not yet been paid their salaries for the previous month were called out on demonstrative strike by their labor association, and routine services for the final Board session, ranging from microphones and the moving of furniture, to the stencilling and distribution of proposed resolutions, were affected.

A Look Back

There can be little doubt that despite the chronic financial problems, the Dori administration marked the beginning of Technion's Golden Era. It was an era marked by explosive increase in student enrollment, in faculty appointments and in campus construction. The following table summarizes some of that growth:

	1951–52	1964–65
Number of Faculties/Departments	7	14
Full-time academic staff	71	303
Undergraduate student body	966	3,005
Graduate School enrollment	–	926
Diplomas awarded	172	422

The most vital developments were not merely quantitative. The entire academic framework was changed, and the struggling technical institute in Haifa began to assume the forms of a progressive, Western-oriented university. The fortunate presence of Prof. Sydney Goldstein at the outset made possible the fixing of academic procedures which stood the test of time during periods of broad development, providing as they did the elasticity for change as required. Dori's awareness of his own lack of experience in the academic world facilitated the introduction of academic reforms. Under the constitutional structure of the Technion the President of the Institute served both as Chairman of the Council, which is the administrative body, and as Chairman of the Senate, the academic forum—unlike other institutions where the respective chief officers are the President and a Rector. In his person the President of the Technion was able to bridge the relations between the Council and the Senate, and for the most part was able to avoid the kind of conflicts which occasionally take place between a President and a Rector elsewhere.

Missing the guidance and advice of Prof. Goldstein, Dori set up an unofficial "kitchen cabinet" with which he consulted on frequent occasions. Principal members of this group were Prof. Rahel Shalon, Prof. David Ginsburg, Prof. Nathan Rosen and Prof. William Resnick, all of them members of the Senate Steering Committee.

Although Dori was an engineer by training, his background had been almost exclusively military. His manner sometimes seemed brusque and rigid, a fact which the students exploited to explain their conflicts with him. Those who were able to penetrate his shell found that he could be warm, friendly, even gentle. He and his wife, Badana, were frequently cordial hosts to visiting Technion delegations from abroad. With the exception of those close to him, he was not at his best in human relations. He seldom praised his staff for their achievements, but in the absence of criticism, they knew he was pleased. He did know how to "sound off" in no uncertain terms against incompetence or neglect, and sounds of his wrath could at times be heard in the corridors. He disliked the concept of "publicity" and only with difficulty succumbed to the argument that fundraising had to be accompanied by appropriate promotional activities. He lived modestly, and if he occasionally brought sandwiches to the office it was as much a matter of his personal austerity as his health diet.

It is true that in administration he thought in military terms. He believed in discipline and in chain of command. He was accustomed to giving an order, and expected to have his instructions followed, though in academic matters he deferred to competence. The various executive personnel at the Institute were supposed to know their duties and to execute them. The President refrained from advising or interfering. Some felt that he took too little interest in what was going on in the various departments and offices, but he expected everyone to give of his best. He was impatient with excuses or alibis for failure. If an executive did not perform as he should, he was transferred. In academic affairs he was impatient with mediocrity. He wanted the Technion's graduates and the Technion's achievements to be superior, and when lack of funds did not always make this possible, he suffered personally.

His belief in delegation of authority led him to expand the team of Vice Presidents, each with his own clearly defined area of operations. The new executive position for management of the financial and administrative affairs of the Institute, which he labelled *Amarcal*, evolved into the position of Vice President for Administration and Finance. He was obviously a strong personality, and when he presented issues before the Council, seeking its guidance, members sometimes had the impression that he had already made his decision, and expected endorsement. This he did not always receive.

The ill health which had plagued him during the final years of his military service returned to bother him at the Technion. It was necessary for him to take occasional leaves for rest and recuperation including one lasting a year. The strains and pressures of the post were enormous, and sapped both his physical strength and his spirit. Retired from the Presidency, he continued as a member of the Council and took an active part in its deliberations for some years.

In 1967, only a few weeks after the Six-Day War, the Technion conferred on Dori its Honorary Degree of Doctor of Technical Sciences, "in recognition of his historic contribution toward the advancement of the Technion and the enhancing of its reputation through the expansion of existing academic units and the establishment of new ones, and for his impressive accomplishments in the construction of Technion City on Mount Carmel."

Conferment of Honorary Doctorate on Gen. Yaakov Dori. At right, Prof. Sydney Goldstein. 1967.

In accepting the honor he repeated what was in effect the essence of his legacy to the Institute, the demand that the Technion inscribe on its banner the single word, "Excellence".

He died in 1972. The broad four-lane road leading into the Technion City Campus from the main gate was named for Yaakov Dori.

* * *

Albert Einstein Institute of Physics.

CHAPTER ELEVEN *Consolidation*

The Teaching of Architecture

Alexander Goldberg assumed the leadership of a large, flourishing institute. His problems were many, not all of them financial. One major controversy, which had begun under his predecessor, was in the Faculty of Architecture and Town Planning.

For several years tensions had been growing in the Faculty as a result of differences between teachers following two schools of thought. The conflict had erupted into the open with publication of a letter in the press, signed by 13 architects, some of them persons of distinction in their field, charging that they could no longer continue teaching at the Technion under the system then in effect. The curriculum and method of teaching pursued by the Dean of the Faculty, Prof. Aviah Hashimshony, supported by a majority of his staff, emphasized the objective teaching of principles and planning, and students were expected to utilize the knowledge thus gained in their architectural projects. Subjects dealing with environment and sociology were also introduced, encouragement given to research, and the postgraduate studies broadened.

The other group, headed by Prof. Alfred Mansfeld, favored greater emphasis on the studio method, whereby students worked in close asssociation with an architect teacher from whom they could imbibe inspiration and ideas. The technical and sociological subjects were integrated with the design project, and the students thereby learned that an architectural project should be the result of team work with other specialists. The difference was one which had already arisen in university schools of architecture in other countries as well. Wittingly or not, the clash was further complicated by the natural competition between teachers who also maintained an outside architectural practice in addition to their academic program. Temperament of the architects and personality conflicts exacerbated the situation.

According to Technion's academic regulations the curriculum is determined by the Faculty Council, and here a majority of the teachers favored the program of the Dean, Prof. Hashimshony. The administration, undoubtedly influenced by the reputations of the dissenters, sought to have the program changed.

The Senate, seeking a compromise solution which would give satisfaction to both groups, proposed dividing the Faculty into two sections, each following its own teaching method. This compromise was opposed by most of the faculty members and by the students as well. The students pointed out that they would be compelled to choose between the two methods. Such choice would identify them with one or the other of the two groups, and this could have undesirable professional repercussions in later life.

Supreme Court Justice Moshe Landau, a member of the Board of Governors, was prevailed upon to offer his good offices in seeking a solution acceptable to both sides and also in the best interest of the Technion. He met with those involved, drew up a detailed plan to bridge the difficulties of the current year, and outlined moderate steps for eventual regularization of the affairs of the Faculty. The proposals were acceptable to both sides only in part, and the crisis deepened. In the meantime, the Faculty appeared to be operating outside the full control of the Technion authorities. As a result, the Senate in January, 1966, formally voted to oust Hashimshony, and Prof. Yosef Karni, of the Faculty of Civil Engineering, was appointed neutral Dean. He failed to receive the cooperation of many of the teachers.

Unfortunately the situation was even further complicated by a matter involving one young teacher in the Faculty who was exceedingly popular among the students, both personally and as a teacher. He was a candidate for promotion and tenure in April, 1966. The nominations committee and the Senate refused to approve promotion on the grounds that he did not possess the necessary academic qualities. This meant, in effect, that his services were being terminated. Under normal conditions the decision might not have received any special attention as there were similar cases every year. However, this young teacher had made himself one of the leading spokesmen of the Hashimshony group, and the cry went up that he was being "purged" because of his role in the controversy.

The students rallied to his defense charging that the administration was showing bias against the Hashimshony group, and defending the teacher's abilities as an instructor. In reply it was pointed out that popularity is not necessarily a criterion for determining academic standards, nor could the students intervene in academic appointments. Nevertheless, the students entered the fray with all the enthusiasm and zeal which young people can bring to a popular cause.

During May the situation in the Faculty deteriorated rapidly. Teachers went on strike demanding that the authority of the Hashimshony group be recognized. The students staged several demonstrations in favor of their popular instructor and agitated against the attempts of the authorities to impose a compromise solution.

In June the Board of Governors dealt with the problem. A neutral committee, headed by Prof. Sydney Goldstein, studied the situation in depth and came up with a new detailed plan which could lead to restoration of peace in the Faculty and at the same time reassert the authority of the Senate as supreme in academic affairs. Based on the Goldstein committee recommendations and subsequent action by the Senate, the following steps were taken:

1. The teachers agreed to call off their strike.

2. It was agreed that a new Dean would be selected in October, and in the meantime Prof. Hashimshony would act as Dean in matters affecting the current academic year, while Prof. Karni would act in matters involving preparation for the forthcoming academic year.

3. A neutral committee set up by the Senate and headed by Prof. Frank Herbstein, of the Chemistry Department, proposed a curriculum for the Faculty which called for a fusion of the two teaching systems. The fused curriculum was approved by both the Council of the Faculty and by the Senate.

4. The Senate agreed with observations of the Goldstein Committee that the regulations were unnecessarily severe toward teachers who were informed in April that they would not be given tenure, thus leaving them little time to arrange for their employment elsewhere in the ensuing academic year. The regulations were amended to give such teachers a one-year severance notice, and such regulations were made retroactive so as to apply to those teachers who had not been given tenure in April. These included the popular young architect.

The dispute made good copy for the press, and those involved did not hesitate to provide the journalists with their versions. The journalistic presentation made it appear as if the Technion were completely disorganized, and there were calls for the new President to resign.

Implementation of the compromise plan did not come easily. The Faculty of Architecture and Town Planning could not agree on a Dean, and Prof. Shklarsky, who was at the time Vice President for Academic Affairs, offered, as a neutral party, to take on the Deanship as well. This proposal was accepted by the Senate, and as of January 1, 1967, he headed the Faculty. What was originally conceived as a stopgap measure extended for six years, during which time Prof. Shklarsky diplomatically returned the Faculty to peaceful conditions, enabling it to continue its function as the sole source of domestically trained architects in Israel. The popular but controversial young architect made his mark, rose to senior rank, and within 15 years, as Prof. Avraham Wachman, was elected Dean of the Faculty. Historic justice, indeed.

An Unusual Building

Still another problem inherited from the previous administration also had an architectural connection — the Danciger Building. A committee of experts advised that the Technion, as an institution training architects, should encourage individual creative expression on the part of its faculty members and students. The committee did not agree that there were

Danciger Mechanical Engineering Building. 1967.

functional drawbacks to the proposed building, though it conceded that construction would be more expensive because of the unusual design. The Technion yielded, and agreed to the construction of one building which was to serve as the prototype for five buildings of similar design.

To save costs during the construction, Technion authorities instructed that certain changes be made in the skylight windows. Later it was discovered that someone had broken into the half-finished structure and had smashed the concrete on the changed section. The police were called and it was alleged that the "vandal" had been none other than Zvi Hecker, who had taken over since Prof. Alfred Neumann had left the country and settled in Canada. The matter went to court as a conflict between an architect seeking to protect the integrity of his artistic creation, and the party which commissioned the building, seeking to reduce costs and emphasize utilitarian aspects.

There were differences of opinion also regarding Hecker's insistence that specific panels be painted in colorful (some critics said "lurid") hues. When invited guests, including representatives of the donors, arrived for the official dedication of the structure in October, 1966, they found that Hecker had already daubed colorful splashes of paint on selected walls to show what he had in mind. The matter dragged through the courts and eventually compromises were reached. Hecker insisted on the painting, and under directives from the court the Technion yielded. The work was

"Daddy, what does it say?" Six-year-old Ilan Levin examines his father's diploma. 1967.

executed at the end of 1971, more than five years after dedication of the building. However, the option for four more buildings of the same design was allowed to lapse.

The Danciger Building became one of the tourist attractions of Technion City. Some of the occupants still maintain that they suffer from gloomy lighting, heat in summer, chill in winter, poor acoustics, and inconvenience because of the corners and angles created by the serrated walls. Everyone agrees that architecturally the Danciger Building is indeed unusual.

Physical and Academic Growth

There was much more to Goldberg's first term than controversies. The momentum of physical development of the campus continued unabated. As facilities were made available at Technion City, space was vacated in the premises of the old campus on Hadar Hacarmel, and late in 1965 the Faculty of Architecture and Town Planning, which had been sharing the new Aeronautics Building with its unwilling host, moved back down to the Hadar and eventually took over almost the entire original Baerwald building.

New facilities completed and dedicated at Technion City during this period included the Shine Student Union Building, the Goldsmith Biology Building, the Brody and San Francisco Agricultural Engineering Buildings, the Taub Computer Center, the Ullmann Teaching Center, the Isidor Goldberg Electronics Center, the Kennedy Leigh and Finkelstein units of International House Student Dormitories and the Spertus Auditorium.

Statistics reflected the numerical increase at the Technion as well. Enrollment of undergraduate students for the academic year 1965–66 rose by 10% over the preceding year, to reach 3290, and continued to rise. The growth in the Graduate School was even greater, as the following table shows:

School Year	Students for B.Sc.	Candidates for M.Sc.	Candidates for D.Sc.	Total Graduate School
1965–66	3290	957	167	1124
1966–67	3644	1159	207	1366
1967–68	3723	1297	227	1524
1968–69	4000	1427	270	1697
1969–70	4256	1671	356	2027

There was also comparable increase in academic staff.

The growth was not left to chance. Obviously there were financial considerations both in the construction of physical facilities and in the enlargement of staff, but the development of the Technion was based on a Master Plan drawn up by Prof. David Ginsburg at the original request of Gen. Dori and submitted in draft form to the administration in October, 1965. Dynamics of natural growth in the country made changes necessary in the forecasts from year to year, but at least there was a reasoned master program to guide the planners.

Ginsburg's basic assumption was that the Development Program should provide for a student body of 4,800 undergraduates and 1,200 graduate students in the ensuing five years. As can be seen from the figures above, the actual totals fell short in undergraduate enrollment, but exceeded the forecast by far in the Graduate School.

Ginsburg recommended that priorities should be given to those Faculties and Departments where there was need for recruitment of additional qualified staff. A detailed analysis of construction needs and anticipated growth in each academic unit then followed.

Monseigneur Hakim, Bishop of Galilee, congratulates Horace W. Goldsmith on the dedication of the Goldsmith Biology Building. 1966.

The emphasis on the need for superior academic staff was echoed by Prof. Sydney Goldstein. Absolved from his Vice Presidential vows, as he put it, he felt free to come for four months as Visiting Professor. Addressing the Board in 1964, he described Technion's need for good teachers, admitting that in some cases the kind of people being sought simply did not exist. However, he volunteered to set up a special program in the United States to recruit and train talented, imaginative academic personnel, especially prepared for posts at the Technion. Considerable effort was put into the project for several years, but the results were on the whole disappointing. The Technion did benefit considerably from the presence of visiting academic personnel who lectured or conducted research for periods of anywhere from two weeks to a full year. The Oscar and Ethel Salenger Guest House on the campus provided comfortable, though limited quarters for their accommodation.

In 1965 Prof. Shalon was elected Senior Vice President (for Academic Affairs) and Prof. Ginsburg was elected a Vice President. Both resigned a year later because of differences of opinion with the President. In 1966 Prof. Pinhas Naor was elected Senior Vice President and Prof. Elisha Shklarsky Vice President for Academic Affairs. In 1969 Prof. Shklarsky, Senior Vice President; Prof. Haim Finkel, Vice President for Academic Affairs; in 1970, Prof. Joseph Hagin, Vice President for Research; in 1972, Prof. Jacob Bear, Vice President for Academic Affairs; Prof. David Hasson, Vice President for Research. Mr. Yosef Ami, first elected in 1965, continued to serve as Vice President for Administration and Finance until 1973.

The first few years of the Goldberg administration were later referred to

Shine Student Union Building.

as years of consolidation. The President recognized the growing trend in inter-disciplinary activity and sought Senate approval for academic units which crossed the traditional Faculty lines. His point of view was not always accepted immediately, but time proved him right. Academic development included establishment of the Departments of Biomedical Engineering (1968), Computer Science, Applied Mathematics, and the Solid State Institute (1969). Advanced programs were inaugurated in materials engineering. An undergraduate program in economics was introduced in 1968. A Road Safety Center was set up in cooperation with the Ministry of Transport. A step which had an important positive effect on Technion's scientific program was the installation in 1968 of an IBM 360/50 computer, at a rental cost of a quarter of a million dollars a year. In retrospect it can be seen as a vital step in the growth of the Technion, but the decision was not easily reached. A six-member committee of Senate members considered the relative merits of CDC and IBM equipment and the arguments raged for three months. One of the participants commented later that as much emotion was aroused in the arguments over choice of a computer as might have been displayed in a decision to get married. It was a daring step at the time but a necessary one, and two years later the equipment was replaced by an even more sophisticated model.

Even the establishment of the Department of Computer Science did not come easily. Goldberg foresaw the coming explosion in computer usage and calculated that within the near future Israel's economy would require from 50 to 100 computer scientists a year. He made the point that the lightning calculations made possible by this instrument would free the engineer from arduous and time-consuming computations and provide him with precious time to think and concentrate on fundamentals. He finally obtained Senate approval for opening of a separate department, not merely an adjunct to one of the existing Faculties.

A further boost to the humanities education of the students was given in 1968 with the establishment of the Bern Dibner Library of General Studies.

The pre-university level of education at the Technion was reorganized, and the Technical High School and Technical Junior College were united under one administration. The School for Senior Technicians, founded in 1962 and operated jointly by the Technion and the Ministry of Labor, moved to the new campus at this time, thus further broadening the scope of the technical education provided at Technion City. The new School, and others like it elsewhere in Israel, conferred the title *Handesai*, variously translated as Senior Technician or Practical Engineer. This led to conflict with engineers (*mehandess* in Hebrew), who charged that the new title tended to blur the differences between two categories of personnel, one with academic, the other with technical training. The *Vaad Menahel* formally noted (in March, 1969) that the title *Handesai* did cause confusion regarding the status and qualifications of a fully trained and accredited engineer, but the term entered popular usage nevertheless.

Nuclear engineering was again put on the agenda. The Yiftach program, which had been shelved in 1963–64, was revised and further attempts were made to increase the Technion's involvement in this field, but they made little headway. Goldberg opposed the spending of a large amount of money (the sum of $8,000,000 had been mentioned) for the acquisition of equipment which in large part already existed at the Weizmann Institute, and for the construction of a new building to house the department. He pointed out that the Israel Atomic Energy Commission was already operating two research reactors, and duplication of facilities would be wasteful. General Dori disagreed, and asserted that if the same cautious and penurious attitude had prevailed in 1950–52, the Technion would never have established its Aeronautical Engineering Department. Goldberg denied there was any basis for comparison, since in aeronautics the Technion had the field completely to itself. Following discussions with the Government, with the Weizmann Institute and with Tel Aviv University, it was decided to limit Technion's involvement in the nuclear field. The site that had been set aside for a Nuclear Engineering Center, on the upper slopes of the campus, remained vacant for years. A high retaining wall supported what was in effect an empty lot. The subject of nuclear engineering was to come up again later.

One of the rapidly growing Faculties was Aeronautical Engineering. By 1968 it had over 300 students, was carrying out more than IL 1,000,000 worth of research a year and had opened its seventh wind tunnel in a set of facilities which ranged from subsonic to hypersonic.

The larger Faculties had in the meantime been undergoing organic reorganization, initiated by Prof. Goldstein as part of his reforms, and extended in the years that followed. By 1970, for example, Civil Engineering offered Options, after the first two years of study, in Structures, Hydraulic Engineering, Sanitary Engineering, Soil and Highway Engineering, Geodesy, Building Sciences and Construction Methods, Municipal Engineering, and Transportation and Traffic Engineering and had an attached Department of Mineral Engineering. Mechanical Engineering offered Options in Planning and Production, Power and Heat and had a Department of Materials Engineering. The Faculty of Electrical Engineering provided choices in Electric Power, Controls, and Electronics and Communications. In the decade that followed additional fields of specialization were added.

New Deficits

Even during a time of academic and physical consolidation the budgets rose, in part because of increase in numbers of students and staff, and in part because of rising costs in the country. The deficits continued to mount as well. In June, 1965, the Board adopted an unbalanced operating budget which envisaged a shortfall of over IL 7,000,000 and noted that the development or construction budget was left with a deficit of over IL 6,000,000 as of 30 September that year. The burden which had been too heavy for Dori to bear was not any easier for Goldberg.

Fund-raising overseas became more and more difficult, especially in the United States. There the leadership of the United Jewish Appeal, pressed by the Israel Government to increase its income, maintained that it was difficult to do so when scores of independent Israel institutions, large and small, were engaged in separate and sometimes competing campaigns for the same dollar. Again and again through the years the suggestion had been advanced, in various forms, that all independent campaigns be discontinued, and the various institutions receive allocation from a central fund-raising agency.

The institutions pointed out that most of what they raised was over and above what donors gave to the U.J.A., and actually constituted added dollars for Israel. U.J.A. pressures continued, and the operations of the various support societies were sharply curtailed. Permissible campaign periods were limited to the least desirable parts of the year, and leading Israel Government personalities were requested not to lend their moral support to the "competitive" campaigns, lest this reduce U.J.A. income. American Jewish communities were, on the whole, well disciplined, and any organization which sought to embark on a private campaign quickly felt the weight of local disapproval from those whose support they needed. Off and on, with varying degrees of severity, this situation continued for years.

In other countries the Keren Hayesod imposed similar limitations, always with the threat that independent fund-raising would result in reduction of Government subsidies.

Another threat to the fund-raising program in the U.S. was the fear that the various organizations in America raising money for Israel might lose their tax-exempt status. The Internal Revenue Service drew up a set of guidelines to ensure that agencies collecting for overseas causes adhered strictly to the law, and the American Technion Society did indeed meticulously observe the regulations. At one stage it was proposed that the Society should explore the possibility of retaining title to buildings constructed at Technion City with the aid of funds from America. A later suggestion was to register the Technion, or one or more of its departments, under the Board of Regents of New York State, and thereby qualify for financial help given to institutions chartered in the United States. All feelers in this direction were rejected, both because of the desire to retain the identity of the Technion as an Israel institution, and also because of the fear that in the event of overseas affiliation or control, the Government of Israel could conceivably wash its hands of all responsibility for budgets, and leave such responsibility to the American Society.

Fund-raising in Israel was also difficult. A proposal was made to institute a voluntary levy on industry for all use of steel in construction, bathtubs, metal accessories, etc., similar to the levy on cement. It was never implemented.

The Technion sought help from the Government, and Minister of Finance Pinhas Sapir agreed that he would cover one-third of the deficit if the Technion societies would raise funds for the balance. After further negotiations he raised his offer to 40%. The agreement was not consummated, and six months later, in January, 1966, the President called a special meeting of the *Vaad Menahel* to deal with the emergency situation. The system of hand-to-mouth operations, and "the running around to extinguish brush fires" could not continue, he said. The Government had

Inside of the atmosphere chamber utilized in the Metallurgy Laboratory for determining the shelf-life of canned products. 1966.

345

approved salary rises, but the money to pay them was not available. Unless the Government would come through with massive financial help, the Council felt, there would be no alternative but to hand the Technion over to the Government and let it take full responsibility for the institution. A strong threatening letter in this spirit was sent off to Prime Minister Eshkol, who refused to receive a Technion delegation, and referred it back to the Minister of Finance.

The situation was no better by the time of the 1966 Board meeting. The operating budget adopted there left a shortfall of IL4,000,000, which brought the accumulated deficit up to over IL17,000,000, and it was noted that the development deficit was approaching IL19,000,000.

Across-the-board salary and wage increases approved by the Government throughout the economy had again left academic staff members with the feeling that their salaries had proportionately been eroded, and a new set of demands was made. In the latter part of 1966, representatives of all the institutions of higher learning met with the Prime Minister and requested, among other things, approval for a rise in the academic salary scale. The response was blunt: If any institution paid extra benefits over and above the agreed scale, such payments would be deducted from the Government grants to the institutions.

The talk of deficits led public and press to question whether the universities were indeed being operated efficiently and economically. There was no sense of complacency at the Technion, and members of the administration repeatedly called on the Institute to make sure that its own house was in order. Accordingly, a committee headed by Dan Tolkowsky was set up to examine executive procedures and to assist the administration in finding ways and means of raising efficiency and reducing expenditure.

The Government approved increases in tuition fees by IL100, whereupon another of the periodic student strikes broke out.

It was in this atmosphere that Prof. Sydney Goldstein accepted the chairmanship of the Board of Governors in place of Justice Landau, and in sharp contrast to his previous attitude toward administrative matters, undertook to effectuate a drastic reorganization in Technion's financing operations. He drew up a many-pronged financial program, and after preparatory meetings held a confrontation session with Finance Minister Sapir at the Essex House in New York in January, 1967. He was flanked by Technion President Goldberg, Maurice Rosen, President of the American Technion Society, Col. Elyachar, Honorary President of the A.T.S., and D. Lou Harris, President of the Canadian Technion Society. He spelled out his program in detail, and gained Sapir's approval in principle. At the same time Goldstein warned that the Technion should not delude itself into expecting that the Government would or should do more than the figure finally agreed upon. "To expect more would be impractical, and to accept more would be dangerous if we are not willing that the Government should exercise complete control," he stated.

A special meeting of Board members was held in April. A subsequent session with Sapir in Jerusalem resulted in a meeting of minds, and the Board session in July, 1967, confirmed the details of what the Board Chairman referred to as the Essex House Plan, but which should more appropriately be known as the Goldstein Plan.

The basis for the program was the bold assertion of a new principle:

"We resolve now that *come what may* there shall, now or in the foreseeable future, be no plans or budgets passed by the Board of Governors that contain built-in deficits; that the Technion debts must be paid or consolidated and its financial affairs put in order. . . (without) delay.

"A. No new debts shall be assumed by the Technion, although old debts may be renewed.

"B. No contract for construction shall be entered into until there is money to pay the contractor.

"C. No building shall be built until provision has been made in the operational budget for the operating expenses of the activities for which the building was erected."

The Board then approved a series of sweeping resolutions designed to pull the Institute out of its difficulties and ensure smooth operations. These included:

1. The launching of a worldwide seven-year appeal for $50,000,000 for the Technion's research and teaching activities. Half of the amount would be set aside to constitute a permanent endowment fund, the income of which would be used for the same purposes. Implementation of this program was tied to a request for a free hand in overseas fund-raising, or at least a relaxation of the freeze which had been imposed by the United Jewish Appeal in the U.S. and by the Keren Hayesod in other countries. The Government of Israel was asked to help in obtaining such freedom from fund-raising constraint. As a matter of fact continuation of the freeze, and even intensification of the restraints, aborted this ambitious program at the outset.

2. The budgeted development costs for the 18-month period April 1, 1967, to September 30, 1968, totalling IL 9 million, would be covered 1/3 by the Government, and 2/3 from a U.S. Government AID Fund loan, interest and amortization of which would be borne by the American Technion Society.*

---

* The U.S. loan had been approved in 1964, and the A.T.S. had repaid $250,000 of principal plus

3. The Government's participation in the operating budget for the current year (1967–68) was set at 70%, a new peak.

4. The accumulated deficit, which had reached IL 30,600,000 by March 31, 1967, was to be eliminated by a series of Government payments and loans, chief among them a long-term loan in the amount of IL 13,700,000 to be guaranteed by the Government and repaid by the Technion as part of its budget in ensuing years. Details remained to be worked out, but this agreement gave additional meaning to the designation of this period as a time of consolidation.

The Technion had become big business indeed, and Goldberg found it necessary to overhaul and revise many of the financial practices which had been suitable for a small institution. In 1967 Nessyahu entered semi-retirement and became part-time financial adviser. Yosef Ami, newly appointed Vice President for Administration and Finance, was entrusted with modernization of the financial structure. Isaac Nissan was named Assistant to the Treasurer, and then Treasurer.

Gradually the Technion transferred its inventory controls and its payroll to computers. Scientific budgetary control was introduced.

In a sense the Technion's financial problems were not unique; other institutions of higher learning faced similar difficulties, some even more acute, and they too sought relief and assistance from the Government. The Ministry of Finance was the obvious source of money, but who was responsible for the general supervision of the country's universities? Respective Ministers of Education had to a greater or lesser degree avoided involvement in university finances, and regarded their own sphere of operation as ending with the secondary schools. The result was near chaos beyond that stage. There were no regulations governing higher education, and anyone who wished could open a "university", register students and offer degrees. The number of institutions did multiply, and almost all eventually came to the Government seeking allocations to help bail them out of their financial difficulties.

In January, 1965, the Government had appointed a 10-member committee, headed by Zeev Sharef, to look into the problem and to recommend what action should be taken to advance, coordinate, guide and plan higher education in the country.

The Sharef Committee submitted its report in October, 1965. It noted that with all due respect for the principles of academic freedom and maximum autonomy of the institutions of higher learning, some form of

interest. The remainder of the principal plus unpaid accrued interest as of that date were "waived and forgiven" under the provisions of Section III of the Foreign Assistance Appropriations Act enacted by Congress as of June 30, 1976.

Government supervision was required. The Committee was not in favor of a Ministry of Higher Education, and called instead for establishment of a Higher Education Authority to prepare long-range master plans, to establish principles for allocation of state funds to the universities, and to be responsible for distribution of these funds. The Authority would also set up regulations governing the opening of new universities or new faculties in existing institutions.

The recommendation was on the whole acceptable to the other institutions, but the Technion was in a minority, holding out for a full Ministry of Higher Learning and Science. At its meeting in June, 1966, the Board of Governors "firmly" pressed its recommendation that such a Ministry be established, and again in 1968 the Board declared its opposition to the Bill for an Authority on Higher Education in the form proposed.

The Six-Day War

Hovering ominously over all was the threatening security situation in the country. Egyptian troops were being massed in the Sinai, and Nasser openly proclaimed his plans to invade and drive the Israelis into the sea. When the Egyptian ruler and Jordan's King Hussein signed their pact for the destruction of Israel, the world expected the worst. Tension was so great that the annual meeting of the Board of Governors, scheduled for June 15, 1967, was postponed *sine die*. Graduation exercises due to be held on May 28 were also deferred.

The story of the Six-Day War and of Israel's lightning strike which completely reversed both the military and political situation is a matter of historic record. As was to be expected, the Technion's role in the victory was very considerable, though as Justice Landau pointed out at the Board meeting which assembled on July 9, 1967, security restrictions prevented the telling of the whole story.

In a special message to the Board, General Moshe Dayan, Minister of Defense, wrote: "... Every field of our existence bears the stamp of the Israel Institute of Technology: education, science, technology, agriculture, industry, national services, the physical construction of our cities and villages, etc. Moreover, the contributions made by the Technion in strengthening the security of our State are, if anything, even more important. Our scientists in Defense Research and Development and the best of our officers in the Air Force, the Tank Corps, the Navy, Communications, Artillery and the Engineering Corps are almost all of them Technion graduates. Our achievements when we faced vastly superior numbers and overcame them so decisively are due to the training and education of our soldiers and officers in which the Technion had so prominent a share. ..."

New type of gas mask for children invented by Eng. F. Zawistowski, of the Faculty of Aeronautical Engineering, during the Six-Day War. 1967.

In the days of mobilization immediately preceding the war some 80% of the Technion family was called into national service. All the physical and human resources of the Institute were placed on a war footing. Laboratories, research facilities and experts in such fields as aeronautics, electrical, mechanical and chemical engineering, all had their roles. A Technion Emergency Committee was set up with Prof. David Ginsburg as Coordinator.

Attempts were made to keep classes going wherever possible, and sometimes academic pursuits and active service were intertwined. The story was told of Captain A, pilot of a Mirage plane, who was a student of electronics at the Technion. He completed his class on Friday just in time to catch transportation to the airfield for his usual practice flight. Sunday morning he was back at his desk in the Fischbach Building, in preparation for his final examinations. What his classmates did not know was that his weekend had been spent in more than just training. The Syrian air force had attacked kibbutzim in the north and Israel planes rose to meet them. That was the day Israel shot down six Migs, and one went to the credit of Captain A.

There were stories and anecdotes galore. One which attracted wide attention, because it typified the spirit of Technion's students, was of a

Staff members prepare camouflage nets for Technion installations in time of stress. 1967.

letter dispatched to the Technion Library by Avivi Adari, a first-year student in Chemical Engineering. He wrote : "On June 6, 1967, during the battle in the Sinai , the tank in which I was fighting was destroyed as the result of a direct hit by an Egyptian shell. All my belongings in the tank were burned. This included, among other items, the book on Calculus by Bacon which had been loaned to me by the Library at the beginning of the year. I regret that I shall be unable to return it. I should be grateful if the Library would release me from the obligation of paying for the missing book."

One of the most frustrated members of the Technion staff was Mrs. Yael Rom. During the Sinai War 11 years earlier, she had been a pilot with the Israel Air Force and had been in the thick of the action. With the outbreak of hostilities she volunteered but was told that as a mother of three children she should find another outlet for her energies. Zealously she organized a group of Technion women to help prepare camouflage nets. Much later she was to serve as adviser to under-privileged students.

At memorial exercises the Technion mourned the death of three teachers, Dr. Moshe Ben Sira, Senior Lecturer in Physics, Dr. Raphael Mokady, Senior Lecturer in Agricultural Engineering, and Yuval Levi, Instructor in Mathematics, as well as ten students and many graduates.

Goldberg noted that the Institute, not for the first time in its history, had been thrust into a special role. "Technion is fundamentally a University for the training of personnel for peace-time pursuits," he said, "but the realities of our circumstances in the Middle East have imposed additional responsibilities upon us as well."

Immediately following the war, proposals were made to set up a Center for Agricultural Development to promote the development of the West Bank, including special courses for Arab students from that area, not too dissimilar from what had been done for Afro-Asian youth.

Technion volunteers were organized to help in the reconstruction of numerous settlements and villages that had been damaged in the fighting in the north. In view of reports that Hebrew University buildings in Jerusalem had been damaged by Jordanian shelling, Goldberg informed the President of the University, Eliahu Elath, that Technion was prepared to place engineering and technological resources at his disposal for any emergency repairs or reconstruction that might be required.

Although the Institute was not in need of "endorsements" it was encouraging to have outspoken support in high places, and considerable publicity was given to a statement issued by David Ben Gurion:

"The Technion—Israel Institute of Technology is one of the cornerstones of Israel's development. Its laboratories and workshops, its scientific and technical staff, are constantly at the disposal of industry,

Foreign students plant trees on campus in memory of members of the Technion family who fell in the Six-Day War. 1967.

Government and the Defense Forces of the nation. Even more important to the State is the steady supply of highly trained engineers and men and women of science who are educated at the Technion.

"The measure of a nation's security may be gauged in terms of its ability to keep abreast of the skills of technology. It is for that reason that we value so highly the Technion, where the youth of Israel obtain their education in engineering and the fields of applied science.

"We stand too at the threshold of great new developments in the fields of atomic energy for peaceful purposes, and we look to Technion to make available the steady supply of trained manpower which is vital to supplement all our other efforts."

Without Discrimination

Almost throughout its entire history the Technion has been free of internal disputes based on political or religious issues. In latter years the Student Association formally barred political activity on campus, and candidates for various positions in the Association campaigned on the basis of personal popularity, or campus issues, rather than under the flags of leading national political parties as at some other institutions in the country. Among the faculty, too, there were activists on both the extreme right and left of Israel's political spectrum, but they ignored these differences when engaged as colleagues in their mutual interests in science or technology.

The question of discrimination at the Technion had been raised occasionally. The Constitution of the Institute is quite clear on the subject. Article 5 states: "The Institute shall be open to all without regard to race, religion, nationality or sex, and shall not discriminate in its activities in any way against any person on grounds of race, religion, nationality or sex."

This policy had been rigidly adhered to throughout Technion's history, though the authorities have sometimes been confronted by problems. If scholarship funds are earmarked exclusively for award to students of Sephardi birth or background, does this constitute discrimination against Ashkenazi students? The considered conclusion has always been that special assistance provided for under-privileged or culturally deprived students in an attempt to help them raise their level does not constitute discrimination nor a selective process against anyone.

The possibility of religious discrimination has sometimes been raised. In 1961 a public agency in Israel offered the Technion a scholarship fund for awards to religious students. In its reply the Technion stated it could not accept such a fund since it struck at a basic principle of Technion's

Ohel Aharon Synagogue, Technion City. 1969.

policy. The reply continued: "Furthermore, the Technion considers it unwise, educationally, to do anything which will segregate religious and non-reiigious students from each other in the study of science and engineering. . . . Were some agency to offer the Technion a large sum of money for the award of scholarships to non-religious students only, we should in similar manner find such an offer contrary to the principles of our Institute. . . ."

In April, 1969, an attractive synagogue, Ohel Aharon, was dedicated on the campus, gift of Mr. and Mrs. Ludwig Jesselson and family of New York. Successive rabbis at Ohel Aharon have been: Rabbi Michael Keller (1969–71), Rabbi H. A. Miller (1971–75), Rabbi Aharon Shear-Yashuv (1975–80), Rabbi Eliahu Zini (1980–   ).

The number of Arab students at the Technion has always been small. The level of high school education in the Arab sector is not uniform, and many Arab candidates for admission find it difficult to compete in the entrance examinations. Security considerations, which limit employment

Minister of Finance, Pinhas Sapir, at dedication of Ohel Aharon, the campus synagogue. 1969.

of Arabs in certain of the large national defense industries which are science-based, also discourage some applicants from studying in a field where their opportunities may be limited. Since most Arab students come from outside Haifa, the majority have been allocated places in the campus dormitories.

Relations With a Neighbor

On other fronts inter-mural relationships were more complicated. By 1965 the Technion awoke to the fact that the University Institute of Haifa had become a fait accompli. Since the Municipality of Haifa had gone ahead on its own, there was no choice but to cooperate with the new school with a view to preventing unnecessary duplication. The 1965 Board meeting went further. It recorded the view that "ultimately the Technion and the Haifa University Institute ought to be integrated into one single institute of higher learning."

A year later the Board returned to the subject again. The Technion was obviously unhappy at the development, but as one speaker noted, a "point of no return" had been reached. It was resolved to seek the greatest possible measure of cooperation between the two institutions in order to prevent duplication of courses and facilities, "with a view to eventual unification."

At a time when the Technion was facing enormous financial difficulties, the fact that the Government was making large sums available for construction of the new institution on the upper slopes of the Carmel, almost adjacent to Technion City, rankled. The principle of unification of the two schools was repeated again and again and the Prime Minister was reminded of this goal, but little real progress was made.

Following protracted negotiations and meetings with Haifa Mayor Abba Khoushy, a joint announcement was issued by the two institutions, setting forth proposals for cooperation. Plans were projected for joint use of the area lying between the two campuses, with a view to creating one joint university campus in the future. On the academic level proposals were reviewed for Technion students to pursue studies in the humanities at the University Institute, and for the other students to study mathematics at the Technion. Cooperation was also proposed in teacher training as well as studies in economics and biology.

At the July, 1967, Board meeting the Technion noted with satisfaction the development of closer ties, and again the ultimate plan to have one campus was recorded by Prof. Pinhas Naor, Senior Vice President. Resolutions on the need for effective cooperation between the two institutions were passed almost every year, but the call for unification of the two was heard less often. Serious plans were discussed for joint establishment and operation of an Institute of Biology.

Identification of the Technion as a "university" had a magic appeal, and Prof. Naor formally proposed that the name be changed to "Technion — Technological University of Israel." More than a year later the name committee reported, and the *Vaad Menahel* decided against the change. Again in 1972 it was proposed that the name be changed to Israel Technological University, but this too was rejected.

Goldberg brought the subject of Haifa University again to the fore in his annual report in 1969. He asked the Board to give consideration to the merging of the Haifa University College and the Technion over a three-year period, thus converting the Institute into a general university.

The Board followed his lead and called not only for cooperation but also for "ultimate merger".

The time for true merger had passed, however. During its early years the municipal institution had its own ambitions, and despite lip service to amalgamation, went ahead with its own independent development. Later, when the financial burden of the University began to weigh heavily on the municipal budget and additional assistance was sought, the Technion had already begun to have second thoughts. The topic was a popular one on campus and in academic circles. The arguments against the merger, as seen from within the Technion, primarily by the academic personnel, were:

1. As the Israel Institute of Technology, the Technion was unique, one of its kind in the country. A merger would transform it into a Haifa University, third in the country after the larger institutions in Jerusalem and Tel Aviv.

2. The merged institution would not have the compactness and efficiency of the modest-sized Technion, and economical administration would be difficult.

3. During its long years of existence the Technion had established a worldwide reputation for its high standards of excellence. A merger with the Haifa Institute, in its then early stages of academic development, could result only in a watering down of the standards.

Early in 1970 Minister of Education Yigal Allon appointed a committee headed by Justice Haim Cohn to look into the matter of unification. Technion representatives testified before the committee, but insisted that Government intervention in Technion's internal affairs was unacceptable.

When Haifa Mayor Moshe Flieman told Technion's Board in June, 1970, that the Municipal Council had authorized negotiations leading to academic and administrative unification, he was already too late. The Senate took up the matter officially, and adopted a resolution which sounded contradictory but did indeed give clear expression to the views held by the majority: "The Senate is of the opinion that the merger of the Haifa University Institute with the Technion will be of advantage to both institutes when opportune conditions are created." The concluding operative clause effectively closed the door. In the subsequent discussion by the Board, Justice Landau likened a merger to harnessing a race horse and a work horse together. The Board adopted the Senate's qualifying clause, but also urged that steps be taken to accelerate achievement of the "opportune conditions". However, serious talk of merger disappeared from the agenda.

A further example of how new educational institutions proliferated was to be found in the discussions regarding textile engineering. In 1968 the Ministry of Commerce and Industry explored the possibility of setting up a program in that field at the Technion to meet the needs for technologically trained personnel in the textile industry. Costs of the program were to be covered one half by the Government, and one half by a levy on the textile industry. The following year a committee of international experts studied the national needs in the field and recommended that a school for the purpose be set up within an existing academic framework. The matter dragged on, but despite Technion's interest, another, separate institution was ultimately established in Tel Aviv.

Medical School

In the field of medicine the developments were different. Prof. David Erlik, for many years a staff member of the Hadassah Hospital in Jerusalem, and since 1948 head of the Surgery Department at the Rambam Government Hospital in Haifa, conceived the idea of establishing a Medical School in that city. The Rambam Hospital had since 1963 been associated with the Medical School in Jerusalem, and provided instruction for fifth-year students. Rambam physicians who instructed the students had academic standing and appointment from Jerusalem. However, when Prof. Erlik proposed an expansion of the program in Haifa the Jerusalem authorities did not approve. At a time when funds were difficult to obtain and existing institutions could not meet their operating budgets, the idea of a separate Medical School to serve the northern part of the country seemed only a dream. Erlik refused to be daunted. In 1968 he asked the Mayor of Haifa, Abba Khoushy, for a municipal grant to enable him to prepare curricular material, and also sought his political influence in procuring the necessary Government approval. The Haifa Municipal Council did approve the grant, but Abba Khoushy died in March, 1969.

The project was brought to the attention of the Technion, and at its 1969 meeting the Board of Governors cautiously resolved that if the proper authorities approved the establishment of a third medical school in the country, it should by all means be set up in Haifa. The Board expressed Technion's desire to cooperate enthusiastically in the realization of this goal, but nothing was said about any direct relationship with the new School.

The creation of a full-fledged Medical School out of nothing was little short of a miracle, but Prof. Erlik achieved that miracle. With the cooperation of the Municipality, he rented the facilities of an old monastery occupied by the Order of *Les Petites Soeures des Pauvres*, directly adjacent to the Rambam Hospital. Only two elderly nuns were in residence, and the Church was willing to turn the building into a source of income. Israel's national lottery, *Mifal Hapayis*, made available funds which were used for complete renovation of the structure. In the meantime, Prof. Erlik obtained temporary use of facilities in the Rambam Nurses' School.

One of Erlik's principal arguments in favor of the opening of another Medical School was the fact that the two existing institutions in Israel were able to absorb only a fraction of the qualified applicants who presented themselves. The disappointed and frustrated candidates for the most part went abroad. It was reported that in Bologna, Italy, alone there were some 500 Israelis studying medicine, with many more in other medical schools in Italy and elsewhere.

Since he had the facilities of a large teaching hospital already available, Prof. Erlik proposed to commence the academic program with a fourth-year medical course, where the emphasis was already on the clinical studies. Haifa's Mayor, Moshe Flieman, was of very great help. He persuaded the City Council to make an appropriation toward the first year's budget. The Government gave its consent, but on condition that no funds would be allocated from the national budget. Nothing daunted, Prof. Erlik raised some IL 600,000 from private individuals and funds in the Haifa area. With the cooperation of Yosef Teicher, he obtained contributions of $300,000 from the United States. Encouragement and valuable political support came from Yigal Allon.

Prof. Erlik issued an invitation to Israeli students who had already completed three years of medical studies abroad, to return home and begin their fourth year in the new Haifa Medical School. From the many applications received, 42 were approved and the school came into being in the fall of 1969. The intention was to accept a similar fourth-year class the following year, and thus build up the upper classes before first-year students were accepted, and this was indeed done.

Municipal expectations were great, and public references to the Medical School made it appear that it would be part of the Haifa University

Institute. However, Prof. Erlik and his colleagues quickly realized that a young school had no standing whatever in the critical academic world. The infant had to be associated with a university, preferably one which already had status and reputation.

Technion's President had been interested in the School from the outset. When a national committee had been set up to determine if there was need for another Medical School in Israel, Goldberg made a strong case for the project, noting also that the necessary academic base already existed in Haifa. In a second endorsement Goldberg called attention to special facilities which the Technion possessed, and which could be useful in the study of medicine. He had special reference to biomedical engineering, an inter-disciplinary program which had grown out of pioneering work done by young staff members in electrical engineering, hydraulics and mechanics. Research in the application of engineering to medical problems multiplied at the Technion, and the program reached the attention of Julius Silver in New York. He made a series of princely gifts through the American Technion Society, and in 1969 the Julius Silver Institute of Biomedical Engineering was established at the Technion. In the years that followed it expanded and was eventually housed in its own building at Technion City.

Goldberg's ambitions for the Medical School were discouraged by some of his colleagues, who were afraid of new, heavy financial burdens. In March, 1969, the Technion Council would only go so far as to say that if the Government did approve the establishment of a third Medical School, then it should by all means be in Haifa, but the Technion should not be expected to undertake any financial obligations. It expressed guarded approval of the project and had no objection if three representatives of the Senate were to serve on the over-all supervisory body together with representatives of other universities, but it repeated the warning that the Technion would undertake no obligation to provide money or facilities for this purpose.

Prof. Erlik had by then had opportunity to gauge the situation and had come to the conclusion that if the Medical School was seeking status it would be far better off with the world-famous Technion than with the Haifa institute which itself was still in its very early stages. Besides, the Technion already had Government authorization pursuant to the Law of the Council for Higher Education to confer the full range of academic degrees, an authorization which the Haifa institution did not yet have. Almost imperceptibly a momentum was begun, and the case for absorption of the Medical School by the Technion was presented and reiterated. By the time of the 1970 Board meeting, the Academic Development Committee recommended that the administration prepare a detailed plan for the merger of the Medical School with the Technion both administratively and academically.

A year later the first step was taken and the academic merger became a reality, with Prof. Erlik as first Dean of Technion's newest Faculty. The warning against financial involvement was repeated. A special Government grant had been promised for a period of five years, and the objection to financial responsibility of the Technion was limited to that period. In the meantime, the support from the Municipality dwindled away. In its early stages the school had borne the name of Abba Khoushy, but with the withdrawal of the city from participation in the budget the name was eventually dropped. By 1973 the Technion extended its full administrative as well as academic control over the School.

Advocates of the program were frequently asked how they could justify the association of a medical school with an institute of technology, a relationship which at the time seemed strange. Prof. Erlik's explanations quickly became a part of the standard exposition of the Technion's program. In recent years medicine had become more and more dependent on the tools provided by electronics and even by hydraulics and mechanical engineering among other technologies. Use of the computer had opened up new horizons in medical diagnostics. In view of these developments, it was only natural that many practicing physicians should seek information about the new fields, and about the equipment which was being created. The association of the new Medical School with the Technion made it possible for the young doctors-in-training to obtain a broad background in the new technological and scientific areas as part of their fundamental training in medicine. The rationale made sense, and the experience in the years that followed more than confirmed the original concept.

The Technion's relations with another educational institution in the southern part of the country took a different direction. The trend toward municipal or regional universities had increased, and in 1963 an Institute for Higher Education in the Negev was established in Beersheba. It assumed the name of University of the Negev, and later Ben Gurion University. Its initial program was in the humanities, social and natural sciences and engineering. The proposal to teach engineering in Beersheba naturally raised many questions at the Technion. Suggestions had already been voiced that perhaps there was need for a second Technion in Israel. General Dori, discussing the matter before Technion's Board in 1965, declared that he saw no need for a second Technion, at least in the immediate future, and certainly not in the Tel Aviv area. When the time did come, he added, the school should be in "the capital of the Negev".

The Board had before it a formal request from Beersheba asking for assistance in setting up the engineering curriculum but the Technion deferred decision until the Council for Higher Education gave its approval.

Technion academic personnel had from the beginning provided assistance in the teaching of postgraduate courses at Beersheba. Within the year the Technion extended more comprehensive help, including drawing up of curricula, general supervision of studies, selection of students, organization of teaching laboratories, and provision of qualified lecturers from Haifa for a complete freshman program in basic engineering. Similar help was provided in the following year, and in 1969 a formal agreement was signed between the two institutions providing for academic supervision by the Technion over the Beersheba engineering program.

In 1970 the Technion granted first degrees in mechanical and chemical engineering at Beersheba. The Hebrew University and the Weizmann Institute assumed similar responsibility for other departments. In addition, Technion's Prof. Haim Hanani served as Rector of the University from 1969 to 1973; Prof. Nathan Rosen was Dean of the Faculty of Engineering, 1969–71, and was succeeded in that post by Prof. Meir Hanin, 1972–73. In June, 1972, the University of the Negev was recognized by the Council for Higher Education as an independent institution. Technion, as one of the "foster parents", sent an appropriate message of congratulations.

Fund-Raising

No matter what the academic innovations or the fluctuating relationships with other institutions, of one thing the Technion was always sure—consistent and continuing financial problems. The brave plan drawn up by Sydney Goldstein in 1967 for reorganization of the Institute's finances and enthusiastically adopted by the Board, was implemented only in part. The Government did increase its participation in the budget, and the accumulated obligations were indeed consolidated conveniently. However, the outbreak of the Six-Day War and the subsequent financial problems of the Government put a halt to the proposed $50,000,0000 world campaign even before it got started, with the result that despite the firm wording of the 1967 resolutions, new deficits began to loom.

The pressures had eased with receipt of the AID loan, and Government participation in the budget had risen from 68.5% to 70% but operating costs had also risen sharply. In opening the 1969/70 academic year the Technion authorities expressed the hope that Government participation would go up to 80%, and a campaign was begun to acquaint Cabinet members with the Technion's role in the country and to seek their support. In the absence of encouraging news from Jerusalem, budgets were cut to such an extent that Deans and Heads of Departments threatened to resign.

There was good will and readiness to help on the part of Technion Societies in various countries, especially in the United States, but the severe restrictions imposed on fund-raising crippled Technion's ambitious plans to mobilize mass support. By 1968 it already appeared that there would be a considerable shortfall. The Government had offered $600,000 as compensation for Technion's suspension of major fund-raising, but this was $2,000,000 short of the amount needed to balance the budget. The Technion had pledged not to incur any new deficits but under these circumstances, the Board warned, it faced the following alternatives: 1. The Government should lift the ban on campaigning and permit Technion's friends to come to its help, or 2. The Government should release the Institute from its non-deficit pledge, or 3. The Technion would have to reduce its activities drastically, including the closing down of some Departments.

Prof. Ludwig Erhard, former German Chancellor, left, received by Alexander Goldberg and Prof. Rahel Shalon. 1967.

It was small consolation that the other institutions of higher learning were in the same position. Leaders of the U.J.A. in America continued to insist that campaigns for the universities diverted contributions which would otherwise have gone into the general funds, and the financial authorities in Jerusalem accepted this point of view. The U.J.A. leadership returned to an old suggestion. They proposed that all of the country's educational institutions merge their fund-raising operations, fold them within the broad U.J.A. campaign, and each receive a proportionate share of the income. The idea was not new. It was welcomed by the new, younger schools, which had not yet established any significant campaign operations overseas, and therefore preferred a fixed, guaranteed income from the U.J.A. The older, larger universities, like the Technion and the Hebrew University, were opposed. They pointed out that each institution had its own circle of friends, and each had its own unique appeal which produced liberal assistance over and above what donors gave to the general funds. This unique appeal would be lost if everything were to be solicited for one big "pot". The threat of unified fund-raising hung heavy over the universities for some years, but was never implemented.

In one geographic area the Technion had a special problem of its own. Israel's acceptance of restitution and reparations from Germany, and the establishment of diplomatic relations with that country gradually altered the public feelings toward the new Germany. German cars were a frequent sight on Israel's roads, German heavy equipment became an important element in Israel's land and sea transportation, and German consumer goods began to appear on the local market. Some of the institutions, with the Weizmann Institute in the forefront, were quick to establish their own relations with that country. Academic contacts quickly developed into relations which produced generous help for research and other projects in Israel. A long-standing policy in the Technion to have no dealings whatsoever with the country which had been responsible for the Holocaust was put to severe test. The moving spirit behind the ban was Supreme Court Justice Moshe Landau, who had been the presiding judge in the trial of Adolf Eichmann, and had in 1969 been returned as Chairman of the Board of Governors. He steadfastly opposed any Technion activity in Germany.

Others argued that the atmosphere had changed, and that some Israeli institutions were already profiting from such relations. They proposed that the Institute be permitted to accept research grants, so long as it was clear that the funds did not come from anyone with previous Nazi associations. At its meeting in June, 1970, the Board took a stand on the matter in a resolution euphemistically headed "Determination of Terms of Reference for Acceptance of Funds from Europe." The decision provided that the Technion "will have no relations with any organizations or individuals which have a Nazi history, irrespective of the present legal constitution of such organizations. The Technion will not receive any gifts which can be interpreted as being conscience money. . . ." On the other hand, the road was left open for acceptance of other funds, and the Vice President for Research was made responsible for carrying out the resolution on the basis of his judgment. In the years that followed research contracts were indeed received from Germany, academic relations were established with German universities and individual faculty members pursued contacts with colleagues in that country, but outright fund-raising or establishment of groups of Technion supporters came only many years later.

In December, 1968, the State Comptroller's Office submitted a report on inspections which it had carried out at the Technion in 1967 and 1968. As an institution receiving Government funds, the Technion was subject to such inspection. The 88-page report (63 pages in English) covered all major administrative matters at the Technion including conditions of employment, assistance to students, development planning, finances and accounts. The report was an objective statement of facts. In accordance with its usual procedure the Comptroller's Office neither praised nor commended anything. If things were going well and in proper order, that was only to be expected. The report did call attention to relatively minor departures from what it felt were proper procedures in such matters as additional emoluments to some employees for overtime, hiring procedures, and insufficient concern in the investment policy for protection of the real value of endowment funds. The Comptroller also commented adversely on the bidding procedures for construction of new buildings, at

Typical student dormitory building, along August Lane at Technion City.

the same time noting the Technion's explanation that it was sometimes found advantageous to give the contract to a bidder who may not have submitted the lowest bid, but offered credit arrangements at a time when the Institute was hard pressed for funds.

Despite press accounts which sought to sensationalize the report, the Technion was pleased with the findings. A committee was appointed to study the Comptroller's observations and to rectify deficiencies where these existed.

In a begrudging tribute to Technion's program of student aid, the report noted that "in actuality every outstanding student can study without having to worry about his livelihood. . . ." In its final words, the report commented that while teaching facilities at the Technion should continue to be expanded, it appeared that the growing demand for engineering studies could not be fully satisfied at the Haifa campus of the Technion alone. The inference that perhaps a second Technion should be set up in Israel had already been rejected by the Institute's Council, which had urged that instead the Government should provide the funds needed to expand the facilities of the existing school.

ings of the Committee for the Examination of the Needs for Expanding Engineering Education in 1969–1980, headed by Prof. Don Patinkin of the Hebrew University. Though the figures produced by the Patinkin Committee did not agree fully with those of the Horev Committee, the basic conclusion was the same, namely, that Israel faced a shortage of engineers in the years ahead. Among the major recommendations of the Patinkin Committee were: 1. Institution of afternoon and summer courses at the Technion. 2. Permission to the Technion to proceed with its building program, which would enable the admission of more students. 3. Development of a program of engineering studies in Beersheba and in the Tel Aviv area.

There was no lack of committees. When Yigal Allon became Minister of Education he willingly assumed responsibility for higher education in Israel, in contrast to his predecessors who had limited their concern to pre-university education. In 1970 Allon appointed a new committee to recommend to the Government criteria for budgetary allocations. Prof. Pinhas Naor was to head this committee, but he was tragically killed in 1970 in an airplane accident over Europe, and the chairmanship was taken over by Prof. Rahel Shalon. The Shalon Committee recommendations led in 1974 to the creation of the University Planning and Grants Committee within the Council for Higher Education.

During the years there had been many other reports and manpower surveys. Valuable studies had been made for the Ministry of Labor by Prof. Eli Ginzberg. An Economic Planning Authority existed in the Prime Minister's office, but as Prof. Patinkin pointed out on one occasion, "to set up a committee is one thing, and to pay attention to the proposals it makes is another."

*   *   *

Agricultural Engineering complex: San Francisco Wing, Samuel Fryer Research Tower and Samuel Brody Wing.

Taking Stock CHAPTER TWELVE

Innovations

Alexander Goldberg's four-year term as President expired in 1969. After some initial hesitation he agreed to stand for re-election and in June of that year was unanimously elected by the Board for a second term. He continued with the academic and administrative developments which he had inaugurated, among them several which were innovations at the Technion.

Through the years, the differences of opinion between the students and the administrative authorities, sometimes erupting into open confrontations, were based either on financial problems, primarily tuition fees, or on matters directly related to the academic program. Solutions for the financial problems were for the most part in the hands of the Government, and therefore beyond the direct control of the Technion. Academic questions, on the other hand, were purely domestic in nature, and it was felt that suitable forums should be established at which student grievances could be heard lest they develop into major issues. Accordingly, joint faculty-student committees were set up within each academic unit to serve as standing bodies for consideration of such matters as might be brought to their attention. It was significant that in many cases the committees met infrequently due to lack of agendas. Almost simultaneously a parallel committee was established on a campus-wide basis to deal with broader aspects of the student-faculty relationship. Based on the deliberations of this committee, the Senate set up its own permanent committee for examining and improving teaching methods.

Another innovation, common in the United States but radical for the Technion, affected the entire structure of requirements for degrees. The Technion's curriculum had gradually been broadened and deepened in accordance with the introduction of new disciplines and new approaches in the academic world generally. However, there had been few changes in the rigid, monolithic system which determined in advance the precise list of courses which each student was required to complete for degrees in the various Faculties/Departments. A committee headed by Prof. Asher Peres studied the matter for several years, and its proposals for inauguration of a flexible credit system were adopted by the Senate. The new system afforded opportunity for students to choose from designated elective courses, as well as to acquire additional credit points by completing courses during summer semesters. The new method also enabled the acceptance of students at the beginning of the second semester in the Spring, since completion of academic requirements for graduation was now related to the number of credits for which students qualified. The credit system went into effect in October, 1971, on a staggered year by year basis. There were inevitable teething pains, and the students did not

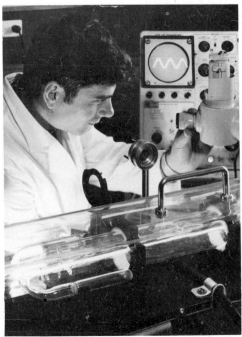

A graduate research worker in the Department of Chemistry with a helium neon laser, built by the Technion, to align a giant pulsed ruby laser used for photochemical research. 1970.

understand a system which sometimes left big gaps in their daily schedule. Eventually the difficulties were overcome. The administrative adjustment to the new program also provided for maximum use of existing laboratories and equipment for parallel courses during evening and summer hours, a system which was known as the David Kohn Plan in recognition of the staff member who had worked out the details.

The Faculty of Medicine, truly an innovation for an institute of technology, continued its careful growth. In 1971 it accepted its first class of freshman students in addition to the upper classmen already enrolled, and additional preclinical and clinical departments were opened. Clinical activity was carried out in close cooperation with the Rambam and Rothschild Hospitals in Haifa. The princely generosity of Baruch Rappaport and his family, from Geneva, was during the ensuing decade to transform the Faculty of Medicine into a major factor on the Israel medical scene.

Technion's library had grown slowly and steadily since it opened in 1924/25 with a collection of some 2,000 books. By 1952 it contained about 32,000 volumes, thanks to gifts from a large number of donors in many countries. Dr. Rivka Feiner, who had been director of the library since 1927, retired from that position in 1954 and was succeeded by Dr. Tunia Gladstein.

Because of space problems the library was for the most part compelled to limit its collection very narrowly to literature dealing with science, technology or those other areas of direct interest to the Technion. Specialized collections of interest were the Baerwald Collection, donated by the architect of Technion's first building, dealing largely with architecture of the Middle East; the Leon Moisseiff Collection, gift of the famous American bridge builder, containing many of his plans; the Rachel-Leah Weizmann Collection of bound periodicals on chemistry, presented by Dr. Chaim Weizmann; the Karl Herz library of books largely in the humanities and social sciences; and the Meyer Gold Library of American Know-How. Just prior to the outbreak of the Second World War, tuition fees deposited to Technion's account with the Keren Hayesod in Vienna by prospective students, were used for acquisition of a large number of books that were shipped safely to Haifa.

By 1970 the library had more than 80,000 books and about 75,000 bound volumes of periodicals. It also subscribed to more than 5,000 scientific, technical and professional journals from all parts of the world. Thousands of textbooks on its shelves were available for loan to students.

A turning point in the history of the library was its entry in 1965 into the new building bearing the name of Col. J. R. Elyachar. The growth of the campus, however, and the widespread location of the various academic units, led to a proliferation of Faculty and Departmental libraries, 18 in all.

Canada Building, in the Chemistry Department. 1971.

The relative advantage of a central collection, as against decentralized libraries, was a perennial subject for consideration.

Dr. Gladstein died in August, 1970, and for some years thereafter much of the administrative burden was borne by Aharon Kutten who had first joined the library staff in 1945.

Applied Mathematics, which had been such an exotic subject in the early days of Prof. Sydney Goldstein, was institutionalized at the Technion in 1962 and emerged as an independent Department in 1969. Undergraduate studies were launched in 1971. Four years later the staff and program of the Department were folded into the Departments of Mathematics and Computer Science.

By 1970 the Technion was comprised of 20 Faculties and Departments. The multiplication of academic units, increase in number of staff and addition of physical facilities to house these activities created new strains. The need for careful planning of future growth led to appointment of a new Development Committee headed by Prof. Elisha Shklarsky. The committee came to the conclusion that maintenance of the Technion's high academic standards required enlargement of academic staff, especially in certain units where demand for admission was high, or existing staff was insufficient to meet the demands made upon it. The feeling was that the science Faculties had forged ahead at the expense of the engineering Faculties. One estimate had it that if the recommendations were implemented, 450 new academic staff members would have to be added in the following five years. Along with this, the committee indicated the need to expand the administrative and technical services, and this was estimated to require about 400 new employees. More students and more staff members inevitably meant more space requirements. The Vice President for Administration and Finance estimated that the committee's proposals added up to 60,000 to 80,000 additional square meters of building. Neither the Shklarsky Committee nor the Senate, which put its stamp of approval on the major recommendations in 1971, dealt with the financial implications of the program.

The Board of Governors warmly approved the findings of the committee but decided that implementation must await the availability of financial resources. The primary need, the Board decided, was to consolidate the recent rapid academic growth. President Goldberg returned to the subject of a possible reorganization of the Technion structure, which had been initiated almost ten years earlier by Gen. Dori and Prof. Ginsburg. Instead of 20 separate Faculties and Departments, each with its parallel internal administration, he urged consideration of a regrouping of the academic units into several larger units, which could possibly be called Schools. Thus, the various science Departments would be grouped administratively into one School of Science and the engineering Faculties into a School of Engineering; there would be separate Schools of Architecture and Medicine. It was later suggested that the latter embrace all the life sciences. The discussion continued, but never got off the ground.

Growth of the Institute, and the consequent complexities of administration, led to a suggestion that the functions and responsibilities of the Office of the President should be studied, and a joint committee of the Council and the Senate was set up for that purpose.

On one specific development there was all-around agreement: the need to upgrade the computer facilities, and an IBM 370/165 was installed. This was later enlarged to a 370/168, and various improvements were added in the years that followed.

A new era in administration opened in June, 1971, when Mr. Evelyn de Rothschild, of London, the distinguished international banker, was elected chairman of the Board of Governors. The body now consisted of 77 members from Israel, and 57 members from overseas, with additional alternate members, from eight countries. In 1968 a new category of membership had been established, according to which members were chosen exclusively on the basis of academic distinction, and their ability to contribute to the academic development of the Institute.

Student Life

In 1969 student life on the campus took on a new dimension with the opening of the Shine Student Union Building. For the first time the student body had adequate facilities for their varied extra-curricular and social activities, for their student supply shop, and for the offices of the Student Association. The building also housed the offices of the Dean of

Evelyn de Rothschild (1932–   ).

370

Students, a post which had since 1964 been occupied by professors. Successive Student Deans were Prof. David Ginsburg, 1964; Prof. Elisha Shklarsky, 1965–66; Prof. Gdalyah Wiseman, 1967–68; Prof. Samuel Sideman, 1969–70; Prof. Brian Silver, 1971–72; Prof. Abraham Rosen, 1973–74; Prof. Zvi Dori, 1975–76; Prof. Amos Komornik, 1977–78; Prof. Raphael Sivan, 1979–80; Assoc. Prof. Ram Sagi, 1981–

Working in close cooperation with the Student Association, the Dean of Students dealt with a wide variety of personal matters affecting students both individually and collectively. These included financial, social and housing problems, difficulties created by army reserve duty, scholarship and loan assistance and various campus services to students such as the restaurants, dormitories, etc. A special mental health program was inaugurated in 1970, providing psychological counselling and psychotherapy programs. The unusual stresses to which the students were subjected made such assistance necessary, and in subsequent years the program was expanded with the establishment of the Philip and Frances Fried Student Counselling Center.

A student program of a different sort entirely was that offered by the International Association for the Exchange of Students for Technical Experience (IAESTE). This program, which had been operating at the Technion for many years, provided practical training overseas for students of the higher courses, usually during summer vacation, and looked after foreign students who came to Israel for similar employment. Several hundred students were involved each year. Largely responsible for the firm establishment and successful operation of IAESTE at Technion was Yehuda Eden, a staff member since 1951, and between 1957 and 1975 Registrar of Students.

A program of physical education was compulsory for all first-year students. Comprehensive sport activities were conducted by the Student Association, financed in part by annual Technion allocations to the Association budget. A student music program, carried on within the framework of the Department of General Studies, was given impetus when Dalia Atlas joined the Technion staff in 1971 and became permanent conductor of the orchestra and choral group.

One of the more unusual programs of the Student Association was its venture into the field of book publishing. The historic insistence that Hebrew be the language of instruction at the Technion was a matter of record, but the unavailability of textbooks in that language made implementation difficult. Obviously no publisher could afford to produce books which would have an extremely limited market. The cost per volume of limited editions would have been an excessive burden on student budgets. Consideration had been given to translation into Hebrew of clas-

Technion's coveted sports trophy, the Henshel Cup, rotates annually. Here it is received by the Department of Industrial Engineering. 1972.

sical texts from English and other languages, but this would have involved expensive payment of copyright fees. The Student Association undertook the printing of Hebrew lecture notes early in its history, but by 1952, when the undergraduate enrollment was still only a little over 1,000, the project emerged as a "business" and has continued to develop ever since. The first volumes were based on the lectures of faculty members, some of whom provided the texts of their lecture notes, edited and annotated. In some cases, where a lecturer was unwilling to yield up his notes, the Student Association was not averse to assigning one or more bright students to take copious notes during class. These were drawn up in proper form and then reproduced for publication and sale to the students of the following year. Most of the teachers were happy to cooperate, and even received a nominal payment.

The early texts were for the most part mimeographed or offset, but bound in hard covers so that aesthetically they created a good impression. In the first ten years of this program the Association published over 140 titles in the entire range of subjects taught at the Institute.

New immigrant students had been a feature of the Technion ever since the beginning of classes in 1924. For many years the new immigrants were a majority of the student body, and their common problems were treated within the normal framework of the operations of the Technion. Gradually the number of native-born Israelis, or at least those who had received their prior education in the country, increased proportionally. By 1959 native-born Israelis were 54% of the student body; the following year the percentage rose to 56.8%, and the percentage continued to rise from year to year. New immigrants found themselves out of the mainstream of student life, and a special adviser was appointed to guide them. It was also found necessary to set up a special preparatory course for immigrant candidates who were lacking in basic academic prerequisites and also required special tutoring in Hebrew. The first course was tried experimentally in 1960 and it grew through the years, giving special attention to the respective waves of immigration from various parts of the world.

The need to prepare candidates for admission was felt by other categories of young people as well, and the program was therefore expanded into a Center for Pre-Academic Studies. The Center dealt also with prospective students who were disadvantaged because of social or economic circumstances, and with veterans of military service who felt the need to refresh their knowledge of mathematics and physics after the long interval which had elapsed since they had graduated from high school. Courses for gifted and science-oriented high school youth were also encompassed within the Center.

On the whole, relations between the student body and the academic and administrative authorities were good, this at a time when campus revolts were taking place throughout the world. In 1969, on President Goldberg's initiative, the Technion constitution was amended to provide for membership on the Board of Governors of two representatives of the Student Association.

Good relations did not prevent student protest where financial matters were involved. A short protest strike in November, 1971, was caused by the raising of prices in the student restaurant. In May, 1972, announcement of increase in tuition fees, in accordance with the Government's economic policy, again put the students out on strike. They sent a delegation to the Minister of Education who told them that this was the business of the institutions involved. In December of the same year when the national Student Association called a strike again over the question of tuition fees, the Technion students did not join out of anger that their own action some six months earlier had not received support from their colleagues in the other institutions. Similar protests were repeated periodically on the university scene. Though they could not arrest the rise in tuition fees, they sometimes did manage to put a brake on the increase so that university fees gradually lagged behind, and were soon exceeded by the cost of attendance at kindergarten.

"Strike" was not a welcome word in the Technion vocabulary, but one of a different kind was held by the students in the Summer of 1973 when

Israel postage stamp issued for the Technion Jubilee. 1973.

they staged a symbolic hunger strike through which they identified themselves with the cause of Russian Jewry.

Student enrollment continued to grow, and the Technion began to feel the influx of new immigrants from the Soviet Union. By 1972 more than 400 Russian Jewish youths were at Technion, either in regular classes or in the Pre-Academic Center.

Major new buildings completed on the campus by the close of the Goldberg administration were the Detroit Mechanical Engineering Laboratories, the Sobell Chemical Engineering Building, the Hassan-Washington Pilot Plant, the Maurice and Ruben Rosen Solid State Institute Building, the Morris and Sylvia Taub Computer Center, the Henry and Marilyn Taub Computer Science Building, the Werksman Building (Pittsburgh) in Physics and the Van Arsdale Wing in Electrical Engineering.

Student housing included Casa Argentina, the Rosalinde and Arthur Gilbert Hostel, the Rose Bender Hostel and the three residence units for junior staff, contributed by David Rose, Barnett Shine and Michael Kennedy Leigh. More than 15 additional buildings were either under construction or on the drawing boards.

In 1972 the campus Master Plan was again subjected to its periodic review. Once more the planners called for appropriate development of the large area south of the campus, and contiguous to both the Technion and Haifa University.

The idea was to locate there central facilities which could be used jointly by both institutions, but no progress was made.

The annual financial reports were almost monotonous in their repetitiousness — only the figures changed from year to year. The maintenance budget rose steadily as a result of the physical growth of the campus and the expansion of staff. The Government grant was increased, but never in proportion to the rise in expenses. Funds from overseas were for the most part for capital growth. The result was once more the creation of an operating deficit. Staff members agitated for higher salaries, and students resisted attempts to raise tuition fees, the income from which remained constantly well under 10% of the total budget. There were always unusual or unexpected expenses. During this period the security situation in the country made necessary a large increase in annual expenditure for guards and other campus precautions.

Nevertheless, the Technion was on solid ground economically. Its deficit was always "manageable". Gone were the days when a Technion messenger had to hurry to Jerusalem to receive the monthly payments from the Government. Salaries were paid promptly on the first of each month. A beginning was made in the building up of internal endowment funds, of which only the income was used for designated purposes.

Well into his second four-year term, President Goldberg could survey a flourishing institution, developing academically and expanding physically.

Hebrew Revived

The fiftieth anniversary of the beginning of classes at the Technion was to occur in December, 1974, and plans were commenced by the Public Relations Department well over a year earlier for a comprehensive program of festivities and special events to mark the historic occasion. The list included an ambitious exhibition to depict dramatically Technion's achievements, the striking of a Government commemorative medal and the issuance of another Technion postage stamp, the writing of a full-scale history of the Technion, academic symposia and other events. Prime Minister Golda Meir agreed to serve as Honorary Chairman of the Jubilee Celebration Committee and plans were drawn up for the launching of a $50,000,000 Jubilee Campaign. Some of the plans were indeed implemented, but others were postponed indefinitely because of the Yom Kippur War in October, 1973.

The celebrations were to be linked also to a serious stock-taking, academically and historically. The Technion had travelled a long and difficult road during its first half century. It had overcome what seemed at the time insurmountable obstacles, and had emerged as a major scientific and educational factor in the life of the State of Israel.

The question of language of instruction, it will be recalled, had been the bone of contention which had prevented the launching of the study program in 1913 and 1914. The proponents of Hebrew had lost and then won their battle at the time, but the language still had to prove itself in use. Despite the pioneering work of Eliezer Ben Yehuda, who compiled the monumental 17-volume Hebrew dictionary, with many of his neologisms, Hebrew was not yet able to meet the requirements for daily use, or for technical/scientific terminology. Ben Yehuda and his colleagues in 1918 established the *Vaad Ha-Lashon Ha-Ivrit* (Committee of the Hebrew Language), much later transformed by law into the Academy for the Hebrew Language, which took upon itself the task of guiding the development of the ancient tongue into a medium of expression fully suitable for use in the twentieth century. This body was vested by the Knesset with legal authority for the growth of the language, including the coining of new words as necessary. The Central Committee for Technological Terminology set up jointly in 1943 by the *Vaad Ha-Lashon*, then by the Academy and the Technion, made its headquarters at the Technion, and for many years, under the dedicated direction of Prof. Shragga Irmay, helped guide the development of the language largely through the production of some 25 multilingual dictionaries of science and technology, containing over 20,000 terms.

No less important than the word in print was the actual infiltration of the new terminology into everyday speech through its use in the teaching process.

The task of the innovators was made easier by the fact that Hebrew was still in flux, not yet rigidly crystallized, and hence free from conservative traditions and popular misusage that affected other tongues.

In normal usage Hebrew has long since caught up with science. New discoveries, which necessitate the coining of fresh words, create terminological problems for all languages, and the Hebrew Academy meets the problem by dipping into the rich treasure house of classical and medieval literature to provide the terms required. Thus it is that aeronautical engineering, nuclear physics and computer technology can, without difficulty, be taught in the tongue of the prophets.

Typical of the use of root words from the Bible, is the adaptation from that source of a word which has entered the modern vocabulary as "electricity". The first chapter of the Book of Ezekiel describes the prophet's vision of a great storm and of fire characterized by obvious displays of energy. Out of the midst of the fire, according to the original Hebrew text, is a manifestation of *hashmal*, a mysterious word, on the meaning of which not all scholars had been able to agree. The Septuagint translates it by the Greek word *elektron*. What better than to adapt this word for the modern concept of electricity? And so with appropriate prefixes and suffixes, *hashmal* has been made a natural part of modern Hebrew speech not only as electricity, but also as electrician, to electrocute, etc.

To be sure, there has also been some compromise to international usage. Thus algebraic formulae, many chemical elements and compounds, and most specialized scientific concepts follow in Hebrew approximately their international equivalents, such as mathematics, physics, telephone, telegraph, autobus, etc.

The Hebrew alphabet made a penetration into international scientific usage in a different way through the pioneering work of one of the Technion's staff members. In developing the mathematical basis for his discoveries in Rheology (for which he was awarded the Israel Prize in 1958), Prof. Markus Reiner used letters of the Hebrew alphabet in his formulae in addition to Greek and Latin letters. Prof. C. Truesdell of Johns Hopkins University, in an extensive essay on Rheology in the Encyclopedia of Physics, utilized Reiner's formulae, and the *aleph, bet, gimmel* were given first international scientific usage. At a 1962 Symposium on Second Order Effects in Elasticity, Plasticity and Fluid Dynamics held at the Technion, the complicated formulae ran out of both Latin and Greek letters, and again the Hebrew letters were put into use. Truly, the Technion had come a long way since the 1913 language confrontation.

There were other conspicuous indications of growth and development as well. As we have seen, the institution had long since spilled over and out of its mid-city campus, and the pine-clad slopes of Technion City had by the 1970's already begun to assume the appearance of a small metropolis, even to the extent of its own Technion City *(Kiryat Hatechnion)* postmark. Neither Prof. Klein nor any of the other early planners of the campus had been able to foresee the transportation explosion, which continued to make necessary the widening of roads and the provision of adequate parking space.

The location in Haifa meant that the Technion was not in the mainstream of tourist traffic, which concentrated largely around Tel Aviv and Jerusalem. Visitors to the beautiful Carmel city either passed through quickly to see the panoramic view from the top of the mountain, or at best remained for only one night. Despite these constrictions, the Technion became a mecca for tourists, constituting, with the Bahai Shrine and Panorama Road, the principal attractions of the city. Indeed the welcoming of visitors became an important "industry" of the Institute, and provisions were made for the opening of a Visitors' Reception Center on campus, thus institutionalizing what had always been a Technion policy.

Staff members were charged with the responsibility of representing the Technion in this vital public relations program. During the 1950's Moshe Shevelov, in Tel Aviv, had served as the Technion's emissary to tourists. In the quarter century that followed thousands of visitors made their first acquaintanceship with the Technion with the helpful and personable guidance of the trained staff. The records show that many who later became warm and generous friends of the Technion first saw the campus in the company of veteran guides and hosts like Otto Stiefel, Janine Bousso and Gedalia Anoushi, who for many were identified with Technion no less than the students and the distinguished professors. Another veteran staff member who was closely associated with Technion's work was Mrs. Rivka Watson who for 20 years (1951–1971) until her retirement developed and coordinated the administration of the scholarship program. She was succeeded by Mrs. Lucy Gottesman.

Public relations extended also to the media of communications, and the Technion made its presence and its achievements known to an ever-increasing degree. In 1970 Morton Dolinsky was named Director of Public Relations.

Technion had become a "big business", and there was much to publicize. Visitors used to ask why there was need for such a large campus, much of which, especially in the early days, was open, green area. Relative to the size of the country, the campus city seemed enormous. The answer given was that just as the founding fathers 50 years earlier had envisioned the needs of the future, so the present generation had to bear in mind possible requirements beyond the immediate present. The words of Winston Churchill were recalled, to the effect that Technion would some day fill a function "to the lasting benefit of all the peoples of the Middle East."

Research

Increase in size was more than merely physical, and was reflected in several spheres of operation outside the primary function of educating and training students. One of these areas was covered by the Technion Research and Development Foundation. As the Institute approached its fiftieth year, the Foundation (known in Hebrew as the *Mosad*, for short) could report an impressive record of achievement. Its functions were threefold: to obtain and administer research projects commissioned from outside the Technion; to manage a number of laboratories and testing stations which provided services to domestic industry; and to hold and manage patents, licensing arrangements, etc., for the Technion and for staff members. In these operations the Foundation was and has continued to be self-supporting.

The comprehensive program of testing and other services was carried out in the following units administered by the Foundation: Elias and Bertha Fife Building Materials Testing Laboratory, Hydraulic Testing Laboratory, Soil and Road Testing Laboratory, Geodetic Research Station, Israel Institute of Metals, Prof. A. Ilioff Chemical Testing Laboratory, Road Safety Center, Industrial Psychology Laboratory and Scientific Editing Office. In 1973 the turnover for both the industrial service units and for commissioned research and consultation amounted to IL 29,308,000.

In addition to the work of the Foundation, a comprehensive research program was pursued within the general framework of the Technion. Some 1800 approved research projects were being carried out by staff members in 1973, not including some 1100 research theses being executed by students in the Graduate School. The major research centers functioning at the Technion were: Center for Water Resources Research, David T. Siegel Hydrodynamics and Hydraulic Engineering Laboratory, Leonard and Diane Sherman Environmental Engineering Research Center, Coastal and Ocean Engineering Research Center, Bruner Institute of Transportation, Mineral Engineering Research Center, Michel Polak Building Research Station, Center for Urban and Regional Studies, Stone Technology Center, Center for Research on Machining Processes and Machine

Tools, Chemical Engineering Research Center, Laboratory for the Technology of Plastic Materials, Food Industries Research and Development Station, Solid State Institute, Julius Silver Institute of Biomedical Engineering Sciences, Mechanics Research Center, Rural Building Research Center, Soils and Fertilizers Laboratory, Aeronautical Research Center, Center for Research in Management Sciences, and the Center for the Study of Man at Work.

Neither the bare statistics nor the listing of the physical facilities provide a sufficient picture of the significance of the research work carried out at the Technion. In scope it was related to practically every aspect of industrial, agricultural and defense needs. Even beyond the studies in Technion's laboratories, the research executed by the Institute's graduates constituted a natural extension of Technion's vital contributions to the nation as a whole. The hundreds of pages listing the research projects of the period are an index of that participation in the advancement of the frontiers of human knowledge. Only by way of illustration, mention may be made of new, coordinated efforts at the Technion to increase research in energy, progress in development of an artificial kidney, and Technion's role in the production of Israel's new Short Take-Off and Landing (STOL) plane, the Arava. The increased funds made available for research were augmented also by a growing trend in research sponsorship from Germany.

Ganna and Anatol Josepho Industrial Research Center.

There were disappointments as well. Considerable publicity was given to a "breakthrough" in water desalination — a technological improvement conceived by Prof. Abraham Kogan which, it was hoped, would enable the production of fresh water at a cost about 30% cheaper than any other known method. Several press conferences were held to announce the new method, and generous financial help was provided by David Rose, of New York. Following success in the laboratory stages a pilot plant was later built on the premises of the Haifa power station of the Electric Corp. The Kogan–Rose project successfully produced 50,000 gallons of water a day, but the spiralling rise in the cost of oil made the process, though feasible, uneconomical. Much later the project was discontinued.

Another area of activity, often eclipsed by the university-level teaching and research, was that occupied by the Extension Division. Since its establishment in 1956, the Division had been broadening its function of providing educational and training services to the general public. The program was carried out through local branches in major cities, in addition to Haifa, and included refresher and retraining courses for graduates, technical and technological training courses for both professionals and non-professionals, symposia and seminars, summer schools, science circles for youth, pre-academic preparatory courses, and various other programs

Towing channel in the Siegel Hydraulics Laboratory. 1970.

designed to serve public needs. Close to 10,000 individuals were enrolled in the various programs in 1973.

The "junior" branches of the Technion had likewise built firm foundations. In 1969 the secondary school program had been reorganized and the Technical High School amalgamated with the Junior Technical College, operating under the latter name, *Bosmat* in Hebrew. Mr. Gershon Ahroni, who had been Principal since 1939, retired in 1969 and after an interval was succeeded by Tuvia Toren who served for only three years until he was killed in the Yom Kippur War. He was followed by Arieh Gur. Conducting its program both on the old campus at Hadar Hacarmel and at Technion City, the Junior Technical College had an enrollment of over 1700. Over and above its numbers, however, it served as a model and pioneer for the entire vocational school system in Israel.

The National School for Senior Technicians *(Handesaim)* in 1973 completed ten years of operations under both the Technion and the Ministry of Labor. More than 2200 students were enrolled. Gershon Har'el was Principal.

President Goldberg's last year in office coincided with the 25th anniversary of the establishment of the State of Israel. The year, which began in a festive mood, was to end on the tragic note of sudden war.

For the Technion it was a "normal" year, with its share of problems and

Henry Ford II (center) on a visit to the Mechanical Engineering Laboratories, with Max Fisher (left), Prof. Ehud Lenz and Alexander Goldberg. 1972.

achievements. In May, 1973, the instructors and assistants had conducted a strike based on salary complaints, but otherwise the labor front was peaceful. Salary and wage scales were determined as part of the national collective agreement with both academic and administrative personnel. At the Technion both sides displayed patience and good will. At other institutions it was claimed that professors at the Technion received extra emoluments for research projects. This was seen as a way of increasing income outside the salary freeze, and similar demands were advanced elsewhere.

The first summer semester, a direct result of the new credit system, opened in July, 1972. The rise in the number of immigrants to the country was also reflected on the Technion campus. During the academic year 1972–73 there were 220 new arrivals in the preparatory courses, 80 new immigrants in the Graduate School and more than 1,000 undergraduate students who could be classified as recent immigrants. During the preceding two years some 85 newly arrived teachers and research workers had received appointments at the Technion. At the Medical School a ten-month course in dentistry was opened for dentists from the Soviet Union, to raise their standards to the level required in Israel.

It was cause for considerable pride at the Technion when in 1972 still another faculty member was awarded the much-coveted Israel Prize for his scientific work. This was Prof. David Ginsburg. Previous faculty members who had received the award were Prof. Franz Ollendorff, 1954;

Prof. Manfred Aschner, 1956; Prof. Markus Reiner, 1958; Prof. Alfred Mansfeld, 1966; Prof. Yitzhak Danziger, 1968. In 1976 the award was given also to Prof. Yosef Rom.

New Finance Procedures

The freeze on overseas fund-raising had not been altogether lifted, but new procedures in Government financing made life much easier. Gone were the days of individual institutional haggling with the Ministry of Finance regarding allocations. Now all financial arrangements were made through the Budgeting and Planning Committee of the Council for Higher Education. The Committee presumably established criteria for such allocations. The grants were made on a global basis to each institution, without designating specific budgeted items. This prevented interference in the internal operations of the institutions. However, the Council retained the right to approve the opening of new departments (to prevent unnecessary duplication), the setting of quotas of new students to be accepted and the building of new physical facilities. This gave it, in effect, de facto control over the growth and development of the institutions.

There were changes, too, in the methods of fund-raising. As capital growth and expansion continued, the need became greater for maintenance funds. The techniques of fund-raising had emphasized the dedication of physical facilities, and the placing of commemorative plaques. Would donors be interested in contributing to an operating budget? The problem was not unique to the Technion, and new types of appeal were developed. These included the establishment of permanent, named endowment funds to support academic chairs, research, scholarships, acquisition of books, etc. The change in direction was not easy, but the realities of the institution's financial needs provided no alternative.

Technion's friends overseas were, for the most part, not found wanting. New records of support were achieved. Chief among the sources of aid was the American Technion Society. This organization continued to enjoy dedicated and competent leadership under a succession of presidents: Jacob R. Sensibar (1960–62); B. Sumner Gruzen (1963–64); Maurice Rosen (1965–68); Lawrence Schacht (1969); Jacob W. Ullmann (1970–71); Laurence A. Tisch (1972–73).

Large contributions began to increase, among them a million dollars or more each from Alexander Konoff, for the Junior Technical College, from Harry DeJur for student housing, from Maurice Amado for a Mathematics building and from Eugene Ferkauf. A similar amount given in 1972 by Leo Harvey, of Los Angeles, established the Harvey Prize Fund. A year later, on the initiative of Samuel Neaman, over three million dollars were raised for what later became the Samuel Neaman Institute for Advanced Studies in Science and Technology. In 1973 Andre Meyer announced a contribution of five million dollars for the construction of the Bella Meyer Advanced Technology Center.

The annual dinner in New York and other cities continued to pay tribute to distinguished personalities. In 1972 the American Technion Society inaugurated its Albert Einstein Award to mark special services to the Society and to the Technion. First recipient was Abraham Tulin.

The A.T.S. Women's Division, established in the 1950's, grew steadily under the leadership of Mrs. Herman Leffert. In 1971 Mrs. Rose Herrmann became President and in addition to its program of scholarships, student aid, book funds and dormitories, the Women's Division adopted a special project at the Technion in support of Medical Engineering, under the chairmanship of Mrs. Pearl Milch.

The President of the Technion and various staff members paid periodic visits to the U.S. and assisted both in fund-raising and in general promotion of Technion affairs. The Society also carried on a widespread public relations program, and for more than a dozen years conducted annual conferences on science and technology which attracted significant participation.

Israel President Zalman Shazar (left) with Leo Harvey on the occasion of the first Harvey Prize Awards. 1972.

The name of the Technion also turned up in a most unexpected place — among the lists of race horses. Mrs. Joan Arbuse, a loyal friend of the Technion as well as a devotee of the sport, registered the name for a two-year-old colt in her stable. Technion (the horse) began its racing career in 1973 but never lived up to expectations. Its owner found ways, other than derby purses, of providing generous aid to the Technion.

In an attempt to streamline its operations, the American Technion Society in 1970 sold its building at the prestigious 1000 Fifth Avenue address, and moved into rented offices in midtown Manhattan, at 271 Madison Avenue.

There was activity in other countries as well. Having completed Canada Building in the Department of Chemistry, the Canadian Technion Society undertook to construct student dormitories. D. Lou Harris, one of the founders of the organization and its national president for many years, died in 1972 and was succeeded by Louis Lockshin. A new dormitory project was named for Harris. A Canadian Women's Division was founded and headed for many years by Mrs. Lilian Mendelssohn. As of 1973 the Technion, jointly with the Hebrew University, benefited from a Lady Davis Fellowship Trust, eventually to reach $5,000,000, which was established by the Eldee Foundation of Montreal, through the good offices of Louis M. Bloomfield, Q.C., and Bernard M. Bloomfield. Sinai Leichter served as Director of the Canadian Society.

The British Technion Society maintained groups in London and in the provinces. Edward Rosen, chairman, died in 1966 and S. J. Birn served as acting chairman. Sir Leon Bagrit was President of the Society and Victor Mishcon, Honorary President. Birn and Barnett Shine were Joint Treasurers. In 1969 Evelyn de Rothschild was elected chairman of the Executive Committee. Academic chairs and buildings on the campus bear the names of such generous donors as Shine, Michael Kennedy Leigh, Michael Sobell, the Wolfson Foundation, the Charles Wolfson Charitable Trust and Mrs. Lily Tobias.

Dr. Jacobo Isler was the foremost leader of the Argentine Technion Society, which raised funds for a student hostel. In South Africa loyal and devoted service was rendered by Gus Osrin for many years. The major gift was received from Morris Mauerberger of Capetown. Stanley Kaplan assumed the leadership of the Society. The dominant personality in Mexico was Max Shein.

Other Technion support groups existed in Australia, Brazil, France, the Netherlands, Uruguay and Venezuela.

Valiant efforts were made also by the Israel Technion Society to raise the level of participation in support of the Institute. Aharon Goldstein and Dan Tolkowsky headed the effort for several years, and were succeeded by Alexander Goldberg when he retired from the Presidency. In 1969 Shimon Halpern retired as Director of the Society and was succeeded by Jonathan Aival. Income increased from IL 1,760,000 in 1971 to IL 2,526,000 in 1972 to IL 5,160,000 in 1973. In the latter year the first two buildings were dedicated of a complex to house a science-based industries center, gift of industrialist Aaron Gutwirth, who had pledged to put up 20 buildings for that purpose. Efforts continued to activate the Alumni Association, but the results still fell far short of what similar organizations accomplished in the U.S. for American universities.

Goldberg's Summary

Goldberg faced his last meeting of the Board of Governors in his capacity as President in 1973, and presented a summary of the achievements during his administration. The record was marked by sound business management and inauguration of new academic fields. He repeated his call for a reorganization of the organic structure of the Technion into three or four large schools, each with a degree of autonomy, to replace the 20 or more individual Faculties and Departments all responsible directly to the President. Such a decentralization, he felt, would reduce bureaucratization and introduce a greater flexibility into decision making, thus also easing the path to academic innovations. He also noted the need for accurate forecasting of the nation's manpower needs, to enable the Technion to plan its own operations properly. He was neither the first nor the last to express such need.

The statistics of his final report were impressive. The Technion had an enrollment of 5,756 undergraduate students and 2,631 students in the Graduate School. At the latest graduation exercises 940 Bachelor degrees, 200 Master degrees and 86 doctorates were awarded.

It was a quiet Board meeting, free from major controversies. It mourned the passing during the preceding year of General Dori, Abraham Tulin and D. Lou Harris, among others. The venerable Arthur Blok, Technion's first head and an annual participant, was present for what turned out to be his last Board meeting.

With regard to business affairs, the Board resolved to implement and complete the financial and administrative merger with the Medical School, in the wake of the academic merger which had already been completed. The setting up of the Samuel Neaman Institute of Advanced Studies in Science and Technology was approved. It was noted with satisfaction that the financial operations of the current year were in the black. For 1973–74 the Board proposed a balanced operating budget of IL 124,500,000 and Development Budget of IL 20,000,000.

In his summation upon adjourning the annual session, Justice Landau said he felt there were hard times ahead: "The Technion enters this stormy period in good order, and it will need the continued help of all those who have helped in the past to assist the new President steer the ship of Technion in these stormy waters." He spoke more truly than even he realized at the time.

The eight years of Goldberg administration were marked, at first, by consolidation after the difficult preceding years, and then by solid achievements. In sharp contrast to his predecessor, who had created a reputation for military style administration, Goldberg was much more informal and easy-going. The fact that almost everyone called him "Sasha" was an indication of his affable manner. Having served for a number of years as Chairman of the Board of Governors he knew the Technion well, and there were no surprises for him. His major achievement was perhaps the systemization of the Institute's administrative procedures along the lines of a modern business or industry. The results were quickly obvious, and though financial crises still occurred, there was none of the hand-to-mouth emergency which had characterized some of the preceding years. Goldberg gave full credit for this to his Vice President for Administration and Finance, Yosef Ami.

Rightly he acknowledged pride in the academic highlights of his two terms: the inauguration of the Biomedical Engineering and Computer

Research on the strength of stiffened cylindrical shells commissioned at the Aeronautical Engineering Department by the U.S. Air Force. The studies, carried out by Prof. Josef Singer and his associates, were applied in Apollo 14's flight to the moon. 1973.

Morris and Sylvia Taub Computer Center.

Science Departments, the absorption of the Medical School, the expansion of work in applied mathematics and in solid state. The credit point system was another innovation which he encouraged and inaugurated.

His relations with the student body were cordial, and it will be recalled that it was he who had initiated student representation on the Board of Governors. Because he had been raised and educated in London, English was his mother tongue and this of course enabled him to communicate freely with supporters abroad. He was in frequent demand as a speaker at Technion functions in the United States and elsewhere, and was responsible in no small measure for the rise in the pace and extent of contributions.

He also initiated informal joint consultations among the Presidents of Israel's universities, which quickly become institutionalized as a Committee, meeting periodically for discussion of matters of common interest.

His retirement from the Presidency by no means meant his withdrawal from Technion affairs. He assumed the chairmanship of the Israel Technion Society and quickly threw himself into promotion of its activities, seeking increased public support for the institution from domestic industry and industrialists.

President Goldberg's second four-year term was due to expire on September 30, 1973. Since he had already made it clear that he would not stand for a further term, the Board, at its June, 1972, meeting, set up a search committee headed by Justice Moshe Landau. By October of that year the committee had already decided to recommend the election of Major General (res.) Amos Horev as the next President of the Technion. After the usual consultation with the Senate, the members of the Board were polled by mail and Horev was elected without opposition. His term of office was to begin October 1, 1973.

The new President came to his post from the Defense establishment where he was serving as Chief Scientist of the Ministry of Defense. His long military career had included command positions as Chief of the Ordnance Corps and Chief of Logistics. He had also founded the Department of Research and Development in the military General Staff. General Horev was 49 when elected President of the Technion.

On Sunday, October 1, 1973, the new President was officially inaugurated at an impressive ceremony held in the Churchill Auditorium and attended by Minister of Education Yigal Allon, among others. In his speech of acceptance, Horev reviewed the mission of the Technion in the past and in the future, and noted that Israel was still standing on the very edge of a great industrial revolution.

The new President settled down in his office. At the end of the same week, on Saturday, October 6, Yom Kippur Day, the Egyptian and Syrian armies suddenly attacked Israel. The Technion President was called back into active military service, and the Technion as a whole, like the entire country, went on a war footing.

An historic epoch had come to an end. A new era was to begin.

*   *   *

Senate House–Sir Louis Sterling Memorial.

Epilogue

Gen. (Res.) Amos Horev (1924–   ).

Our narration ceases with 1973, but the story continues nonetheless. In the years that have followed, the institution has continued to flourish. It has increased in size, broadened its scope, and enhanced its prestige in the academic world. The full dramatic story of the ensuing period, the appraisal of the Technion's achievements, even the unique and anecdotal aspects of its existence, must await the pen of the next historian who will pick up the tale where it has been temporarily halted here. Yet even in succinct form, some notice must be taken of what has transpired during the subsequent years. The Technion has drawn upon the rich traditions left by those who built and developed it, while adding fresh spirit and renewed vision. Continuity has been assured by sound organization and loyal personnel; new directions have been provided by wise and farsighted leadership.

After the Yom Kippur War there was a five-month delay until classes resumed under Amos Horev, the Technion's first Sabra President. He was particularly qualified for the post, not only because of his military and administrative background and experience, but also because he was an engineer, holding two degrees from M.I.T. He served two full terms of four years each, and though he asked to be released, he was prevailed upon to continue for a ninth year (1973–1982) as the Search Committee named by the Board of Governors sought a suitable successor.

The signing of the peace treaty with Egypt was the major political development of the period. No matter what may happen thereafter, it has set in motion a progression of events whose effect on the country and on the Technion can be measured only in long-range terms and from the proper historical perspective.

As the history of the period unrolls, the Technion continues its dynamic existence according to the forms and procedures which had been developed through the years. The Board of Governors continues to meet every year, and new classes of freshmen are accepted annually. Harvey Prizes continue to be awarded. Student life on the campus flourishes. A thousand or more engineers and scientists have been graduated each year, adding to the pool of skilled manpower in the country. The general industrial growth of the country, especially in electronics and other science-based industries, may be traced directly to the basic facilities provided by the Technion, and the personnel which it has trained.

In 1980 Uzia Galil, a Technion graduate, former member of the staff and head of Elron, a major complex of electronic industries in Israel, was elected Chairman of the Board, succeeding Evelyn de Rothschild. This in itself was symbolic of new trends.

Uzia Galil (1925–   ).

Giant solar collector, planned and designed by faculty members, and used for solar research at the Technion Energy Center. 1980.

Academic Growth

In the academic area a mere listing of programs cannot do justice to the growth during this period. Brief mention must be made, however, of some significant developments. One is the intensified research in various fields of energy, and the resolve of President Horev to project the importance of nuclear power for Israel. An immediate result has been a program to provide better facilities on campus for studies and research in nuclear science. A new program has also been launched in marine engineering and pioneer work undertaken in industrial robots.

The Faculty of Medicine has more than lived up to expectations and has blazed new paths in medical education as an outcome of the pioneering relationship between medicine and technology.

Perhaps the fastest growing department at the Technion is Computer Science, which in 1980 was elevated to the rank of a full Faculty. Between 1973 and 1981 it quadrupled the number of students enrolled. Staff and space have had to be increased accordingly.

Left to right: Prof. and Mrs. David Erlik, Julius Silver, David Rose. 1976.

The computer equipment has been further upgraded and now includes four major computers: three of these, an IBM 370/168, an IBM 4341 and a VAX 11/780 for teaching and research, and an additional IBM 4341 for administrative requirements. Microcomputers were installed as data entry equipment to replace old key-punches used by the students.

Much of the technological and scientific progress in recent years has resulted from close cooperation and interchange between various disciplines of human knowledge; the academic program at the Technion has reflected such development in the establishment of a number of inter-disciplinary units. Perhaps the first, on a major scale, was the Institute of Biomedical Engineering, inaugurated in 1969. Generous help from Julius Silver, of New York, and the initiative of staff members have led to a broad program of teaching and research in biological and medical sciences including such fields of specialization as biomechanics (tissue mechanics, artificial organs and prostheses), bioelectronics, medical physics and bio-materials. The studies draw upon staff and facilities of Electrical Engineering and Electronics, Physics, Biology, Medicine, Computer Science, Chemical Engineering and others.

The Solid State Institute was set up in 1976 to carry on both fundamental and applied research, but with an eye to having an impact on local industry. Among its chief activities have been ion implantation, surface analyses, laser spectroscopies and crystal growing. The Institute, which is housed in the Maurice and Ruben Rosen Building, includes staff members from Physics, Electrical Engineering, Materials Engineering and Chemistry.

The Transportation Research Institute was established in 1977 and draws its members from Civil, Mechanical and Agricultural Engineering and from Architecture and Town Planning. Problems of traffic and vehicle engineering and planning of public transportation fall within its scope. The Institute also embraces a Road Safety Center, first set up in 1966 in collaboration with the Ministry of Transport. A helpful source of assistance is the Henry Ford II Fund for Transportation Research set up through the American Technion Society.

The Samuel Neaman Institute for Advanced Studies in Science and Technology was established at the Technion in 1978 for the purpose of assisting in the search for solutions to national problems in the fields of economic, scientific and social developments in the State of Israel, the raising of the standard of living of its citizens and the search for methods of facilitating Israel's integration into the Middle East. Thus far the Institute has dealt with such subjects as rehabilitation of slum areas, vocational education, national energy resources policy, national policy for water resources and minerals, low altitude aviation, application of mathematics to industry and others. There are about 100 researchers involved in the various projects, representing most of the disciplines at the Technion.

The Materials Engineering unit, first inaugurated in 1968, and attached to the Faculty of Mechanical Engineering, was in 1981 recognized by the Senate as an independent Department. Although it offers graduate degrees only, the Department provides instruction to well over a thousand students a year from other units.

Courses in biology had been given in various academic units until 1971, when the Department of Biology was established as an independent unit. It has grown appreciably since then. A broad teaching program places emphasis on cell biology, biochemistry and physiology. The Department also offers courses for students from the Faculty of Medicine and other Faculties as well. Both experimental and theoretical research cover a wide field including physiological cell biology, genetics, biochemistry, bioenergetics and molecular biology.

A shortage of properly qualified teachers of technological subjects in the country's secondary schools had led in 1959 to the setting up of a special teacher training program as part of the Department of General Studies. In 1964 this became an independent Department, and in 1978 was given its new name: the Department of Education in Technology and Science. A year later it moved into its own quarters in the George and Beatrice Sherman Building. Though the Department fills an important need in the educational framework in the country, there has been a

Crystal-growing in the Rosen Solid State Institute.

391

decrease in the number of students choosing this field (at Israel's teacher training seminaries as well) possibly because of the decline in the status of teachers and a negative popular image of the profession. The Department at the Technion has added new scope with special programs dealing with both disadvantaged and highly talented children.

For the sake of academic and administrative efficiency, several of the smaller academic units have been discontinued as separate entities, and their activities continued within the framework of the larger Faculties. Among these have been the Departments of Mechanics, Applied Mathematics and Mining Engineering.

While commissioned and sponsored research at the Technion continue to show an encouraging increase both financially and numerically, the true measure of the broad scope and extent of research at the Institute is not to be found in these terms alone. Practically every member of the academic staff is engaged in research in one form or another, and the close to 2,000 students enrolled in the Graduate School are working on research projects as part of the requirements for their Master or Doctor degrees. In every major field of science, engineering and technology there is an ongoing program of research, both pure and applied, in which such

Control room of the Tark Turbo and Jet Engine Research Laboratory in the Faculties of Aeronautical and Mechanical Engineering.

Heavy ion accelerator, the only machine of its kind in Israel, for implantation of ions in target materials, usually silicon. It is located in the Rosen Solid State Institute Building.

distinctions are becoming more and more blurred. The consequences are reflected both in the industrial development of the country, and in the Technion's standing in the academic world.

For years the Technion has enjoyed close friendly relations with many of the leading universities in the United States, and academic exchanges on both the official and the personal plane have been frequent. Similar relations have been created with leading European institutions but on a more formal basis. Agreements spelling out such relationships have been signed with the Eidgenössische Technische Hochschule (ETH) in Zurich, the Ecole Nationale Supérieure des Mines in Paris, the Rijksuniversiteit in Ghent, the Rheinisch-Westfalische Technische Hochschule in Aachen and the Technische Hochschule of the University of Karlsruhe.

Technion's library, housed in the central Elyachar Library Building at Technion City and in 22 departmental libraries, has some 375,000 books and some 225,000 bound periodicals, for a total of 600,000 volumes.

A radical change in the degree requirements was instituted by decision of the Senate and the Board in 1978. Other universities in the country had been awarding their first degree (B.A.) after three years of study, whereas

Technion required four years for its B.Sc. The additional year of study was reflected in higher standards to be sure, but many students seeking university degrees preferred to save the extra year and enrolled at the other institutions. This was especially true of studies in the basic sciences, which were offered at all the universities. The new decision, taken only after lengthy deliberation, and even then with some hesitation, provides that students of chemistry, physics, mathematics, biology and computer science can obtain a B.A. at the Technion, after three years of study. The B.Sc. is still available after four years, leading to opportunity for an M.Sc. in the respective subjects. The innovation is being watched carefully to ascertain if the change has indeed been justified. Admission statistics have shown an appreciable increase in registration only in computer science, mathematics and physics.

Changes in the industrial complexion of the country have been reflected in the popularity of the various Faculties and Departments. The classical subjects, such as electrical, civil and mechanical engineering, continue to draw large enrollments, but newer disciplines have risen in popularity. The following table provides an illuminating picture of the undergraduate enrollment at the Technion in 1981 in the top ten Faculties:

Electrical Engineering	1213
Civil Engineering	646
Industrial Engineering and Management	642
Computer Science	610
Architecture and Town Planning	603
Mechanical Engineering	504
Medicine	352
Aeronautical Engineering	314
Agricultural Engineering	205
Chemical Engineering	147

The spurt in immigration from the Soviet Union, especially during the early part of this period, was reflected in a marked increase in the number of students and faculty members from that country. Award of an earned doctorate in applied mathematics served to dramatize one aspect of the situation. Anatol Mendelevich Galperin had conducted his research and prepared his thesis in Russia and submitted it to the examiners of the Central Economic-Mathematical Institute of the U.S.S.R. in 1972. He and his wife had previously applied for permission to leave for Israel. Although

Anatol Mendelevich Galperin receives his doctorate in applied mathematics. 1976.

the thesis was favorably received, the degree was not granted, and the couple were dismissed from their jobs. By a circuitous and secret route the research material and thesis reached Israel, and the Technion set up a Board of Examiners to judge the doctoral thesis in the usual way except for the oral examination, which was waived. The degree was authorized in absentia, and Galperin, after a long wait for his exit visa, arrived in Haifa in time to receive the diploma in person in June, 1976.

There has been a levelling off in student enrollment, as other institutions in the country now also offer studies in engineering, but 72% of all the students of technology and engineering in Israel are enrolled at the Technion.

The President has repeatedly expressed the desirability, for both administrative and academic reasons, of structural reorganization of the Technion along lines suggested by both his predecessors, possibly to replace the 20 existing Faculties and Departments with five or six comprehensive Schools.

In recent years the Technion has been engaged in an ongoing constructive study of its operations, both academic and administrative. In 1976 the

Scale model of an airport being tested in the Lipson Wind Tunnel.

Senate approved a system of periodic inspection of the various academic units by commissions which include outstanding personalities from abroad. These inspections are intended to assess the goals of the units and the efficacy of execution, and to provide an evaluation of achievements. A majority of the major Faculties and Departments have already undergone such intensive study, leading to inauguration of changes where deemed necessary.

At the same time there has been critical study of the operational practices of the Senate on the one hand, and of the program and deliberations of the Council, on the other. Both have resulted in improved operations but the spirit of self-analysis continues. Smooth and efficient functioning of the Senate, which by 1982 had grown to over 150 members, is made possible by the work of its Steering Committee, a small executive group of its members, which systematically and thoroughly prepares most material for discussion and decision by the Senate.

During this period President Horev enjoyed the cooperation of a competent Cabinet of Vice Presidents, chosen by the Board of Governors, with the approval of the Senate, as provided by the Institute's Statutes. Vice President for Academic Affairs was Prof. Jacob Bear (1974–1976); Prof. Moshe Zakai (1976–78); Prof. Jacob Ziv (1978–81); Prof. Ephraim Kehat (1982–   ). Vice President for Development: Prof. Avraham Ginzberg (1973–76); Prof. Paul Singer (1976–80); Prof. Yakov Eckstein (1980–   ). Vice President for Research: Prof. David Hasson (1973–76); Prof. Zeki Berk (1976–79); Prof. Ehud Lenz (1979–   ). Vice President President for Administration and Finance: Yosef Ami continued until 1974; Menahem Zehavi (1974–79); Isaac Nissan (1979–   ).

Few Controversies

The nine years of the Horev administration was the first prolonged period in the history of the Technion not marked by major controversies. It could well be known as the Period of Placid Progress, though some clouds began to gather in the latter years. Initial clashes on the campus between Arab and Jewish nationalists were cause for concern. Unlike other campuses, the Technion had hitherto been free of any such problems, perhaps because of the relatively small number of Arab students, but a sharpening of political views at both extremes led for the first time to physical confrontation.

Another issue had to do with the future of the old building on the Hadar Hacarmel campus. The Board of Governors had long since decided to gather all academic units on the one Technion City campus, with the exception of the Faculty of Medicine which, because of the nature of its studies, had to be located near a hospital. Accordingly, plans were approved for adequate building space. The Faculty of Architecture and Town Planning, however, having taken over almost all of the original Baerwald building, gave notice that it preferred to remain where it was, rather than move to the new facilities being built for it at Technion City. A public relations campaign was launched by members of the faculty and students, charging that if they vacated the historic structure the Technion administration would have it torn down, and the site would be handed over for commercial exploitation. Repeated denials by Gen. Horev were followed up by firm decision of the Board in 1980 that the Baerwald "building and gardens are an historic landmark which must be preserved as a national monument marking the creation of the Technion and the early days of Jewish settlement in Haifa." The Board further endorsed the stand that "the building and gardens shall be retained as a vital physical facility which can continue to play an important role in advancing Technion's broad program in the fields of education, technology and science."

During his incumbency the President was asked to establish his residence in Haifa. The cost of building a President's Residence on the campus was found to be too great and he preferred to commute from his home in Ramat Hasharon.

Labor relations have, on the whole, been excellent. Salary and wage controversies erupted from time to time within the context of the general economic situation in the country, but good will on both sides, and appropriate financial assistance, usually from the Government, helped to bridge the difficulties. The academic staff fought for and obtained special grants to finance their research projects. Through the years there has been an increase in the number of organizations representing the Technion staff at various levels. Today there are the Faculty Association, the Employees' Association, largely of clerical, administrative and maintenance personnel, the Organization of Engineers Employed by the Technion, and the Organization of *Handesaim* and Technicians Employed by the Technion.

During this period the Student Association has had the benefit of responsible and dedicated leadership, elected from within the ranks of the student body. As a result the Association has been able to conduct successfully a broad program of social, cultural, athletic, recreational and other extra-curricular activities.

Campus Growth

The Horev period has been marked by a dramatic spurt in the physical development of the campus, as well as by a broadening and deepening of

Technion City. 1979.

the academic program. During the years 1973–1982, the built-up campus area almost doubled from 140,000 square meters (1,507,000 sq. feet) of floor space to 260,000 square meters (2,799,000 sq. feet), comprising both Technion City and the separate facilities of the Faculty of Medicine. More than 20 buildings have been added to the campus as the Technion approaches a levelling off in its construction program. The new units include the DeJur Materials Engineering Center, the Food Engineering Pilot Plant, additional units in the Gutwirth Science-Based Industries Center, the Kranzberg Industrial and Management Engineering Wing, the Lady Davis Mechanical-Aeronautical Engineering Center, the Petrie Pavilion in Mathematics, the Penner Wing in Civil Engineering, the Leonard and Diane Sherman Environmental Engineering Center, the George and Beatrice Sherman Science and Technology Education Center, the Julius Silver Biomedical Engineering Institute, the Stein Food Engineering Building, the Werksman Physics Building, Australia House–Coastal and Marine Engineering Research Institute, the Ivanier Laboratory for Welding and Casting Research, the Natovich Orthopedic and Rehabilitation Research Center, the Carasso Wing for Self-Study, the Fried Student Counselling Center, the Gruss Swimming Pool, the Mexico, Shein, Kessel and France Student Hostels, and additional units in Canada Village.

Storage and warehousing facilities are an urgent requirement of a large institution. Lack of adequate space at the Hadar campus had in the early 1950's led to the utilization of supplementary storage space in an old factory building in the Haifa Bay area. Dampness and exposure there caused considerable damage to historical files. Later the basement of the new Churchill Auditorium was utilized, and in 1977 very ample facilities were provided at Technion City in a Central Supplies building. Head of the Central Stores from 1951 until his retirement in 1977 was Dov Czopp.

The crowning glory of the past decade has been the development of the Faculty of Medicine, both in the construction of the magnificent B. Rappaport Family Medical Sciences Building, and no less important, in the laying of the foundations for the extensive program of research in medicine and the life sciences to take place there.

Considerable attention has been given to the physical appearance of the campus of Technion City, and a master plan for gardening and landscaping, drawn up by Prof. Ronald T. Lovinger, is being implemented. In addition to tastefully landscaped open areas which had been created in the past, like the Sherman Family Forum, Schacht Park and the Riesman Plaza, new green belts have been planted including August Lane, Federmann Park and, at the heart of the campus, Kislak Park.

The peak in the construction program has been reached and passed. The only top priorities which remain are to meet the need for additional housing on campus for accommodation of all students who come from outside the immediate Haifa area, and to provide proper facilities for a faculty club.

The financial needs of the Institute have not diminished; on the contrary, the growth in physical facilities has resulted in increase in the operating budget. There has been a running battle with inflation, and again a cycle of efforts to avoid deficits by allocations from the Government, contributions from Technion Societies, and tight paring of expenses while at the same time maintaining all necessary priorities. The achievement of a balanced budget, especially in recent years, without any sacrifice in standards, is little short of a miracle.

Fresh emphasis in fund-raising has been placed on endowment funds and there has been considerable success in establishment of such funds for academic chairs, research, fellowships and scholarships. A list of such funds now operating at the Technion appears in the Appendix.

New moratoriums on fund-raising were imposed following the Yom Kippur War, but these difficulties were surmounted. Emanuel Shimoni was in 1977 named overall Director of Public Affairs, and new impetus was given to activities in this field. David Friedlander is Director of Public Relations.

Simplified one-hand typewriter keyboard developed in the Faculty of Industrial Engineering and Management, for use with computer terminal. Each letter is typed by a combination of strokes. 1980.

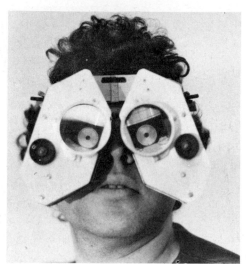

An improved infrared device for night vision developed by Dr. Itzhak Hadani of the Biomedical Engineering Department. 1981.

As usual, the American Technion Society has been the major bulwark of support. Its successive Presidents have been Laurence A. Tisch (1972–73), Henry Taub (1974–76), Samuel Neaman (1977), Alexander Hassan (1978–79), Theodor H. Krengel (1980–81) and Jacob Goldman (1982–   ). President of the Women's Division is Mrs. Pearl Milch. Executive Vice Presidents of the Society following Charles Scher have been Louis Levitan, who died in office in 1979, and Saul Seigel.

The Americans have continued their massive aid for the physical building of the campus as well as the intensification of research, provision of scholarship funds and the many other avenues of support.

When the Churchill Auditorium at Technion City was gutted by fire in 1979, the British Technion Society mobilized additional resources which made possible reconstruction of an improved building with additional facilities. Officers were Edgar Astaire, Chairman; Lord Mishcon, Hon. President; Sir Isaac Wolfson, Bt., Hon. Life President.

Under the leadership of Bernard M. Bloomfield, the Canadian Technion Society added to the student dormitory units of Canada Village on the campus. The Lady Davis Aeronautical and Mechanical Engineering Center has been completed and work started on the Canada Nuclear Engineering Institute. Eugene Stearns has continued as Chairman of the Board. Nicholas Munk retired as treasurer after some three decades of service.

The Society in France, Le Groupement des Amis du Technion, has continued under the effective leadership of Jean Paul Elkann and Josef Vaturi, and has established the Pierre Mendes France Chair in Economics.

Valuable assistance has been received from South Africa, primarily from the Mauerberger Foundation, through the efforts of Solm Yach.

Dedicated work by the late Harry Castle over many years in Sydney provided a firm foundation for a Technion group in that city. Parallel activity has taken place in Melbourne. Their joint project has become Australia House–Coastal and Marine Engineering Research Institute.

The Mexican Technion Society has continued to enjoy the leadership of Max Shein, and its principal project, among others, has been the Enrico Berman–Max Shein Energy Fund.

Similar groups exist in Argentina, Brazil, Uruguay and Venezuela, and new Technion support societies have been organized in Switzerland, The Netherlands and Germany.

The Future

What of the future? The Technion is not an end in itself. It survived against all odds because it fulfilled a need. It is difficult to conceive of Israel as we

In the laboratory for low temperature physics. Technion scientists have managed to reach a temperature of within a few parts of a thousand of absolute zero, minus 273.16 degrees, Celsius.

know it today, had it not been for the skilled manpower made available by the Technion and the services rendered by its facilities to both the civilian economy and the defense establishment. So long as those needs exist, and the Technion remains true to its vital mission, the institution will flourish.

Indeed, informed expectations are that the needs will mount. The anticipated growth of sophisticated, science-based industries is already increasing the demand for precisely the kind of personnel trained at the Technion. Figures drawn up in 1980 revealed that the demand for engineers in industry and other economic sectors, including replacements for retirement, amounted annually to 2,200. This number was met by the supply composed of domestic graduates, Israelis returning to the country, and immigrant engineers. The chart shows a mounting demand curve which during the decade thereafter will call for addition of almost 21,000 new engineers to the work force. If the present number of domestic graduates and returnees remains static, less than 13,000 trained men and women will be available, leaving a very serious manpower deficit. Even an

B. Rappaport Family Medical Sciences Building.

erin, Sarah. *Dr. A. Biram and His Reali School: Tradition and Experimentation in Education.* Jerusalem, 1970.

oel Hatzair. Jaffa, successive issues, 1912–1914.

sverein der deutschen Juden. *Gedenkschrift für Paul Nathan, Jahresbericht für 1926.* Berlin, 1927.

usalem Post. "Appeal in Sitte Case Dismissed" (July 5, 961).

ish Technical Bulletin. New York, 1923.

ch, Frederick H. *Palestine Diary.* London, 1938.

ng, Simcha. *The Mighty Warrior: the Life Story of Menahem Ussishkin.* New York, 1965.

dau, Dr. Yitzhak. *B'Einei Yedidim.* Jerusalem, 1968.

honenu La-Am: Popular Pamphlets in Matters of Language, Elul 1979 (Scientific Secretariat of the Hebrew Language Academy). Jerusalem, 1979.

vin, Schmaryahu. *The Arena.* New York, 1932.

ichro Shel Shmuel Pewsner. Haifa, 1940.

oz, Moshe, ed. *Studies on Palestine During the Ottoman Period.* Jerusalem, 1975.

eyer, Isidor S., ed. *Early History of Zionism in America.* New York, 1958.

chaeli, Y. *Doar Hayom* (January 21, 1926).

or, Mordecai. "Hasabta Shel Haknesset," *Yediot Ahronot* (August 25, 1978).

— "K'sheYerushalayim Hashka Batechnion," *Maariv* (September 23, 1977).

athan, Paul. *Palästina und palästinensischer Zionismus.* Berlin, 1914.

esher, Aryeh. "Yerushat Hamillionim Shel Swope," *Haaretz* (December 24, 1959).

— "Hamachloket B'Fakulta L'Architecktura Batechnion," *Haaretz* (April 3, 1966).

w York Times (October 2, 1910) (3:3).

— (August 23, 1913) (6:8).

— (January 18, 1914) III, 3:6.

— (January 20, 1914) 3:7.

(February 1, 1914) III, 3:1.

(July 6, 1914) 7:5.

February 15, 1915) 7:5.

December 28, 1918) 17:3.

October 5, 1920) 1:3.

Rabinowicz, Oscar K. *Fifty Years of Zionism: A Historical Analysis of Dr. Weizmann's Trial and Error.* London, 1950.

Razsvet. "A Dirty Trick" (April 5, 1915).

Rinott, Moshe. *Hevrat Ha-Ezra L'Yehudei Germania B'-Yetzira Ub'Maavak.* Jerusalem, 1971.

Sacher, Harry, ed. *Zionism and the Jewish Future.* London, 1916.

Sefer Hayovel Shel Histadrut Hamorim 1923–1928. Jerusalem, 1929.

Stern, Gabriel. "L'Zichro Shel Yehuda Magnes." *Al Hamishmar* (October 25, 1968).

Szakowski, Zosa. "Conflicts in the Alliance Israélite Universelle and the Founding of the Anglo-Jewish Association, The Vienna Allianz and the Hilfsverein," *Jewish Social Studies,* No. 1–2, 1957.

Technion Souvenir Journal. American Society for the Advancement of the Hebrew Institute of Technology, N.Y. (1942, 1944).

Technion Yearbook (1945–1968), American Technion Society, N.Y.

Thon, Jacob. "Jewish Schools in Palestine," *Zionist Work in Palestine,* ed. Israel Cohen. New York, 1912.

Toldot Hatechnion B'Reshito (1908–1925). Haifa, 1953.

Universal Jewish Encyclopedia, Vol. V. p. 360. New York, 1941.

Urofsky, Melvin I. *American Zionism from Herzl to the Holocaust.* New York, 1975.

Weizmann, Chaim. *Trial and Error.* New York, 1949.

Wilbushewitz, Gedalyahu. "Hayesh Tzorech Batechnikum?" *Haaretz* (July 27, 1923).

Yellin, Aviezer. "Al Hayovel Shel Milchemet Hasafot," *Hahinuch* (1963–64).

— — "Mardut," *Sefer Hayovel Shel Histadrut Hamorim.* Jerusalem 1939.

A corner of Australia House–Coastal and Marine Engineering Research Institute.

immigration of 5,000 engineers during the ten-year period, at the optimistic rate of 500 per year, will still leave industry short of the personnel required. Obviously, only an increase of domestically trained manpower will meet the situation.

Fortunately, population statistics indicate that the number of high school graduates in Israel qualified for university studies, which reached a total of 17,450 in 1980–81, will continue to rise, reaching 22,980 in 1990, and 25,580 in 1994. This will result in increased demands upon the Technion which will be relieved only to a degree by the expansion of technological/engineering/scientific studies at Ben Gurion University of the Negev or at other institutions.

This is truly a case of history repeating itself. It may be recalled that in 1928 an "expert" had after thorough research determined that in the foreseeable future the country would have need for no more than four or five engineers a year. Even the surveys made during the 1950's and 60's, which many felt exaggerated the needs, eventually fell short of reality. Historical development has a dynamism of its own, often unpredictable.

Sources

This generation, and that which follows, may be expected to bear their responsibilities to the Technion no less than those which preceded them. In his report to the Board of Governors in 1980 General Horev paid tribute to those whom he credited with the successful operation of the Institute, each in his own field, each with his specific functions. He termed them the "Technion family", working cooperatively toward a common goal: "The diligent student body, the learned members of the faculty, the loyal administrative and technical employees, the faithful alumni, and the generous friends and supporters in Israel and abroad."

Their reward is the ability to implement Technion's program and fulfill the tasks imposed on the Institute, in continuation of a long and distinguished tradition.

* * *

An epilogue is usually the added and concluding part of a work. In this case there is no conclusion. Our epilogue serves as a bridge from the history of the past, over the present, in continuity into the as yet unrecorded future.

* * *

Unpublished sources include: Minutes of the Technion *Vaad Menahel, Moetzet Hamorim (Haprofessorim)*, Senate and Board of Governors; Protocols of the Provisional Technion Committee in Haifa; Protocols of the Palestine Executive of the Zionist Organization; Minutes of the Executive Committee and Board of Directors of the American Technion Society; periodic Technion reports to the Zionist Congresses; Asher Mallah papers and Memoirs of Gedalyahu Wilbushewitz in the Technion Archives; the letters of Jacob H. Schiff and others in the American Jewish Archives; collections of letters and other unpublished files in the Zionist Archives, Jerusalem; and most especially the correspondence, financial and academic files in the Technion Archives.

Various publications of the Technion, including *Technion* Magazine, *Bet Hatechnion, Yediot Agudat Dorshei Hatechnion*, Technion annual *Catalogue*, Technion City Master Plan (1971–   ), Annual Report of the President to the Board of Governors (1952–   ), *Technion News Overseas Bulletin* (1951–71), *Yediot Hatechnion, Inside Technion* and others.

Personal interviews with principals concerned.

CHAPTERS I–III:

Adler, Cyrus. *Jacob H. Schiff: His Life and Letters*. Vol   New York, 1929.

Ahad Ha-Am. *Essays, Letters, Memoirs*. Translated f   the Hebrew. Ed. Leon Simon. Oxford, 1946.

— — *Iggrot*. Jerusalem, 1924.

Alsberg, Avraham P. "The Hebrew Technical College," *Z*   (January–February 1950).

Avissar, David. "Hatesisah Harishonah," *Sefer Hay*   *Shel Histadrut Hamorim* (Jerusalem, 1939).

Biska, A. L. "A Report from Manchester," *Hazvi*, Jerusal   (26 Tammuz, 1909).

Buber, Martin, Berthold Feiwel and Chaim Weizmann.   *Sefer Gavoa Yehudi*. Geneva-Lausanne, 1902.

Carmel, Alex. *Toldot Haifa Biyemei Haturkim*. Haifa, 19

Chomsky, Zeev. *Halashon Ha-Ivrit B'Darkei Hitpath*   Jerusalem, 1967.

Cohen, Israel. *The German Attack on the Hebrew Scho*   *in Palestine*. London, 1918.

Cohen, Israel, ed. *Zionist Work in Palestine*. London, 1

Cohn-Reiss, Ephraim. *M'Zichronot Ish Yerushalayim*.   salem, 1933, 1936.

Eliav, Mordecai. *Hayishuv Hayehudi B'Eretz Yisr*   *Hamediniut Hagermanit 1842–1914*. Tel Aviv,

Feder, Ernst. "Paul Nathan and His Work for Eas   and Palestinian Jewry," *Historia Judaica*

— — "Paul Nathan, the Man and His Work," *L*   *tute Yearbook III* (1958).

— — *Politik und Humanitat, Paul Nathan*   Berlin, 1929.

— — *Die Technische Hochschule von H*   *und ihre Entstehung*. Rio de Janei

Friedman, Isaiah. *Germany, Turkey*   *1918*. Oxford, 1977.

Gordon. A. D. *Michtavim U'reshi*   1954.

Gross, David C. "Strange to   (January 22, 1977).

Grunwald, Kurt, "Jewish Sc   Ottoman Palestine," in   *Ottoman Period*, ed.

404

406

CHAPTERS IV–XII and Epilogue:

Alexander-Katz, Dr. E. *The Technion and Its Future.* London, 1948.

Alpert, Carl. *Palestine Between Two Wars.* Washington, D.C., 1944.

Bentwich, Norman and Michael Kisch. *Brigadier Frederick Kisch: Soldier and Zionist.* London, 1966.

Blok, Arthur. "Early Days at the Technion," *Technion Yearbook,* Vol. XI (1952–53).

— — "Those Were the Days," *Technion* Vol. I, No. 1 (June 1965).

Campbell, John C. *Defense of the Middle East.* New York, 1958.

Celler, Emanuel. "Technion and Israel," *Congressional Record* (September 18, 1963).

Davar, May 14, 1929 (On the Badian affair).

Die Neuer Welt (April 24, 1929).

Eshel, Zadok. *Maarachot HaHaganah B'Haifa.* Tel Aviv, 1978.

Haaretz. "Interview with Prof. H. Hanani" (July 30, 1963).

The Hebrew Technical Institute, Haifa: Its Achievements and Its Aims. Haifa, 1938.

Irmay, Shragga. "The Development of Scientific Hebrew," *Technion Yearbook* (1959).

— — "Kaytzad Yotzrim Milim Ivriyot B'Technologia," *Technion Quarterly* (Autumn 1971).

Jerusalem Post. "Unusual Building to House Technion Mechanical Engineering" (October 11, 1966).

Jewish Chronicle, London (August 22, 1924) (February 27, 1925).

Jones-Grossman, Judith. "My Father, Prof. Jeremias Grossman," *Technion* (August 1965).

Kohn, David, ed. *Manpower Planning: Research and Statistics.* Haifa, 1970.

Learsi, Rufus. *Fulfillment: The Epic Story of Zionism.* Cleveland, 1951.

Levin, Nahum. *Maavak Harishonim Al Yeud Hatechnion.* Tel Aviv, 1964.

Mardor, Munya M. *Haganah.* New York, 1964.

Nessyahu, Yehoshua. Text of address prepared for delivery at reception in honor of his 50th anniversary with the Technion, September 29, 1974.

Neufeld-Tzerniak, Zippora. Mizichronot Hastudentit Harishona Batechnion," *Yediot Hatechnion* (June 1965).

New Judaea (September 26, 1924).

— — (February 13, 1925).

— — (February 27, 1925).

— — (March 4, 1927).

— — (August–September 1932).

— — (December 1932).

New York Times "British Rebuffed by Gerard Swope," (June 4, 1948).

— — Gerard Swope, 84, Ex-G.E. Head Dies" (November 21, 1957).

— — "Haifa School Gets Swope Millions" (December 4, 1957).

Palestine Post (February 3, 1943).

— — Editorial (December 15, 1943).

Rappaport, Shimon. "Ha-Adrichal Ha-Akshan Nitzach Et Hamimsad," *Maariv* (January 3, 1972).

Reiner, Markus. "Ten Years of Technological Research in Israel," *Technion Yearbook* (1959).

Ruppin, Arthur. *Pirkei Chayai.* Tel Aviv, 1968.

Salomon, Ya'akov. *B'Darki Sheli.* Jerusalem, 1980.

Shuval, Dov. "Adrichal Neevak Al Raayonotav," *Al Hamishmar* (January 7, 1972).

Singer, Mendel. *Shlomo Kaplansky, Chayav U-Poalo,* Jerusalem, 1971.

Slutsky, Yehuda. *Sefer Toldot Hahagana, Vol. II.* Tel Aviv, 1959.

Tahazit Hapotential Shel Studentim Hadashim B'Mosdot L'Haskala Gvoha L'Tkufat 1980–81 – 1996–97. Jerusalem, 1981.

Tcherniavsky, Aharon. "Zichronot," *Yediot Hatechnion* (June 1965).

Wilmington, Martin W. *The Middle East Supply Centre.* Albany, 1971.

Yaari-Folskin, Y. *Arthur James Balfour, His Life and Speeches.* Tel Aviv, 1930.

Zionist Review. "Interview with Dr. Kaplansky" (June 16, 1950).

Shikun. Housing project.

Slick. Hidden cache of arms and ammunition.

Small Zionist Actions Committee. Executive group of the Zionist General Council which functions between Zionist Congresses.

Solel Boneh. Construction, industrial and contracting arm of the Histadrut.

Spritz. Dilution of wine with soda water.

Tanach. The Hebrew Bible.

Tav Melet. The levy on cement for the benefit of the Technion.

Templars. German Christian group which in the latter part of the 19th century set up settlements in Haifa, Jerusalem and Jaffa.

Tu B'Shevat. Holiday also known as New Year of the Trees, marked by tree-planting.

UIT. Joint fund-raising office of the Hebrew University, Weizmann Institute and Technion in the U.S.

UJA. United Jewish Appeal, comprehensive fund-raising organization of American Jewry.

Ulam Haknesset. Literally, assembly hall.

UPA. United Palestine Appeal of American Jewry.

Vaad Halashon Haivrit. Committee of the Hebrew Language, later became the Academy for the Hebrew Language.

Vaad Leumi. The executive committee of the democratically elected National Assembly of Palestine Jewry.

Vaad Menahel. Literally, Administrative Committee. The name was used for the different executive bodies of the Technion at various times.

Wadi. Land depression, often bearing water during the rainy season.

Wadi Rushmiyah. Land cleft between hills in East Haifa.

Yerida. Emigration out of Israel.

Yishuv. The collective name for the organized Jewish community in Palestine before establishment of the State of Israel.

Yom Kippur. Jewish Holy Day of Atonement.

Zionist Commission in Palestine. The Commission appointed by the British in 1918 to advise on the implementation of the Balfour Declaration.

Zionist Congress. The supreme authority of the World Zionist Organization. The first Congress was held in 1897. At first it met annually, then biennially, and in recent years quadrennially.

Zionist General Council. The Zionist Actions Committee, which exercises authority between Zionist Congresses.

Appendices

APPENDIX A

Major Engineering and Science Buildings

Australia House — Coastal and Marine Engineering Research Institute

Baltimore Aerospace Laboratories

Samuel and Belle Bernstein Building, Chemistry

Anne Borowitz Building, Civil Engineering

Canada Building, Chemistry

Karl Taylor Compton Building, Chemistry

Benjamin Cooper School of Industrial Engineering & Management

Danciger Mechanical Engineering Building

Ralph and Frances DeJur Materials Engineering Center

Detroit Mechanical and Aeronautical Engineering Laboratories

Albert Einstein Institute of Physics

Henry F. Fischbach Electrical Engineering Building

Food Engineering Pilot Plant

Samuel A. Fryer Aeronautical Engineering Building

General Studies — Humanities — Social Sciences Building

Goldsmith Institute of Industrial Microbiology

Miriam and Aaron Gutwirth Science-Based Industries Center

Hassan Washington Chemical Engineering Pilot Plant

Ganna and Anatol Josepho Industrial Research Center

Alexander and Anna Konoff Junior Technical College Building

Lidow Building for Research in Physics

Helen and Morris Mauerberger Soil Engineering Building

Ostrow Building, Aeronautical Engineering

Petrie Pavilion — Mathematics and Applied Mathematics

B. Rappaport Family-Medical Sciences Building

Maurice and Ruben Rosen Solid State Building

George and Beatrice Sherman Education in Technology & Science Center

Leonard and Diane Sherman Environmental Engineering Center

Julius Silver Biomedical Engineering Building

Sobell Chemical Engineering Building

Louis and Bess Stein Food Engineering Building

Henry and Marilyn Taub Computer Science Building

Morris and Sylvia Taub Computer Center

Werksman Building, Pittsburgh Project, Physics

APPENDIX B

Major Research Laboratories and Wings

Charles and Bertha Bender Aeronautical Engineering Lab.

Bernard High Pressure Laboratory

Samuel Brody Wing, Agricultural Engineering

Bruner Institute of Transportation

Buckstein Research Laboratory, Chemistry

Cooperband Shock Research Center

Fenichel Laboratory for Human Psychobiology

Elias and Bertha Fife Materials Testing Laboratory

Samuel Fryer Research Tower, Agricultural Engineering

Isidor Goldberg Electronics Center

Goldstein Packaging Laboratory

Ida (Babe) Goodstein Center for Food Chemistry (Fund for Higher Education in Israel)

Arturo Gruenebaum Metallurgy Center

Isin and Fancia Ivanier Laboratory for Welding & Casting Research

Josepho Family Wing (Materials Engineering Center)

Bessie Koenig Road Safety Center

Kranzberg Industrial Engineering & Management Wing

Kranzberg Laboratory in Electronics

Krumbein Aircraft Structures Laboratories

Lipson Environmental Wind Tunnel, Toronto Project

Esther and Shmuel Margolin Automation & Control Laboratory — Agricultural Engineering

Milwaukee Electrical Testing Laboratory

Hy and Anne Natovich Orthopedic & Rehabilitation Research Center

Albert and Jean Nerkin Microcomputer Laboratory

Edith and Louis Penner Wing, Civil Engineering
Michel Polak Building Research Station
Rich-Rohlik Engineering Laboratory, Mechanical Engineering
Rohlik Energy Laboratory, Mechanical Engineering
San Francisco Wing, Agricultural Engineering
David T. Siegel Hydraulics Laboratory
Jacob and Sallie Simon Acoustics Laboratory

Tark-Recu Turbo & Jet Engine Building
Mendel and Hersz Tenenbaum Research Laboratory, Chem.
Van Arsdale Electrical Engineering Wing
Wedner Physics Wing
Wunsch Memorial Laboratory, Mechanical Engineering
Harry Zekelman Research Laboratory, Chemistry

APPENDIX C

Major Faculty and Student Facilities

Calgary-Edmonton Student Computer Training Center
Moshe and Palomba Carasso Wing for Self-Study, Ullmann Center
Churchill Auditorium
Elyachar Central Library
Philip and Frances Fried Student Counselling Center
Samuel and Isabelle Friedman Auditorium
Horowitz Audio-Visual Center
Kunin-Lunenfeld Academic Preparatory Center
Medical Students' Activity Center — ATS Women's Division

Ohel Aharon — Campus Synagogue
Oscar and Ethel Salenger Guest House
Senate House — Sir Louis Sterling Memorial
Shine Student Union Building
Spertus Auditorium
Sports Facilities
  Pinhas Rasner Sports Ground
  Woolf Senior Sports Center
  Joseph and Caroline Gruss Swimming Pool
Siegfried and Irma Ullmann Teaching Center

APPENDIX D

Student Hostels

Harry DeJur Village
  Casa Mexico, Rosario Castellanos
  La Maison de France — Louis Armand — Student Hostel
  Max and Amparo Shein Student Hostel
  International House Student Hostel
    Finkelstein Building
    Kennedy Leigh Building
  Harry DeJur Student Center
D. Lou Harris Student Dormitories (Canada Village)
  Hyman and Sophie Bolter Building
  Gertrude H. and Harold J. Caster Building

Estelle and Eugene Ferkauf Building
Theodore, Albert and Henry Gildred Building
Dr. Jacob Isler Building
Louis and Sylvia Lockshin Building
Nicholas and Hedy Munk Building
Chris Sharp Cafeteria
Harry and Abe Sherman Building
Noona and Shoua Soffer Building
Eugene and Dr. Anna Stearns Building
Jacob and Esther Stiffel Building
Tachna Building

Joseph and Faye Tanenbaum Lobby
Rifkin Village
 Ahavas Chesed Student Center
 Belchtovski Student Club House
 Rose Bender Student Hostel
 Casa Argentina Student Hostel
 Jerome Loeb DeJur Student Hostel
 Beatrice E. Fischbach Student Hostel
 Eva Frost Student Hostel
 Rosalinde and Arthur Gilbert Student Hostel
 Goldblatt Student Center
 Henrietta and Stuard Hirschman Student Hostel

 Rose Mazer Student Hostel
 Philadelphia Chapter Women's Division Student Hostel
 Rebecca Rose Student Hostel
 Sylvia and Barnett Shine Student Hostel
 Student Cultural Center
 Louis and Rebecca Susman Student Hostel
 Michael and Anna Wix Student Hostel
 Jacob Ziskind Student Hostel
Junior Staff Housing
 Kennedy Leigh Buildings
 Rose-Shine Buildings
 Kessel Building-Mexico Medical Student Dormitories

APPENDIX E

Roads, Parks and Other Facilities

August Lane
Arthur Blok Lane
Chicago Avenue
A. Cutler Gardens
Abraham and Esther Davis Garden
Deutsch Bridge of Learning
Yaakov Dori Road
Federmann Garden
Grossman Garden
Gutwirth Ecological Garden
Kislak Park
Kutz Science Park

Lilienthal Tribune
Munk Garden
Haim Reiskin Promenade
Joseph and Sadie Riesman Plaza
David Rose Avenue
Lawrence and Aleen Schacht Park
Harry and Betty Shapiro Garden
Nate Sherman and Family Forum
David Silbert Square
Technical Services and Storage Center
Abraham Tulin Forest
Joseph and Theresa Wertheimer Water Tower

APPENDIX F

Buildings Under Construction or Being Planned (as of 1981)

Maurice Amado Mathematics Building
Canada Nuclear Engineering Institute
Lady Davis Mechanical-Aeronautical Engineering Center

Meyer Davis Agricultural Engineering Building
Yaakov Dori Faculty Center
Samuel and Isabelle Friedman Nutrition and Food Chemistry
 Building.

Joseph and Caroline Gruss Dormitory
Walter and Ruth Leventhal Dormitory — Philadelphia
Project
Bella Meyer Advanced Technology Center
Miami Education Center

Michigan Dormitory
Palm Beach Married Students' Dormitory
San Francisco Married Students' Dormitory
Rahel and Uriel Shalon Married Students' Dormitory

APPENDIX G
Academic Chairs and Lectureships
Permanent Chairs

Yigal Allon Chair in the Sciences of Man at Work
Bertha Axel (nee Hertz) Chair in Chemistry
Joan Goldberg Arbuse Chair in Electronics
Stephen E. Berger Chair in Aeronautical Engineering
Joseph and Sadie Danciger Chair in Engineering
Lady Davis Chair in Experimental Aerodynamics
Max and Lottie Dresher Chair in Aerospace Performance
& Propulsion
Louis Edelstein Chair in Cancer Research
George Farkas Chair in Computer Science
Carl Fechheimer Chair in Electrical Engineering
Rebecca and Herman Fineberg Chair in Geriatrics
Minnie and Ruben Finkelstein Chair in Aviation & Space
Medicine
William Fondiller Chair in Telecommunications
Pierre Mendes France Chair in Economics
Samuel O. Freedlander Chair in Physiology
Henri Garih Chair in Materials Processing
Henry Goldberg Chair in Biomedical Engineering
Sydney Goldstein Chair in Aeronautical Engineering
Herman Gross Chair in Communications
Arturo Gruenebaum Chair in Mining & Metallurgy
Josef Gruenblat Chair in Production Engineering
Matwei Gunsbourg Chair in Civil Engineering
Shlomo Kaplansky Chair in Agricultural Engineering
Harry Lebensfeld Chair in Industrial Engineering
Levinson Chair in Food Technology
Lidow Chair in Solid State Physics
Lunenfeld-Kunin Chair in Urban Planning
Chair in Management Sciences

Albert and Anne Mansfield Chair in Water Resources
Roy Matas/Winnipeg Chair in Biomedical Engineering
Joseph Meyerhoff Chair in Urban & Regional Planning
Markus Reiner Chair in Mechanics & Rheology
Joseph and Sadie Riesman Chair in Electrical Engineering
Rosenblatt Chair in Mechanical Engineering
Louis and Samuel Seiden Chair in Soil Engineering
Beatrice Sensibar Chair in Environmental Engineering
David T. Siegel Chair in Hydraulics
Philip Slomovitz Chair in Hebrew Language
Sir Michael and Lady Sobell Chair in Electrical &
Electronic Engineering
Sigmund Sommer Chair in Structural Engineering
Gerard Swope Chair in Electrical Engineering
Gerard Swope Chair in Physics
Gerard Swope Chair in Mechanics
Mary Hill Swope Chair in Architecture
L. Shirley Tark Chair in Aircraft Structures
Philip Tobias Chair in Glass and High Temperature
Technology.
Abraham Tulin Chair in Operations Research
Wolfson Chair in Chemical Engineering
Charles Wolfson Chair in Nuclear Sciences
Abel Wolman Chair in Sanitary Engineering
Gideon Zimmerman Chair in Food Engineering

Chairs and Lectureships Occupied by Annual
Invitation

Albert Alberman Visiting Lectureship
Helen and Norman Asher Visiting Lectureship

414

Andre Ballard Chair in Psychiatry

Karl T. Compton Chair

Corob Visiting Lectureship

Lady Davis Fellowships for Visiting Professors

Lena and Ben Fohrman Lectureship in Aeronautical Engineering

Nat Goldman Visiting Lectureship in Energy Research

George Griffith Lectureship in Medicine

Miriam and Aaron Gutwirth Visiting Lectureship

S.B. Harbour Visiting Professorship

Harry Zvi Josselson Lectureship

Kranzberg Visiting Lectureship in Electronics

David and Gertrude Krengel Chair in Metallurgy

Prof. Pinhas Naor Memorial Lectureship

Petrie Chair — Visiting Professorship

Harry D. Pierce Visiting Professorship

Sidney Quitman Lectureship

Sandy and Russell Rosen Lectureship in Chemical Engineering

Harry and Abe Sherman Chair in Engineering

Louis Susman Visiting Professorship in Aeronautical Engineering

Isaac Taylor Visiting Professorship in Energy

Sidney and Beatrice Wolberg Chair in Mechanical Engineering

Joseph Wunsch Annual Lectures

Anna R. Zager Visiting Lectureship in Pediatrics

Felix Zandman Visiting Lectureship in Aeronautical Engineering

Chairs and Lectureships in Process of Establishment (as of 1981)

Irving Boren Chair

Kathleen and Morton Bank Chair in Mathematics

David Erlik Chair in Surgical Sciences (established by Eedis Cooperband)

Dr. Joel Hamburger Chair in Clinical Thyroidology

Harry W. Labov and Charlotte Ullmann Labov Chair

Samuel and Dore Liebman Chair in Medical Research

Bernard Mirochnick Chair in Bioenergetics

Ruben and Tallu Rosen Chair in Solid State Physics

Sherman–Gilbert Chair in Solar Energy

Barnett Shine Chair in Rheumatology

Meyer E. Smirnoff Surgical Lectureship

Gertrude W. and Edward M. Swartz Professorship in Engineering

Stanley Vineberg Memorial Visiting Fellowship

Washington Chapter ATS Chair in Energy Conversion

APPENDIX H

Special Endowment Funds

The Rose and Isak Alcazar Scholarship Fund

Lester Aronberg Prize Fund

Lester Aronberg Amyothropic Lateral Sclerosis Research Fund

August Lane Maintenance Fund

Kathleen and Morton Bank Fund in Mathematics

Barsky Award in Optics

Dario Beraha Fund for Cancer Research

The John and Sara Jean Berg Scholarship Fund

Berman-Shein Solar Energy Fund

Walter Bernstein Scientific Research Fund

Adolfo and Evelyn Blum Medical Research Fund

Edith Blum Foundation Cancer Research Fund

Manuel A. Borinstein Medical School Fund

Canada Village Maintenance Fund

Carasso Self-Study Center Maintenance Fund

Isidore and Theresa Cohen Fund

Caesarea Foundation Fund

Joseph and Sadie Danciger Fund for Scientific Purposes

Women's Division Louis and Pauline Cohen Scholarship Fund

Leon and Ruth Davidoff Library Fund

Davidson Library Fund

Lady Davis Foundation

Lawrence Deutsch Memorial Research Funds

Discount Bank Endowment Fund for Textbooks

Milton and Lillian Edwards Scholarship Fund

Mike Feldman Energy Fund

Henry F. Fischbach Exchange Fellowship Fund

Phillip and Frances Fried Counselling Center Maintenance Fund

Henry Ford II Fund for Transportation Research

The Foundation for Charity Scholarship Fund

Max and Dora Frocht Fund for Research & Teaching Fellowships

Blanche and Norman Ginsburg Research Endowment Fund in Energy Research

Meyer Gold Memorial Library Fund

Alexander Goldberg Fund for Chemical Engineering Education & Research

Bernard Goldberg Fund

Leopold Goldmuntz Fund

Horace W. Goldsmith Microbiology Fund

Leslie Gulton Fund for Medical Research

Henry Gutwirth Fund for Promotion of Research

Julius and Dorothy Harband Scholarship Fund

Ruth Lee Harvey Fellowship Fund

Harvey Prize Fund

Frank J. Iny Scholarship Fund

Isasbest Fund

Louis R. Jabison Memorial Scholarship Fund

Martin and Dorothy Kellner Fund

Julius Kislak Fellowship Fund

Raymond Klein Prize Fund

Jules Kramer Music & Fine Arts Fund

Max Harris and Betty Muriel Kranzberg Fellowship Fund

Ronald Lawrence Neuro-Psycho Pharmacological Research Fund

Kennedy Leigh Fund for Biomedical Engineering

William Levenson Memorial Fund for Medical Engineering

Ralph Levitz Fellowship Fund

Leon Lidow Fund

Mailman Fellowship Fund

Mexico – Israel Energy Fund

Andre Meyer Fund for Operations Research

Ronnie T. Meyer Memorial Scholarship Fund

Miami Energy Research Fund

Bernard and Louis Mirochnick Fellowship Fund

Oil Industry Research Fund in the name of Prof. Arieh Litan (Kwitny)

Seniel Ostrow Research Fund

Pecker Steel Fund

Colonel Asher Pelled Memorial Fund

Raymond Pepp Research Fund

Regina Pollak Fund

Felice and Nathan Ratkin Scholarship Fund

Leo, Johanna, Oskar and Melitta Reichmann Scholarship Fund

Israel and Leon Reiskin Graduate Endowment Fund in Architecture & Town Planning

Herman and Mary Robinson Scholarship Fund

Louis Rogow Aeronautical Research Center Fund

Maurice M. Rosen Solid State Building Maintenance Fund

Carl and Lillian Schustak Student Aid Fund

Joseph Haim Sciaky Memorial Fund

Architect Arthur Shragenheim Fund

Max and Doris L. Starr Ophthalmology Fund

Dr. Max and Brunia Steigman Fund

Martha and Harry Stern Program of Continuing Education

Dr. Jacob and Jeanne Sternberg Scholarship Fund

Morris and Ann Sussman Loan Fund for Graduate Students and Fellows

Helen and John Strykoff Memorial Scholarship Fund

L. Shirley Tark Research and Scholarship Fund

Joseph and Arlene Taub Biological Research Fund

Monte H. and Bertha Tyson Fellowship Fund

Ullmann Center Endowment Fund

Western Region Energy Research Fund

Nahum Wilbush Memorial Fund

Women's Division Medical Engineering Project Fund

Felix Zandman Fund

Max and Ida Zuckerwise and Family Faculty Lounge Maintenance Fund

APPENDIX I

Honorary Degrees Awarded
Doctor of Science in Technology

Dr. William Fondiller	1949
Dr. Shlomo Kaplansky	1950
Dr. Theodore von Karman	1951
Dr. Chaim Weizmann	1952
Dr. P. F. Danel	1952
Sir Ben Lockspeiser	1952
Dr. Walter C. Lowdermilk	1952
Dr. L. A. Richards	1952
Dr. Albert Einstein	1953
Dr. J. Franck	1953
Sir Patrick Abercrombie	1953
Dr. R. J. Forbes	1953
Dr. George Sarton	1953
Dr. Karl Taylor Compton	1954
Dr. J. W. Wunsch	1955
Dr. E. D. Bergman	1955
Dr. Gerard Swope	1957
Dr. F. Julius Fohs	1957
Dr. Abraham Tulin	1957
Dr. M. Novemeysky	1957
Dr. Emanuel Goldberg	1957
Dr. Simha Blass	1958
Dr. Haim Slavin	1958
Dr. Philip Sporn	1960
Dr. Robert L'Hermite	1960
Dr. David Rose	1961
Dr. J. R. Sensibar	1963
Dr. Edward E. Rosen	1966
General Yaakov Dori	1967
Dr. Al Schwimmer	1968
Dr. Manes Pratt	1968
Dr. Aharon Wiener	1971
Dr. Julius Silver	1971
Dr. Harry F. Fischbach	1971
Dr. Arthur Blok	1972
Dr. Alexander Goldberg	1975
Dr. David Laskov	1975

Dr. Bern Dibner	1976
Dr. Uzia Galil	1977
Dr. Maurice M. Rosen	1978
Dr. Bernard M. Bloomfield	1978
Dr. Morris Cohen	1979
Dr. Jehiel R. Elyachar	1979
Dr. Bruce Rappaport	1979
Prof. Nicholas John Hoff	1980
Justice Moshe Landau	1980
Sir Michael Sobell	1980
Dr. Isin Ivanier	1981
Dr. Avraham Suhami	1981
Dr. Jacob Walter Ullmann	1981

Doctor of Science

Prof. Niels Bohr	1958
Dr. Carroll V. Newsom	1958
Dr. F. Houphouet-Boigny	1962
Dr. Harold C. Urey	1962
Dr. I. I. Rabi	1963
Dr. A. Biram	1965
Dr. Robert B. Woodward	1966
Dr. Selman A. Waksman	1966
Prof. Yuval Ne'eman	1966
Lord Rothschild	1968
Prof. Sydney Goldstein	1969
Prof. Eugene Paul Wigner	1971
Prof. M. Schiffer	1972
Prof. J. Wolfowitz	1972
Prof. George B. Dantzig	1973
Prof. Herman F. Mark	1975

Doctor of Architecture

Dr. David Ben Gurion	1962

Honorary Engineer

Mr. Joseph W. Wunsch	1946

Mr. Alexander Konoff	1949
Mr. Elias Fife	1955

Honorary Fellow

Col. J. R. Elyachar	1953
Mr. Arthur Blok	1954
Mr. Max Hecker	1954
Sir Louis Sterling	1956
Sir Isaac Wolfson	1956
Mr. Samuel Fryer	1959
Mr. S. J. Birn	1965
Dr. Jacob Isler	1970
Ing. Aharon Goldstein	1971
Mr. Alexander Whyte	1972
Mr. Barnett Shine	1972
Mr. Leo M. Harvey	1972
Dr. Abel Wolman	1972
Mr. Maurice M. Rosen	1972
Mr. Jacob W. Ullmann	1972
Sen. Jacob K. Javits	1973
Mr. Ludwig Jesselson	1973
Mr. Eliyahu Sacharov	1973
Mr. Victor Tabah	1973
Mr. Samuel M. Bernstein	1975
Mr Horace W. Goldsmith	1975
Mr. Alexander Hassan	1975
Mr. Michael Kennedy Leigh	1975
General Dan Tolkowsky	1975
Mr. Aryeh Carasso	1976
Mr. Avraham Lev	1976
Mr. Leon Lidow	1976
Mr. Joseph Riesman	1976
Mr. Eugene Stearns	1976
Mr. Nathan Goldenberg	1977
Mr. Isaac Taylor	1977
Mr. Lawrence Harvey	1977
Mr. Nate Sherman	1977

Mr. Aaron Gutwirth	1978	Mr. Norman Seiden	1979	Mr. Morley Blankstein	1981
Mr. Max Shein	1978	Mr. L. Shirley Tark	1979	Mrs. Eedis Cooperband	1981
Mrs. Rose Herrmann	1978	Mr. Anatol Josepho	1980	Mr. Zvi Langer	1981
Mr. Yekutiel Federman	1978	Mr. Joseph Meyerhoff	1980	Justice Roy Joseph Matas	1981
Mr. Moshe Bernard (Benno)		Mrs. Pearl Milch	1980	Mr. Louis Stein	1981
Gitter	1979	Mr. Louis Susman	1980	Mrs. Beatrice Sherman	1981
Mr. Josef Gruenblat	1979	Mr. Henry Taub	1980		
Mr. Louis L. Lockshin	1979	Mr. Solm Yach	1980		

APPENDIX J

Harvey Prize Laureates

Prof. Willem J. Kolff	1972	Prof. Saul Lieberman	1976	Prof. Shlomo Dov Goitein	1980
Prof. Claude E. Shannon	1972	Prof. Herman F. Mark	1976	Prof. Michael Rabin	1980
Sir Alan Howard Cottrell	1974	Prof. Seymour Benzer	1977	Prof. Efraim Racker	1980
Prof. Gershon Scholem	1974	Prof. Freeman John Dyson	1977	Prof. Hans W. Kosterlitz	1981
Prof. George Klein	1975	Prof. Bernard Lewis	1978	Prof. James M. Lighthill	1981
Prof. Edward Teller	1975	Prof. Isaak Wahl	1978		

APPENDIX K

Tenured Academic Staff (as of 1 March, 1982)

Faculty of Civil Engineering

Professor
Bear, Jacob
Diskin, Mordechai
Gluck, Jacob
Komornik, Amos
Kott, Yehuda
Livneh, Moshe
Poreh, Michael
Rebhun, Menahem
Reiss, Max
Shamir, Uri

Shelef, Gedaliahu
Shmutter, Benjamin
Soroka, Itzhak
Wachs, Alberto
Winokur, Arnold
Wiseman, Gdalyah
Zeitlen, Joseph G.
Zaslavsky, Aron

Professor Emeritus
Glucklich, Joseph

Irmay, Shragga
Karni, Joseph
Shalon, Rahel
Shklarsky, Elisha
Spira, Ephraim
Yitzhaki, David

Associate Professor
Adin, Moshe
Argaman, Yerachmiel
Craus, Joseph

Frydman, Sam
Gellert, Menachem
Getzler, Zvi
Golecki, Joseph J.
Jaegermann, Chanoch H.
Kirsch, Uri
Lin, Israel
Peer, Shlomo
Peranio, Anthony
Rohrlich, Vera
Rubin, Hillel

Rutenberg, Avigdor V.
Stoch, Leslie
Tene, Yair
Vajda, Michael
Warszawski, Abraham

Teacher A
Lupu, Alexander

Senior Lecturer
Becker, Rafael
Ben-Arroyo, Abraham F.
Bentur, Arnon
Braester, Carol
Hakkert, Shalom A.
Harel, Gershon
Ishai, Ilan
Narkis, Nava
Papo, Haim
Pisanty, Avraham
Ravina, Dan
Rom, Dan
Rosenhouse, Giora
Sharni, Dan
Sheinman, Izhak
Tatsa, Elisha Z.
Uzan, Jacob
Zelikson, Amos
Zimmels, Yoram

Senior Research Fellow
Raphael, Miriam
Rosenthal, Israel

Faculty of Architecture and Town Planning
Professor
Elon, Yochanan
Herbert, Gilbert
Hill, Moshe
Kashtan, Aharon
Wachman, Avraham

Professor Emeritus
Hashimshony, Aviah
Hoenich, Paul K.
Mansfeld, Alfred

Associate Professor
Amir, Shaul
Burt, Michael
Enis, Ruth
Gilead, Shlomo
Havkin, Daniel
Mochly, Josef
Shefer, Daniel
Thau, Avshalom

Senior Lecturer
Alterman, Rachelle
Gat, Daniel
Ilan, Yeshayahu
Law Yone, Hubert
Ne'eman, Eliyahu
Oxman, Robert
Pearlman, Wolf
Peled, Arie
Segal, Yoram
Shaviv, Edna
Tzamir, Igal
Voghera, Renzo B.
Yanai, David
Yelin, Nathaniel

Teacher A
Tibon, Haim

Faculty of Mechanical Engineering
Professor
Bodner, Sol
Gutfinger, Chaim
Hetsroni, Gad
Ishai, Ori
Lenz, Ehud

Pnueli, David
Solan, Alexander
Stotter, Artur
Weill, Roland
Wolberg, John

Associate Professor
Adler, Dan
Ber, Abraham
Blech, Joab J.
Braun, Simon
Dayan, Yehoshua
Grossman, Gershon
Koren, Yoram
Lifshitz, Jacob
Malkin, Stephen
Pessen, David
Rotem, Assa
Shavit, Arthur
Shitzer, Avraham
Tirosh, Jehuda
Yarnitsky, Yeshaya
Zvirin, Yoram

Senior Lecturer
Etsion, Izhak
Gutman, Shaul
Kaplivatski, Yona
Navon, Uri
Rozenau, Philip

Department of Materials Engineering
Professor
Blech, Ilan
Brandon, David
Katz, Dov
Minkoff, Isaac
Rosen, Abraham
Weiss, Ben-Zion
Yahalom, Joseph

Professor Emeritus
Rozeanu, Lou

Associate Professor
Dirnfeld, Shraga F.
Gutmanas, Elazar
Komem, Yigal
Levin, Lev A.
Nadiv, Shmuel
Ron, Moshe
Siegmann, Arnon

Senior Lecturer
Shechtman, Dan

Faculty of Electrical Engineering
Professor
Bar-David, Israel
Bar-Lev, Adir
Erlicki, Michael
Heymann, Michael
Katzenelson, Jacob
Kidron, Izhak
Kohavi, Zvi
Navot, Israel
Raz, Shalom
Schieber, David
Sivan, Raphael
Weiser, Kurt
Zakai, Moshe
Ziv, Jacob

Professor Emeritus
Ben-Uri, Josef
Klein, Nicholas
Madjar, Leon
Naot, Jehuda
Stricker Stefan
Zederbaum, Israel

Associate Professor
Alexandrovitz, Abraham

Arbel, Arie
Inbar, Gideon F.
Kreindler, Eliezer
Malah, David
Margalit, Shlomo
Preminger, Julius
Schoen, Eliezer
Segall, Adrian
Shamir, Joseph
Zeevi, Yehoshua
Zeheb, Ezra

Senior Lecturer
Cory, Haim
Nemirovsky, Yael
Shimony, Uri

Teacher B
Frohner, Alfred

Department of Chemistry
Professor
Ariel, Magda
Ben-Ishai, Dov
Cais, Michael
Dori, Zvi
Folman, Mordechai
Halevi, E. Amitai
Herbstein, Frank
Kimel, Sol
Loewenstein, Aharon
Loewenthal, Eli H.J.
Mandelbaum, Asher
Pauncz, Ruben
Ron, Arza
Rubin, Mordechai
Silver, Brian

Research Professor
Ginsburg, David

Professor Emeritus
Kalugai, Yitzhak

Associate Professor
Bien, Shlomo
Katriel, Jacob
Kohn, David H.
Oref, Izhack
Schmuckler, Gabriella
Speiser, Shammai
Welcman, Nathan

Senior Lecturer
Becker, Dan
Gilboa, Haggai
Gruenwald, Theodor B.
Kozirovski, Yaffa
Vromen, Sander
Yarnitzky, Chaim N.

Department of Biology
Professor
Avi-Dor, Yoram
Ben-Ishai, Ruth
Gershon, David
Nelson, Nathan
Warburg, Michael

Associate Professor
Lifschytz, Eliezer
Manor, Haim
Shalitin, Yechiel
Tal, Moshe

Senior Lecturer
Eytan, Gera D.
Kuhn, Jonathan
Shalitin, Channa

Department of Chemical Engineering
Professor
Hasson, David
Kehat, Ephraim

Narkis, Moshe
Pismen, Leonid
Ram, Arie
Resnick, William
Rigbi, Zvi
Rubin, Eliezer
Sideman, Samuel
Tadmor, Zehev

Associate Professor
Aharoni, Chaim
Lavie, Ram
Nir, Avinoam
Orell, Aluf

Senior Lecturer
Marmur, Abraham
Talmon, Yeshayahu

Department of Food Engineering and Biotechnology

Professor
Berk, Zeki
Mannheim, Chaim H.
Mizrahi, Shimon

Associate Professor
Cogan, Uri
Kopelman, Isaiah
Margalith, Pinhas
Mokady, Shoshana
Ulitzur, Shimon
Yannai, Shmuel

Senior Lecturer
Miltz, Joseph

Department of Physics

Professor
Cohen, Elisha

Dar, Arnon
Eckstein, Yacov
Gilat, Gideon
Goldberg, Jacques J.
Hirsch, Aaron A.
Kalish, Rafael
Kuper, Charles G.
Lipson, Stephen G.
Peres, Asher
Revzen, Michael
Ron, Amiram
Rosner, Baruch
Rudman, Peter
Schulman, Lawrence
Shaviv, Giora
Singer, Paul
Tannhauser, David S.
Zak, Joshua

Visiting Professor
Senitzky, Israel

Associate Professor
Altman, Colman
Ben-Aryeh, Yacob
Beserman, Robert
Brafman, Oren
Eckstein, Shulamith
Ehrenfreund, Eitan
Felsteiner, Joshua
Fibich, Moshe
Fisher, Bertina
Genossar, Jan
Oppenheim, Uri P.
Pratt, Baruch
Rosendorff, Simcha
Shechter, Hanan
Weil, Raoul B.

Senior Lecturer
Ben-Guigui, Lucien
Dado, Shlomo

420

Eilam, Gad
Gronau, Michael
Mann, Adi
Shapiro, Boris

Department of Mathematics

Professor
Aharonov, Dov
Finzi, Arrigo
Lewin, Mordechai
London, David
Marcus, Moshe
Reichaw, Meir
Saphar, Pierre
Schwarz, Binyamin
Zaks, Abraham
Ziegler, Zvi

Permanent Visiting Professor
Erdos, Paul

Professor Emeritus
Bonfiglioli, Luisa
Hanani, Haim
Netanyahu, Elisha

Associate Professor
Atzmon, Aharon
Berman, Avraham
Charit, Yehuda
Fridman, Gideon
Gordon, Yehoram
Lerer, Leonid
Liron, Nadav
Loewy, Raphael
Pinkus, Allan
Rottenberg, Reuven
Sonn, Jack
Srebro, Uri
Steinberg, Jacob

Teacher A
Arwas, Jack
Pollingher, Adolf

Senior Lecturer
Benyamini, Yoav
Boleslavski, Moshe
Chillag, David
Cwikel, Michael
Elias, Uri
Friedkin, Solomon
Horowitz, Charles S.
Lorenz, Dan
Merkine, Lee-Or
Orenstein, Avraham J.
Ran, Abselom

Teacher B
Banai, Abraham
Katz, Moshe

Teacher C
Stossel, Josepha

Department of Computer Science

Professor
Even, Shimon
Ginzburg, Abraham
Lempel, Abraham
Paz, Azaria
Yoeli, Michael

Associate Professor
Israeli, Moshe

Senior Lecturer
Francez, Nissim
Hofri, Micha
Itai, Alon
Kantorowitz, Eliezer

Department of Biomedical Engineering

Associate Professor
Dinnar, Uri
Foux, Amnon
Gath, Isac
Karni, Zvi
Lanir, Yoram
Seliktar, Rahamim
Shalev, Shlomo

Senior Lecturer
Gur, Moshe

Faculty of Agricultural Engineering

Professor
Hagin, Joseph
Kornecki, Aleksander
Seginer, Ido
Zaslavsky, Dan

Professor Emeritus
Finkel, Haim
Orlowski, Samuel

Associate Professor
Avnimelech, Yoram
Karmeli, David
Kimor, Baruch
Naveh, Zev
Nir, Dov
Ravina, Israela
Sagi, Ram
Zur, Benjamin

Teacher A
Nir, Zeev

Senior Lecturer
Amir, Ilan

Benami, Amnon
Galili, Naphtali
Manor, Gedaliahu
Neumann, Peter
Peleg, Kalman
Wolf, Dan

Department of Aeronautical Engineering

Professor
Baruch, Menahem
Gal-Or, Benjamin
Hanin, Meir
Kogan, Abraham
Libai, Avinoam
Merhav, Shmuel
Nissim, Eliahu
Rom, Josef
Singer, Josef
Stavsky, Yehuda
Timnat, Yaakov

Associate Professor
Bar-Itzhack, Itzhack
Berkovits, Avraham
Betser, Abraham A.
Elishakoff, Itzhak
Seginer, Arnan
Shafer, Jerome
Shinar, Josef
Wasserstrom, Eliahu
Weihs, Daniel
Wolfshtein, Micha
Zawistowski, Ferdynand

Senior Lecturer
Burcat, Alexander
Durban, David
Rosen, Aviv
Steinberg, Avraham

Stricker, Josef
Weller, Tanchum

Department of Nuclear Engineering

Professor
Rothenstein, Wolfgang
Shafrir, Naftali H.

Professor Emeritus
Aschner, Fritz

Associate Professor
Notea, Amos
Segal, Yitzhak
Yiftah, Shimon

Senior Lecturer
Elias, Ezra

Faculty of Medicine

Professor
Barzilai, David
Better, Ori
Gutman, David
Haim, Salim
Hershko, Avram
Palti, Yoram
Peyser, Eli
Robinson, Eliezer
Winter, Shimon T.
Youdim, Moussa

Professor Emeritus
Erlik, David
Gellei, Baruch
Peretz, Aron
Valero, Aharon

Associate Professor
Barzilai, Ami

Bental, Ephraim
Brandes, Joseph
Bursztein, Simon
Eliachar, Isaac
Eidelman, Shmuel
Front, Dov
Fry, Michael
Gershon, Harriet
Ginsburg, Haim
Hirshowitz, Bernard
Lichtig, Chaim
Merzbach, David
Neumann, Eliyahu
Paldi, Eytan
Riss, Egon
Rosenberger, Alex
Scharf, Jehuda
Schramek, Alfred
Sharf, Mordehai
Silbermann, Michael
Timor, Ilan

Senior Lecturer
Alroy, Gideon
Ber, Rosalie
Brook, Gerald J.
Coleman, Raymond
Dagan, Daniel
Epstein, Leon Mervy
Finberg, John
Finkelbrand, Sara
Gorin, Erela
Hashmonai, Moshe
Hocherman, Shraga
Laufer, Dov
Lavie, Peretz
Levitan, Emanuel
Ludatscher, Ruth
Naot, Yehudith
Nir, Izhak
Sharf, Benjamin
Zinder, Oren

Teacher B
Lindenbaum, Ella

Faculty of Industrial Engineering and Management

Professor
Avi-Itzhak, Benjamin
Avriel, Mordecai
Dar-El, Ezey Meyer
Epstein, Benjamin
Mannheim, Bilha
Passy, Ury
Rubinovitch, Michael
Yadin, Micha

Associate Professor
Adiri, Igal
Ben-Tal, Aron
Ben Zion, Uri
Feigin, Paul David
Gopher, Daniel
Jacobsen, Chanoch
Maital, Shlomo
Pollatschek, Moshe A.
Rim, Yeshayahu
Roll, Ya'akov
Rosenstein, Eliezer
Rubinstein, Reuven J.
Trifon, Raphael

Senior Lecturer
Cohen, Ayala
Goldberg, Albert I.
Kirschenbaum, Alan B.
Rosenberg, Richard D.
Weissman, Ishay

Department of General Studies

Professor Emeritus
Kurzweil, Zvi
Radday, Yehuda T.

Associate Professor
Barzel, Alexander
Atlas, Dalia

Senior Lecturer
Nameri, Dvora

Teacher B
Ararat, Nissan
Ben-Bassat, Nurith
Katz, Moshe
Mentcher, Ezra
Sasson, Moshe
Sendler, Bernard
Steinberg, Joshua S.

Teacher C
Adelman, David
Bismuth, David
Gottlieb, Maurice
Lipkunsky, Sarah
Rabinovitz, Selma
Rosenbluth, Sally Rena
Shapek-Oren, Rivka
Uzvolk, Benjamin H.

Department of Education in Technology and Science

Professor
Avital, Shmuel
Evyatar, Azriel

Associate Professor
Perlberg, Arye

Senior Lecturer
Finegold, Menachem
Hadar, Nitsa
Lazarowitz, Reuven
Lerman, Noah
Moore, Michael
Waks, Shlomo

APPENDIX L

Board of Governors

(as of February, 1982)

Chairman: Uzia Galil
Deputy Chairmen: Alexander Goldberg, Judge Leonard Rabinowitz
Vice Chairmen: Col. J.R. Elyachar, Maurice M. Rosen, Uriel Shalon, Eugene Stearns, Jacob W. Ullmann
Executive Vice Chairman: Carl Alpert

Members from Israel

Representatives in Official Capacity

President of the Technion: Maj-Gen. (res.) Amos Horev
Vice President for Academic Affairs: Prof. Ephraim Kehat
Vice President for Research: Prof. Ehud Lenz
Vice President for Development: Prof. Yakov Eckstein
Vice President for Admin. and Finance: Isaac Nissan
Dean of Students: Assoc. Prof. Ram Sagi
Dean of the Graduate School: Prof. Zvi Rigbi
Dean of Undergraduate Studies: Prof. Abraham Rosen
Representatives of the Government: Minister of Education and Culture, Minister of Labor, Minister of Transport
Mayor of Haifa: Arie Gurel
Chairman of the Alumni Association: Zvi Langer

Representatives of Groups and Associations

The Jewish Agency: Dr. Raanan Weitz and A. Katz, M.K.
The Alumni Association: Zvi Dvoretzky, Amos Kimchi and Moshe Y. Turetz
The Association of Engineers and Architects: Mordechai Shoshani, Uriel Stock and Michael Walden
The Senate: Prof. E. Amitai Halevi, Prof. Chaim H. Mannheim and Prof. Max Reiss
The Associate Professors: Assoc. Prof. F. Zawistowski
The Senior Lecturers: Dr. Dan Lorenz
The Lecturers: Dr. Edna Ishay
The Instructors and Assistants: Haim Abramovich
The Faculty Association: Assoc. Prof. Arnan Seginer
The Organization of Engineers: Shlomo Goldberg
The Organization of Handesaim and Technicians: Jehoshua Niselevitch
The Employees' Association: Yosef Harel
The Student Assoc.: Aviv Abramovitch and Haim Ben Zion

Members at Large

Avraham Agmon
Yosef Ami
Ariel Amiad
Yitzhak Barnov
Mrs. Sarah Baruchin
Yitzhak Ben Dov
Naftali Blumenthal
Josef Boxenbaum
Aryeh Carasso
Michael Comay
Joseph Creiden
Prof. David Erlik
Yosef Even
Yekutiel Federman
Dror Galezer
Uzia Galil
Yeshayahu Gavish
Yehuda Genossar
Benno Gitter
Dov Givon
Alexander Goldberg
Menahem Gottlieb
Dr. Reuven Hecht
Ernst Japhet
Gershon Kader
Yosef Koen
David Koren
Justice Moshe Landau
Dov Lautman
Avraham Lev
Mordecai Limon
Jonathan Moller
Mark Mosevics
Benny Pelled
Judge L. Rabinowitz
Yaakov Recanati
Yaacov Rechter
Max Reis
Chaim Rubin
Eliyahu Sacharov

Aharon David Sella
Uriel Shalon
Moshe Shamir
Avraham Shavit
Israel Shenkar
David Shoham
Moshe Shohamy
Paul Shulman
Yitzhak Streifler
Dr. A. Suhami
Uriel Tamir
Yitzhak Tcherniavsky
Dan Tolkowsky
Zeev Wertheimer
Zvi Zur

Honorary Life Members
Menahem Bader
Dr. Jacob Bach
Yaacov Ben Sira
Avigdor Bartel
Leon Carasso
Judge Joseph Herbstein
Abraham Klir
Naftali Lipschuetz
Dr. Ernst Ne'eman
Moshe Sanbar
Aharon Wiener
Dr. Naftali Wydra

Academic/Scientific Members
Prof. Malcolm Chaikin
Prof. George B. Dantzig
Prof. George Feher
Prof. Robert Hofstadter
Prof. Joshua Jortner
Prof. George Klein
Sir Claus Moser
Dr. Emanuel R. Piore
Prof. Ascher H. Shapiro
Dr. Harold J. Simon
Prof. Louis D. Smullin
Dr. Myron Tribus

From Overseas
Members from the U.S.A.
Mrs. Joan Arbuse

Samuel M. Bernstein
Marshall Butler
Alan H. Cummings
Lester Deutsch
David Dibner
Col. J.R. Elyachar
Ruben Finkelstein
Benjamin J. Free
Dr. Jacob B. Goldman
Burt I. Harris
Alexander Hassan
Mrs. David W. Herrmann
Lawrence G. Horowitz
Martin Jelin
Ludwig Jesselson
Martin Kellner
Theodore H. Krengel
Leon Lidow
Norman D. Louis
Mitchell J. Marcus
Lester Matz
Mrs. Pearl Milch
Louis Milgrom
Samuel Neaman
Mrs. Dorothy Rautbord
Sam Rich
Dr. Leon Riebman
Maurice N. Rosen
Norman Seiden
Leonard H. Sherman
Julius Silver
Henry Taub
Sam B. Topf
Jacob W. Ullmann
Lewis M. Weston
Dr. Felix Zandman

Alternate Members from the U.S.A.
Louis Avner
James S. Balter
George Berbeco
Stanley Berenzweig
Nathan Berkowitz
Paul Bernstein
Dr. Jules H. Bromberg

Mrs. Eedis Cooperband
Richard Davison
William Davidson
Ben Domont
Jerome Drexler
Melvin Dubin
Alex J. Etkin
Herman Fialkov
Bernard Fife
Murray M. Friedman
Mrs. Gustav Gettenberg
Arthur Gilbert
Edward Goldberg
Dr. I. Ralph Goldman
Salman Grand
Willard Hackerman
Dr. Harold L. Harris
Stanley Hatoff
D. Dan Kahn
Alan Keiser
Mason Lappin
Irvin Larner
David M. Levy
Bernard Mars
Mrs. Joan Callner Miller
Joseph Mitchell
Ernest Nathan
Harry D. Pierce
Robert A. Riesman
Richard J. Schwartz
Irving Shepard
Norton Sherman
Alfred P. Slaner
Morris Sussman
Alfred B. Teton
Sidney Wolberg
Louis A. Zuckerman

Members from Great Britain
Edgar Astaire
Peter Blond
Sidney Corob
Dr. Rodney Grahame
Lord Mishcon
Dr. Isaac Muende

424

Evelyn de Rothschild
Barnett Shine
Lord Sieff of Brimpton
Michael Sorkin
Harold M. Stone

Alternate Members from
Great Britain
Louis N. Harris
Dr. L. Kopelowitz
Maurice B. Links
Simon Susman
Alfred D. Webber
Alex Whyte

Members from Canada
Dr. David Azrieli
Morley Blankstein
Bernard M. Bloomfield
Irving Greenberg
Jack Hahn
Isin Ivanier
Louis L. Lockshin
Justice Roy J. Matas
Eugene Riesman
Eugene Stearns

Alternate Members from Canada
Jack Abugov
Jack Chisvin
Mrs. Frances Cohen
Dr. Richard Goldbloom
Frank Kettner
Archie Micay
Norbert Rand
Lewis Rosenfeld
Harry Sheres
Mrs. Irma Wigdor

Members from South Africa
Prof. Leslie J. Cohen
Maurice Ostroff
Jack Rubenchik
Solm Yach

Alternate Members from
South Africa
Laurie Brazg

Basil Tim Michel
Barney Seidle
Mrs. Estelle Yach

Members from Argentina
Aida de Barenboim
Gregorio Faigon
Mrs. Martha de Wolff

Alternate Member from
Argentina
Dr. Roberto Kohen

Member from Mexico
Max Shein

Member from Australia
Sidney M. Renof

Alternate Member from
Australia
David Faen

Members from France
Jean-Paul Elkann
Joseph Vaturi

Alternate Members from France
Ing. Gen. Robert Munnich
Henri Strosberg

Members from Brazil
Leon Feffer
Maurice Shashoua

Member from Venezuela
George M. Brief

Member from Switzerland
Bruce Rappaport

Member from Spain
Mauricio Hatchwell
Toledano

Member from Hong Kong
Horace Kadoorie

Honorary Life Members
Stephen Berger
Jack N. Berkman

Joseph Berman
Ignacio Blasbalg
Louis M. Bloomfield
David Borowitz
Joseph Bronfman
Benjamin Cooper
Ralph DeJur
Dr. Bern Dibner
Mrs. Anna Tulin Elyachar
Eugene Ferkauf
Henry L. Goldberg
Prof. Sydney Goldstein
Sofia L. de Grinberg
Jordan Gruzen
Homer Harvey
Harry B. Henschel
Dr. Arthur Kantrowitz
Charles Krown
Michael Kennedy Leigh
Justice Sam Lieberman
Louis B. Magil
Joseph Mailman
Joseph Meyerhoff
Alfred Miller
Marco Mitrani
Nicholas M. Munk
Arthur Pascal
Morris Pearlmutter
Ralph Philipson
Sam Posner
Daniel Rose
David Rose
Elihu Rose
Frederick Rose
Prof. Louis Rosenhead
Mrs. Frances Rosenstein
Burton D. Rudnick
Murray Rubien
Lawrence Schacht
Samuel Sebba
Victor Sefton
Max Seltzer
David Silbert
Abe Simkin
Maurice Spertus

Index

Aachen, 393
Aaronsohn, Aaron, 54
Aaronsohn Agricultural Station, 71
Abramowitz, Leo, 91
Abramski, Israel, 262
Academic By-laws and Regulations, 123, 190, 264, 288, 336
academic level, 95–101, 116, 120, 123–124, 127, 138, 139, 143–145, 156, 174, 264, 271, 277–278
Academy for the Hebrew Language, 375, 376
Acre, 14, 15, 23, 24, 25, 97, 189
Actions Committee, see Zionist Actions Committee
Adari, Avivi, 351
Adelson, Charles, 282
Adler, Cyrus, 42, 54
Adler, Roberto, 166
admission to the Technion, 112–113, 160, 171, 172, 176, 193, 195, 224, 229, 242, 295, 364–365
aeronautical engineering, 227, 229, 239–241, 249, 250, 257, 261, 350, 364, 376
Aeronautical Engineering Building, 282, 291, 295, 340
Aeronautical Engineering, Dep't of, 19, 225, 240, 241, 252, 266, 286, 310, 343, 344, 394
Aeronautical Research Center, 379
Aeronautical Research Council of Great Britain, 239, 240
Affuleh, 80
Africa-Asia Aid Program, 318–321
Agency for International Development, see AID
Agranat, Shimon, 300
Agricultural Development Center, 351
agricultural engineering, 83, 180, 194, 216, 228, 239, 241, 254, 267, 319, 351
Agricultural Engineering, Dep't of, 228, 266, 322, 391, 394
agriculture, 83, 96, 148, 192, 216

Ahad Ha-Am, 9, 10, 11, 12, 14, 15, 17, 21, 22, 25, 26, 29, 30, 33, 35, 39, 40, 41, 42, 43, 45, 49, 50, 52, 55, 56, 103, 111, 132
Ahroni, Gershon, 216, 310 380
Ahuza, 189, 268
AID, 316, 347, 360
Air Force, Israel, 236, 250, 351
Aival, Jonathan, 384
Akavia, Avraham, 184, 259
Alabama Polytech, 127
alcohol, production of, 275
Aleinikoff, Michael, 136, 150, 155, 160, 180, 181, 189
Alexander, Samuel, 240
Alexander-Katz, E., 246
Alexandria, 139
Aligarh, Muslim University in, 195
Allenby, Gen. Sir Edmund, 67
Alliance Israélite Universelle, 4, 5, 6, 7, 36, 40, 61, 155
Alliance Schools, 14, 46, 49, 50, 57, 62
Allon, Yigal, 331, 355, 357, 365, 386
Almagia, Roberto, 166
Alpan, Yitzhak, 321
Alperovitz, Yaakov, 141
Alpert, Carl, 274
Alsberg, Avraham P., 273
Altneuland, 14, 105
alumni, 156, 218, 384, 404
Amado, Maurice, 382
Amarcal, 288, 334
American Engineering Society, 148
American Fund for Palestine Institutions, 243
The American Hebrew, 13, 58
American Palestine Fund, 221
American Society for the Advancement of the Hebrew Institute of Technology, see American Technion Society
American Special Cultural Program, 289
American Technion Society, 45, 83, 93, 149, 201, 218–226, 235, 239, 247–249, 252, 257, 269, 282–286, 307, 309, 315, 316, 345, 346, 347, 358,

382, 383, 391, 400; see also Women's Division
Ami, Yosef, 341, 348, 385, 396
Amsler Machine, 176
Amuli, Beda Jonathan, 318–319
Anglo-Jewish Association, 5, 90, 102, 186, 194
Anglo-Palestine Bank, 17, 38, 165
Ankara, 232
Ankara, Middle East Technical University, 321
Anoushi, Gedalia, 274, 377
Anti-Semitism, 6, 61, 233
Anti-Tuberculosis League, 243
Apisdorf, 92
applied mathematics, 240, 241, 386
Applied Mathematics, Dep't of, 342, 369, 392
Arab students, 88, 130, 236, 351, 353–354, 396
Arabic language, 7, 36, 44, 118, 132
Arabs, 18, 23, 24, 27, 29, 48, 55, 88, 95, 115, 117, 138, 181, 182, 185, 186, 199, 201, 231, 232, 233, 255, 351
Arava plane, 379
Arbuse, Joan, 383
architecture and town planning, 83, 97, 108, 113, 115, 120, 125, 154, 167, 178, 190, 191, 215, 227, 236, 259, 263
Architecture and Town Planning, Dep't of, 108, 176, 178, 190, 215, 241, 261, 265, 274, 328, 329, 336–338, 340, 391, 394, 397
Argentina, 246
Argentina, Casa, 374
Argentine Atomic Energy Commission, 304
Argentine Technion Society, 384, 400
Arlosoroff, Chaim, 158, 160
Armon, David, 267, 321
Arnon, Jacob, 271–273, 364
Artisans' High School, see Bosmat
Aschner, Manfred, 382
Ashkenazim, 310, 312, 352

Asquith, Premier, 40
Association of Engineers and Architects, 95, 97, 99, 133, 136, 138, 140, 144, 156, 169, 170, 174, 192, 242, 250, 328
Association of Settlers from Germany and Austria, 198
Association of Technicians 100
Astaire, Edgar, 400
Aswan Dam, 140
Atid, 22
Atlas, Dalia, 371
Atlit, 27
Atomic Energy Commission (Israel), 343
Auerbach, Eliyahu, 22, 29, 37, 52
August Lane, 399
Australia, 125, 384, 400
Australia House, 398, 400
Austria, 5, 59, 82, 125, 198, 215
automotive engineering, 227, 239
Avigdor, Ariel, 120, 139
Avissar, David, 45

Bach, Dr., 197
Badian, Avner, 120, 124, 125, 134, 136, 137, 147, 149, 150, 151, 154, 166, 169, 170
Baerwald, Alexander, 21, 25, 26, 27, 28, 29, 41, 107, 108, 120, 124, 133, 135, 136, 151, 154, 164, 187, 340, 368
Baghdad, 14
Bagrit, Sir Leon, 279, 384
Bahai Shrine, 377
Baharav, Yekutiel, 258, 259
Balfour Declaration, 61, 200
Balfour, Lord, 117
Baltimore, 65, 221
Bar Rav Hay, David, 215
Bar Rav Hay, Meir, 215
Baratz, Yosef, 181, 183, 184, 204, 210, 211, 236
Bardin, Shlomo, 176 177, 178, 196, 199, 216, 221
Barkai, Dora, 274
Barney, Edgar S., 226
Barsky, Yosef, 22, 26, 35, 98, 107, 116, 120, 187
Barth, Aaron, 189, 262

Basle, 1, 133, 151, 157, 158, 239
Bastuni, Rustum, 236
Bat Galim, 116
Bausch lamp, 183
Bear, Jacob, 341, 396
Beckman, Harold E., 282
Beersheba, 67, 331, 359, 365
Beersheba University, see Ben Gurion University of the Negev
Beethoven, Ludwig van, 254
Beigel, Ze'ev, 82
Beilis, Mendel, 6
Beirut, 14, 23, 24, 25, 45, 47
Beirut, American University of, 111
Beitania, 84
Bejarano, Shimon, 262
Belgium, 65, 220, 255
Bender, Charles and Bertha, 291
Bender, Rose, Hostel, 374
Benenson, G., 194
Ben Gera, Yitzhak, 292
Ben Gurion, David, 47, 186, 200, 204, 240, 254, 255, 259, 260, 269, 284, 292, 309, 311, 316, 351
Ben Gurion University of the Negev, 328, 359–360, 403
Benjamin, Abe, 246
Ben-Shabbat, Shmuel, 53
Ben Sira, Moshe, 351
Ben Sira, Yaakob, 262
Ben-Tur, Shimshon, 184, 262
Bentwich, Norman, 137
Ben-Uri, Joseph, 267
Ben Yehuda, Eliezer, 42, 375
Ben Zvi, Yitzhak, 47, 174
Bergman, Ernst David, 266
Berhane, Israela, 320
Berhane, Zawde, 320
Berk, Zeki, 396
Berkson, Isaac B., 151–157, 160, 161, 181
Berligne, Eliyahu, 161, 168, 174, 181, 215, 262
Berlin, 5, 9, 10, 12, 18, 19, 22, 23, 25, 29, 33, 39, 41, 43, 44, 46, 48, 52, 54, 55, 59, 67, 72, 75, 76, 77, 79, 88, 91, 92, 103, 108, 153, 173, 196, 197
Berliner, Emile, 89, 91
Berliner, M., 194

Berlowitz, Max, 103, 125
Berman, Enrico, 400
Bernstein, Belle, 221, 324
Bernstein, Herbert, 267
Bernstein, Samuel, 324
Bet Erdstein, 327
Bet Hakranot, 175, 244, 290
Bet Hataasiyah, 233
Bet Najada, 233–234
Bezalel School, 22, 100
Bialik, Chaim Nachman, 103, 111, 189
Bible, 47, 376
Bina, Baruch, 76
bioelectronics, 390
biology, 354, 355, 390, 391, 394
Biology, Dep't of, 391
biomaterials, 390
biomechanics, 390
biomedical engineering, 358
Biomedical Engineering, Institute of, 342, 358, 385, 390, 398
Biram, Arthur, 41, 51, 52, 76, 79, 81, 85, 93, 102, 103, 107, 109, 123, 129, 133, 152, 155, 161, 176, 177, 221, 262, 270, 288, 316
Birn, S. J., 274, 384
Bloch-Blumenfeld, David, 27
Blok, Arthur, 89, 90, 101, 102, 105–119, 125, 143, 252, 279, 385
blood pipeline, 313–314
Bloomfield, Bernard M., 383, 400
Bloomfield, Louis M., 383
Bnai Brith, Grand Lodge of, 58
Bnai Brith, London, 167
Board of Governors, 10, 84, 90, 91, 92, 94, 101, 102, 107, 110, 118–119, 121, 129, 142, 150, 154, 155, 163, 165, 257, 258, 262, 264, 270, 274, 287–288, 290, 298, 304, 309, 328, 329, 330, 337, 341, 344, 346–347, 349, 354, 355, 357, 362, 370, 373, 384, 385, 386, 388, 396, 404; see also Kuratorium
Board of Regents, 345
Bokharan language, 7
Bologna, 147, 357
Bombay, 218

Bonfiglioli, Luisa, 173
book publishing, 371
Borowitz, Anne, 324
Bosmat, Technical High School, 104, 176–177, 179, 181, 190, 192, 199, 201, 203, 204, 210, 215, 216, 221, 226, 228, 232, 243, 252, 270, 310, 343, 380, 382
Boston, 221, 222
Bousso, Janine, 377
Brainin, Reuben, 132
Brandeis, Louis D., 63, 64, 69, 219, 283
Braude, Nathan, 91
Braverman, Joseph, 310
Brazil, 321, 384, 400
Breuer, Joseph, 96, 136, 151, 152, 154, 159, 160, 166, 180, 181, 191
Brin, Alexander, 222
British Army, 67, 76, 78, 201, 203, 210, 212, 233–234, 255
British Committee for Technical Development, see British Technion Society
British Controller of Enemy Property, 73
British Government, 61, 68, 172, 200, 202, 282–283
British police, 184, 199, 211
British soldiers, 181
British Technion Society, 101, 119, 195–196, 246, 278–280, 384, 400
British warships, 181
Brode, Heinrich, 49
Brodetsky, Selig, 148, 157, 159, 160, 161, 241, 248, 249
Brodie, Israel, 195
Brody Agricultural Engineering Building, 340
Brosh, Nehemiah, 209
Bruck, Zvi Ben Yaakov, 31
Bruner Institute of Transportation, 378
Buber, Martin, 1, 2, 3
Buchenwald, 304
Budgeting and Planning Committee, see Council for Higher Education
Buffalo, N.Y., 221
Building Materials Experimental Institute, 98

Building Materials Laboratory (and Building Research Station), 154, 165, 166, 169, 176, 179, 190, 204, 206, 210, 211, 257, 292, 304, 316, 317, 378
Building Research Station, see Building Materials Laboratory
Bulgaria, 6
Burma, 321
Butzel, Fred M., 222

Cahen, Jules, 289, 321
Cairo, 104, 139, 207, 230
Calculations of Magnetic Fields, 275
California, University of, 127
Cambridge, 144
Canada, 125, 246, 274, 339
Canada Building, 324, 383
Canada Nuclear Engineering Institute, 400
Canada Village, 84, 398, 400
Canadian Technion Society, 246, 346, 383, 400; see also Women's Division, Canadian Technion Society
Cantor, Louis, 83
Cap Pilar, 192
Capetown, 384
Caplan, Louis, 222
Caquot, Albert, 247
Carasso Wing for Self-Study, 398
Cardiff, 280
Carlsbad, 86, 90
Carmeli, David, 321
Carmelite Order, 66
Carmi, Dov, 325
Castle, Harry, 400
Celler, Emanuel, 315–316
Cement and Concrete, 242
cement levy, 317
Central British Fund for German Jewry, 172, 179, 194
Central Bureau for Settlement of German Jews, 198
Central Comm. for Technical Terminology, 375–376
Central Economic-Mathematical Institute of the U.S.S.R., 394
Central Supplies Building, 399
centripetal pump, 303

Century of Progress Fair, Chicago, 218
Ceylon, 321
Chapingo Agricultural College, 321
Chapiro, Jorge, 246
Charlottenburg, 19, 105, 108
Chasen, Philip, 282
chemical engineering, 83, 96, 176, 190, 215, 226, 241, 268, 350, 351, 360, 379
Chemical Engineering, Dep't of, 226, 241, 265, 275, 289, 310, 390, 394
chemical technology, 96
Chemicals and Fertilizers, Ltd., 309
chemistry, 108, 115, 178, 180, 184, 190, 232, 365, 394
Chemistry, Dep't of, 266, 310, 391
chemistry, industrial, 176
chemistry laboratories, 204, 206
Chemnitz, 92
Chicago, 149, 218, 221, 222, 224, 249
Churchill Auditorium, 280, 281, 292, 315, 386, 399, 400
Churchill, Randolph, 281
Churchill, Sara, 292
Churchill, Winston, 246, 280, 281, 292, 378
Ciffrin, Assaf, 239
Ciffron, A., 97
Cincinnati, 221
Citroen, André-Gustave, 167
civil engineering, 83, 96, 98, 115, 125, 146, 190, 191, 215, 242, 261, 275, 320
Civil Engineering, Dep't of, 125, 190, 215, 241, 265, 337, 344, 391, 394, 398
Civil Service Commission, 273
Cleveland, 221
Cluson Steel Works, 204
coastal and ocean engineering, 378, 398, 400; see also marine engineering
Cohen, Ben-Zion, 109
Cohen, Hirsch, 266
Cohen, Isidor, 282
Cohen, Joseph, 279
Cohen, Joseph H., 222

Cohen, Leonard, 89
Cohen, Y., 108
Cohn, B., 157
Cohn, Ephraim, 15, 17, 18, 19, 21, 23, 46, 47, 50, 73, 75
Cohn, Haim, 355
Cohn-Oppenheim, Baroness Julie von, 22
Colle, S. S., 246
Cologne, 43
Colombia Technion Society, 246
Columbia University, 288
Compton, Karl T., 219, 324
Comptroller's Report, 362-363
computer science, 376, 394
Computer Science, Dep't of, 342, 343, 369, 385, 389, 390, 394
computers, 308, 310, 342, 348, 359, 370, 390
Congressional Record, 315-316
Constantinople, 9, 14, 23, 24, 25, 47, 48, 49, 67, 75
Constantinople, University of, 24, 47
constitution, 261-263, 274, 352
construction and public works, see civil engineering
Contractors' Association, 317
Cooper, Benjamin, 221, 224, 262, 292
Cooper, Miriam, 221
Cooper, Simon, 179
Cooperstock, J., 82
Cornell University, 266
cornerstone, 25, 29, 255
cosmic rays, 304, 305
Council, see Vaad Menahel
Council for Higher Education, 263, 328, 358, 359, 365, 382
Council of Professors, see Senate
Cowen, Joseph, 90
credit system, 367, 381, 386
Cyprus, 140, 220, 319
Czechoslovakia, 92, 125, 190, 304
Czechoslovakia Bnei Brith, 179, 194
Czopp, Dov, 399

Damascus, 14, 66, 79, 117, 155, 225

Dan, Hillel, 262
Danciger Building, 328-329, 338-340
Danciger, Dan, Sadie and Joseph, 328
Daniels, Julius, 222
Danzig, 92, 108
Danziger, Yitzhak, 382
Dardanelles, 65
Davis, Lady, Fellowship Trust, 383
Davis, Lady, Mechanical-Aeronautical Engineering Center, 398, 400
Dayan, Moshe, 349
Dayton, 221
Dean of Students, 276, 309, 370-371
Degania, 184
degrees, 174, 191, 232, 242, 247, 265, 297, 360, 367, 393-394
DeJur, Harry, 382
DeJur, Ralph, Materials Engineering Center, 398
dentistry, 381
desalination, 216, 276, 379
Detroit, 58, 135, 198, 221, 222
Detroit Mechanical Engineering Laboratories, 374
Deutsches Landes-Komitee des Technischen Instituts, 93, 125
Dibner, Bern, 274, 285, 343
Dingott, Meir, 108
discrimination, 352-353
Dizengoff, Meir, 77, 111
Dobrzynski, 92
Dolinsky, Morton, 377
Dori, Badana, 333
Dori, Yaakov, 29, 185, 211, 235, 238, 251, 252, 254, 255, 257, 261, 262, 264, 266, 270, 271, 274, 276-278, 286-288, 290-294, 300, 302, 305, 306, 308, 309, 313, 314, 316, 319, 322, 323, 324, 326, 327, 329-335, 341, 343, 344, 359, 364, 370, 385
Dori, Zvi, 371
dormitories, see student dormitories
Dostrovsky, Yaakov, see Dori, Yaakov

Douglas, William O., 247
Dresden, 120
Dultzin, Samuel, 246
Duma, 9
Dumanois, P., 247
Dunnia, Tuvia, 22
Düsseldorf, 92
Dykar, Moshe, 186

Eastern Group Supply Conference, 202
Easton, Pa., 147
Echtman, A., 217
Eckstein, Yakov, 396
Ecole Nationale Supérieure des Mines, 393
economics, 109, 194, 216, 227, 239, 342, 354
Edelman, Joseph, 176, 268
Eden, Yehuda, 371
Eder, Montague David, 89, 90, 106, 119, 126, 142-143
Education in Technology and Science, Dep't of, 391-392, 398; see also teacher training
Egypt, 11, 36, 75, 125, 130, 139, 156, 165, 204, 208, 213, 274, 290, 349, 351, 386, 388
Eichmann, Adolf, 362
Eidgenössische Technische Hochschule, 393
Eilat, 313
Ein Ganim, 100
Ein Hod, 258
Einstein, Albert, 94, 107, 108, 125, 172, 219, 220, 222, 249, 250, 282, 283, 292
Einstein Award, 383
Einstein, Hans, 292
Eisenman, Morris, 186
Ekaterinoslav, University of, 107
El Alamein, 207, 213
Elath, Eliahu, 246, 351
Eldee Foundation, 383
electrical engineering, 83, 96, 176, 190, 215, 261, 350, 358, 364
Electrical Engineering, Dep't of, 241, 265, 344, 390, 391, 394
electrical laboratories and building, 190, 191, 204, 205, 210, 226, 247, 250, 251, 270, 292, 350, 374

electronics, 178, 359, 365, 388, 390

Eliot, Charles W., 54

Elkann, Jean Paul, 400

Elkes, Hugh, 98

Elon, Yohanan, 261, 321

Elron, 388

Elyachar, Jehiel R., 45, 195, 219, 224, 247, 252, 257, 262, 269, 282, 309, 312, 324, 346, 368, 393

Employees' Association, 397

employees, waiver of salaries, 163–164

Engineers' Association, *see* Association of Engineers and Architects

Engineers Employed by the Technion, Organization of, 397

England, *see* Great Britain, British Government

English language, 5, 36, 40, 44, 108, 116, 132, 192, 193, 230, 319

environmental engineering, 176, 398; *see also* sanitary engineering

Epstein, A. K., 222

Eritrea, 204

Erlik, David, 104, 356–359

Erlik, Yaakov, 104, 108, 176

Eshkol, Levi, 309, 317, 331, 346

Essex House, 346

Ethiopia, 320

Ethiopia, Imperial College of Technology, 321

Etkes, Peretz W., 82

Ettingen, Shlomo, 120, 155, 242–243, 251, 254, 256, 266

Euphrates River, 203, 225

evening classes, 129, 165, 176, 207, 242, 252, 275

examinations, 112, 113, 136, 140, 170, 171, 174, 180, 186, 201, 212, 231, 242, 271, 278, 301

Extension Studies, 275, 379

Ezekiel, Book of, 376

Faculty Association, 397

Faculty Club, 308, 399

faculty members, statistics of 132, 190, 268, 302, 333

Fain, Irving Jay, 222

FAO (Food and Agriculture Organization), 267, 289

Federation of American Zionists, 56

Federation of Synagogues, London, 167, 179, 196

Federmann Park, 399

Feiner, Rivka, 368

Feiwel, Berthold, 1, 2, 3, 90

Felman, A. L., 109

Fels, Mary, 83

Fels, Samuel S., 221

Feltenstein, 91

Ferdinand, Archduke, 59

Ferkauf, Eugene, 382

Ferrara, 166

Fife, Elias, 262, 282, 324, 378

Finkel, Haim (Herman), 267, 321, 341

Finkelstein, Alphonse, 35, 39–40, 41, 42, 50, 51, 52, 57, 65

Finkelstein International House Student Dormitory, 340

Finland, 220

Finzi, Arrigo, 268

Finzi, Comm., 166

Fischbach, Beatrice, 324

Fischbach, Harry, 191, 247, 262, 270, 291

fish ponds, 192

Flexner, Bernard, 148

Flieman, Moshe, 355, 357

Fohs, F. Julius, 148, 224

Fohs Foundation, 267

Fondiller, William, 219, 222, 224, 227, 239, 247, 282

food engineering and biotechnology, 275, 310, 398

Food Engineering and Biotechnology, Dep't of, 310

Food Industries Research and Development Station, 379

Ford Foundation, 267

Ford, Henry, 135

Ford, Henry II, 391

France, 6, 28, 40, 112, 127, 166, 167, 168, 190, 278, 384

France, Pierre Mendes, Chair in Economics, 400

France Student Hostel, 398

Franck, James, 282

Frankel, John, 91

Frankfort-am-Main, 12, 54

Franz, Prof., 41

Freeman, Aryeh, 319

Freiburg, 92, 305

Freiman, A. J., 125

French College, 45

French language, 5, 7, 44

French Technion Society, 247, 278, 400

Frenkel, Avraham, 112

Frenkel, Zvi (Gregory), 112, 141

Fried, Philip and Frances, 371, 398

Fried Student Counselling Center, 371

Friedenwald, Jonas, 221

Friedland, Rahel, *see* Shalon, Rahel

Friedland, Uriel, *see* Shalon, Uriel

Friedlander, David, 399

Friedman, Elisha, 71

Friedman, Emanuel, 141

Friedman, Louis, 40

Frocht, Max M., 222

Frost, Charles, 236, 262

Frost, Eva, 221, 292

Fryer, Samuel, 291

Fuad University, 230

Fulbright Program, 289

fund-raising, 166, 180, 250, 344–345, 347, 360–362, 375, 382; *see also* Technion Societies of the respective countries

fund-raising in Israel, *see* Israel Technion Society

fund-raising in Palestine, 165, 189, 217–218; *see also* Israel Technion Society

Galeries Lafayette, 167

Galicia, 6, 120

Galil, Uzia, 388

Galilee, 183

Galperin, Anatol Mendelevich, 394

Gan Shmuel, 100

Garcy, Haim, 28

Gasko, Zeev (Wolf), 141

Gawronsky, Boris, 10

Gaynor, Israel, 282

Geddes, Sir Patrick, 260

Geisler, Mordecai, 268

Geist, Jacob M., 268

Gelfand, Louis, 222

Gelfman, Amnon, 299

General Electric Corp., 83, 282

General Studies, Dep't of, 289, 302, 391; *see also* humanities

Geneva, 2, 199, 330, 368

Geneva University, 1, 33, 107

geodesy, 108, 115

Geodetic Research Station, 378

George V, 149

Georgian language, 7

Geri, Yaakov, 262

German army, 65, 79, 80, 207

German colony, Haifa, 15

German Foreign Ministry, 6, 7, 9, 23, 37, 53, 75, 93

German immigrants, 178, 179, 185, 189

German language, 4, 9, 36, 38–45, 47, 48, 50, 53, 56, 60, 61, 65, 73, 74, 75, 118, 178, 192

Germany, 6 7, 8, 9, 12, 13, 15, 23, 24, 28, 29, 33, 37, 38, 42, 46, 49, 51, 52, 60–64, 66, 67, 75, 79, 91–94, 108, 125, 171, 172, 173, 175, 176, 190, 195, 196–198, 199, 200, 215, 220, 222, 304, 361–362, 379, 400

Germany, fund-raising in, 91–94

Gershenfeld, Dr. Louis, 221

Gerson, Leon, 246

Gestapo, 197

Ghana, 321

Ghent, University of, 255, 393

G.I. Bill of Rights, 224

Gilbert, Rosalinde and Arthur, Hostel, 374

Gilead, Shlomo, 325, 326

Gill, Max, 233

Ginsberg, Asher, *see* Ahad Ha-Am

Ginsberg, William, 219

Ginsburg, Avraham, 52, 103

Ginsburg, David, 267, 288, 306, 307–308, 309, 314, 327, 330, 333, 341, 350, 370, 371, 381

Ginsburg, Shlomo, *see* Shaag, Shlomo

Ginzberg, Avraham, 396

Ginzberg, Eli, 366

Ginzberg, Louis, 180
Givat Ram, 273
Givon, Dov, 158, 159, 208, 262
Gladstein, Tunia, 368, 369
Glikson, Arthur, 259
Glunts, James D., 222
Gold, Meyer, 368
Goldberg, Alexander, 47, 85, 309, 329, 330, 332, 336, 340–343, 344, 345, 348, 351, 355, 358, 367, 370, 373–375, 380, 384–386
Goldberg, Boris, 47, 79, 81, 82, 84, 85, 86, 87
Goldberg, Elias, 85
Goldberg, Harry, 246, 274
Goldberg, Isidor, Electronics Center, 340
Goldberg, Saadia, 108, 130, 147, 184
Goldman, Conrad, 92
Goldman, Harvey H., 222
Goldman, Jacob, 400
Goldmann, Nahum, 248
Goldsmid, O. D'Avigdor, 89, 90
Goldsmith, Horace W., Biology Building, 340
Goldsmith, Lester M., 221
Goldsmith, Samuel A., 222
Goldstein, Aharon, 317, 384
Goldstein, Mrs. Rosa, 274, 276
Goldstein, Sydney, 19, 239–241, 244, 249, 252, 254, 255, 261, 262, 264–265, 266, 268, 270, 271, 274, 286–288, 300, 305, 333, 337, 341, 344, 346–347, 360
Goodman, C. Davis, 246
Gonen, Dan, 321
Gordon, A. D., 53
Gottesman, Lucy, 377
Göttingen, University of, 93, 107
Grabinsky, Isaac, 246
Graduate School, 264, 265, 275, 295, 297–298, 305, 320, 340, 341, 378, 381, 385, 392
Great Britain, 2, 40, 85, 93, 101, 127, 144, 166, 167, 173, 190, 192, 195, 196, 218, 239, 274; see also British Government, British Technion Society

Greece, Union of Jewish Communities, 25
Greece, Zionist Federation of, 25
Green, Louis, 143–144, 148, 152, 364
Greenberg, Max A., 82
Grenoble, Technical School, 82
Gromyko, Andrei, 233
Grossman, Jeremias, 107, 111, 124, 134, 135, 136, 151, 158, 159, 160, 166, 169–171, 179, 268, 288
Groupement Francais des Amis du Technion, see French Technion Society
Gruenebaum, Arturo, 324
Gruenebaum, Heinz, 262
Gruenebaum, Y., 215
Gruss Swimming Pool, 398
Gruzen, B. Sumner, 262, 382
Gur, Arieh, 380
Gurevitch, Gershon, 274
Gut Bridge, 203
Gutwirth, Aaron, 384
Gutwirth Science-Based Industries Center, 384, 398
Gutz, R., 11
Guy, M., 40

Haaretz, 99, 258, 259, 275, 314
Haavara, 176, 197
Haberschaim, Izhak, 173
Hacohen, David, 189, 274
Hadar Habrazim, 181
Hadar Hacarmel, 18, 80, 105, 106, 130, 171, 174, 179, 181, 183, 209, 250, 254, 292, 340, 380, 396, 399
Hadassah, 216
Hadassah Hospital, Haifa, 155
Hadassah Hospital, Jerusalem, 356
Haganah, 28, 117, 131, 181–186, 208–211, 212, 213, 229, 231, 233–234, 255, 287
Hagin, Josef, 267, 341
The Hague, 62
Haifa, 5, 12, 13–16, 19, 22, 23, 25, 26, 27, 28, 29, 35, 37, 42, 44, 46, 47, 48, 51, 52, 53, 54, 57, 59, 61, 65, 66, 68, 73, 76, 77, 78, 79, 80, 84, 85, 87,

88, 94, 95, 97, 100, 102, 103, 105, 128, 129, 130, 131, 145, 159, 162, 163, 165, 168, 169, 172, 181, 183, 185, 186, 189, 190, 206, 224, 228, 233, 236, 251, 258, 261, 273, 310, 327, 354, 356, 357, 358, 359, 377, 397
Haifa-Baghdad Railway, 165
Haifa Labor Council, 84, 242, 253
Haifa University, 327–328, 354–356, 357, 358, 374
Haimovitch, Dov, see Givon, Dov
Haines, James, 82
Hakim, Rafael, 15, 18
Halevi, E. Amitai, 298
Halfon, Rabbi Moshe, 47
Halpern, Benjamin M., 82, 83, 179, 195, 219, 274
Halpern, G., 90
Halpern, Shimon, 384
Hamburg, 12, 92, 152
Hanani, Haim, 311, 313–314, 327, 329, 360
Handesaim and Technicians Employed by the Technion, Organization of, 397
Hanin, Meir, 266, 360
Hanita, 183
Hantke, Arthur, 58
Hapoel, 188
Hardegg, Loytved, 37, 38, 42, 48, 49, 65, 66
Hardy, Charles, 221
Har'el, Gershon, 380
Hari, Hayim, 26, 46
Harris, Ben R., 222
Harris, D. Lou, 246, 274, 346, 383, 385
Hartog, Sir Philip, 180
Harvard University, 54, 287
Harvey, Leo, 382
Harvey Prize Fund, 382, 388
Hashimshony, Aviah, 268, 336, 337, 338
Hashimshony, Yehudah, 236
Hashomer, 27, 28
Hassan, Alexander, 400
Hassan-Washington Pilot Plant, 374
Hassin, Alexander, 115, 130, 141, 262

Hasson, David, 341, 396
Hebrew army, see Jewish Army, Jewish Brigade, Jewish Legion
Hebrew Gymnasium, see Herzlia Gymnasium
Hebrew language, 3–5, 7, 9, 12, 36, 38–41, 43–45, 47–50, 52, 54, 58, 60, 61, 62, 73, 74, 79, 82, 90, 91, 100, 101, 104, 107–108, 110, 112, 116, 123, 130, 147, 184, 192, 193, 224, 242, 371, 375–377
Hebrew Language Committee, 40; see also Vaad Ha-Lashon Ha-Ivrit
Hebrew Teachers' Association, 3, 44, 45, 55, 61–62
Hebrew Technical Institute, 226
Hebrew University, 68, 70, 75, 81, 86, 93, 102, 117, 118–120, 122, 125, 126, 127, 128, 129, 131, 135, 139, 142–144, 147, 148, 152, 153, 156, 158, 164, 167, 174, 176, 180, 194, 195, 196, 211, 216, 229, 235, 237, 238, 240, 241, 248, 249, 255, 267, 270, 271, 272, 283, 293, 300, 327, 351, 360, 361, 364, 365, 383
Hechalutz, 93
Hecker, Max, 41, 51, 52, 76, 81, 84, 85, 87, 88, 89, 91–94, 100, 101, 102, 103, 106, 107, 109, 111, 113, 119, 120–129, 130, 132, 133, 134, 139, 195, 215, 243, 273, 274
Hecker, Zvi, 328, 339
Hedjaz railroad, 8, 14, 66, 72, 79
Heidelberg, University of, 107
Heimann, Hugo, 173, 202
Hemed, 266
Herbstein, Frank H., 298, 338
Herman, Basil, 288
Herrmann, Lotte, 92
Herrmann, Rose, 383
Herz, Karl, 368
Herzer, Hugo, 196, 197
Herzl Club, Vienna, 154, 176
Herzl, Theodor, 1, 5, 14, 53, 105, 157, 189
Herzl, Tomb of, 258

Herzlia Gymnasium, 22, 39, 40, 43, 72, 107, 188, 258
Het, N., 151
Hexter, Maurice, 161
Heyman, Moses D., 224, 235
Higher Education Authority, 349, 364; see also Council for Higher Education
Hilfsverein der deutschen Juden, 5–12, 14, 15, 17, 24, 29, 30, 33, 35, 36, 37, 38, 39, 40, 44, 45, 46, 47, 48, 49, 50, 53, 57, 59, 61, 62, 63, 65, 67, 68, 69, 72–74, 76, 79, 84, 93, 98, 111, 152, 155, 262
Hindes, Matityahu, 254, 262
Hirsch, Giacomo, 166
Hirsch, Sigfried, 194
Hirschman, Stuard and Henrietta, 292
Hirssenberg, 92
Hirst, Lord H., 89, 196
Histadrut, 77, 84, 85, 99, 102, 116, 170, 200, 253, 327
history of the Technion, 187, 243, 273–274, 375
Hitler, Adolf, 171, 172, 187, 208
Holford, William, 324
Holland, see The Netherlands
Hollander, Herman, 262
Homel pogroms, 6
Hong Kong, 218
Horev, Amos, 85, 365, 386–387, 388–389, 396–397, 404
Hornthal, Anna, 197
horse-racing, 383
Horwitz, Aaron B., 321
hostels, see student dormitories
Houphouet-Boigny, Felix, 320
House of Lords, 246
Houston, 224
Hovevei Zion, 10
Huleh, 204
Humanitarian Trust, 279
humanities, 3, 263, 308, 314, 315, 343, 354
Hungary, 190, 288, 290
Hussein, King, 349
hydraulic engineering, 179, 180, 313, 344, 358, 359
hydraulics laboratory, 178, 186, 190, 191, 192, 207, 226, 292, 378

IAESTE, 371
Ilioff, Alexander, 107–108, 124, 151, 184, 190, 378
"illegal" immigration, 198, 199, 208, 209
Illinois Institute of Technology, 223
ILO (International Labor Organization), 289
Imperial College, 180
Import-Export Bank, 250
India, 195
Indianapolis, 221
industrial chemistry, see chemical engineering
industrial engineering, 175, 190, 191, 215, 226
Industrial Engineering and Management, Dep't of, 241, 289, 292, 394
Industrial Psychology Laboratory, 378
industrial relations, 289
inter-disciplinary studies, 342, 390–391
Internal Revenue Service, 345
International Geophysical Year, 304
ionosphere recording station, 304
Iraq, 225
Iraq Petroleum Co., 205, 207
Iraq Sueidan, 236
Irmay, Shragga, 176, 298, 327, 375
Isler, Jacobo, 384
Israel Bonds, 72
Israel Defense Forces, 238, 255, 349, 352
Israel, Government of, 244, 247, 250, 254, 256, 260, 269, 270, 293, 300, 309, 316, 324, 343, 345–349, 355, 360, 361, 364; see also State of Israel
Israel Government Archives, 49
Israel Institute of Industrial Design, 289
Israel Institute of Metals, 289, 378
Israel Prize, 275, 376, 381
Israel Technion Society, 270, 316, 345, 384, 386; see also fund-raising in Palestine

Italy, 166, 168, 169, 208, 239, 268, 357
Italy, Union of Jewish Communities, 179, 194
Itzkovich, Shmuel, 22, 136, 150, 152, 155, 156
Ivanier Laboratory for Welding & Casting Research, 398
Ivory Coast, 320

Jabotinsky, Eri, 321
Jabotinsky, Ze'ev, 181
Jacobovitz, Myra, 131
Jacobson, Israel, see Gaynor, Israel
Jaffa, 5, 14, 23, 27, 35, 39, 40, 43, 48, 61, 67, 70, 74, 82, 85, 95, 97, 188
Jaffa oranges, 71
Jaffe, Hillel, 76, 109
Japan, 4, 147
Java, 218
Jerusalem, 5, 8, 11, 13, 14, 18, 27, 39, 40, 45, 46, 48, 49, 54, 61, 67, 74, 75, 88, 89, 95, 100, 102, 124, 135, 144, 145, 151, 153, 154, 157, 161, 165, 176, 189, 355, 356, 377
Jerusalem Post, 214
Jesselson, Ludwig, 353
Jewish Agency, 145, 149, 151, 153, 154, 155, 157, 158, 160, 161, 163, 165, 167, 168, 171, 172, 173, 174, 178, 179, 186, 188, 189, 192, 193, 198, 199, 215, 216, 230, 237, 243, 244, 248, 257, 262, 270, 317
Jewish army, 201; see also Jewish Legion
Jewish Brigade, 201
Jewish Chronicle, London, 111
Jewish Colonization Association, 84, 90, 186
Jewish Comment, Baltimore, 65
Jewish Exponent, Philadelphia, 65
Jewish Legion, 255
Jewish National Fund, 16, 56, 62–64, 68, 69, 73, 90, 132, 159, 175, 194, 197, 216, 228, 251, 257
Jewish State, 189, 190, 207, 231

The Jewish State, 1
Jewish Technical Bulletin, 83
Johns Hopkins University, 376
Joint Distribution Committee, 186
Jordan, 349
Jordan River, 100, 147
Josepho, Anatol and Ganna, 323
Jubilee Campaign, 375
Judaism, 6, 11, 41, 63
Jüdisches Institut für technische Erziehung in Palestina, see names of the Institute
Judith, 326
Junior Technical College, see Bosmat

Kadoorie estate, 102
Kadoorie, Lawrence, 218
Kahn, Bernhard, 41, 42, 67
Kahn, Milton, 222
Kahn, Rabbi Zadoc, 6
Kaiserman, Nathan, 15, 17–19 22, 24, 25, 29, 52, 187
Kalandia, 100
Kalugai, Yitzhak (Isaac), 83, 108
Kanovitz, M., 108
Kanowitz, Siegfried, 92
Kaplan, Eliezer, 85, 161, 181, 215, 251, 269
Kaplan, Jacob J., 222
Kaplan, Stanley, 384
Kaplansky Chair in Agricultural Engineering, 254
Kaplansky, Shlomo, 27, 43, 123, 145, 158, 159–162, 163–171, 172, 173, 174, 175, 176, 178–180, 181, 186–190, 193, 194–195, 196, 199, 200–202, 211, 213, 214, 215, 216–217, 218, 219, 221, 223, 226–233, 235, 236, 237–239, 240, 241, 242–243, 244, 246, 247–254, 266, 268, 286, 330, 331
Kaplansky Square, 254
Karlsruhe, Hochschule of the University of, 393
Karman, Theodore von, 147, 148, 229, 239, 240–241

Karni, Yosef, 236, 257, 259, 260, 267, 268, 337, 338
Kassab, Alexander, 15, 18
Kassif, Gabriel, 321
Kastein, Josef, 215
Katzen, Bernard, 289
Katznelson, Berl, 159, 160
Kauffmann, Richard, 98
Kaufman, Eitan, 299
Kaufmann, Edgar J., 222
Kay, Leon, 222
Kazmann, Boris, 82
Kehat, Ephraim, 396
Keller, Friedrich, 37
Keller, Michael, 353
Kenya, 308, 309, 318
Keren Hayesod, 73, 80, 81, 84, 88, 89, 92, 121, 123, 130, 134, 142, 149, 174, 194, 197, 198, 345, 347, 368
Kerensky, Alexander, 85, 185
Kessel Student Hostel, 398
Kfar Saba, 100
Kfar Tira, 252
Kharkov, Polytechnicum, 107
Khoushy, Abba, 189, 354, 356, 359
Khreiba, 258
kibbutzim, 176, 185, 350
kindergarten, Hebrew, 3, 7, 14, 49, 74
Kinneret, 100
Kipling, Rudyard, 179
Kirstein, Louis E., 222
Kiryat Anavim, 100
Kiryat Eliyahu, 216
Kisch, Frederick, 102, 123, 125, 149, 179, 189, 207, 222, 223, 226, 228, 232
Kishinev pogroms, 6
Kishon River, 228
Kislak Park, 399
kitchen cabinet, 333
Klebanoff, Jacob, 137, 274
Klein, Alexander, 173, 215, 226, 247, 252, 257, 261, 324, 377
Klir, Avraham, 262
Knesset, 2, 236, 270, 273, 295, 299, 364, 375
Knessiah Rishonah, 2, 3
Kogan, Abraham, 266, 379
Kohn, David, 368
Kohn, L., 89, 101
Kol Hastudent, 314

Komornik, Amos, 371
Königsberg, 92, 93
Konoff, Alexander, 221, 226, 247, 262, 270, 292, 382
Kook, Rabbi Avraham Yitzhak Hacohen, 46
Kopelowitz, U., 215
Korea, 321
Kramer, Abba, 141
Kranzberg Industrial and Management Engineering Wing, 398
Krauze, A., 170
Krengel, Theodore H., 400
Kristallnacht, 196
Krupnick, Hayim A., 109, 215
Kuhlman, Richard von, 60
Kumasi School of Technology, 321
Kupat Tagmulim, see Provident Fund
Kuratorium, 10, 11, 12, 17, 21, 22, 25, 26, 29, 31, 35, 38, 39, 40, 41, 42, 44, 49, 50, 51, 53, 54, 55, 56, 58, 59, 62, 84, 87, 100, 262; *see also* Board of Governors
Kurrein, Max, 172, 173, 190, 191
Kutten, Aharon, 321, 369
Kutzinsky, D., 120

labor, Jewish, 27
Labor Party, 158, 159, 171
Ladino, 7
Ladyjensky, M., 120, 151, 217
Laemel School, 5, 45, 50
Lafayette College, 147
Lagos, University of, 321
Lake Success, 231
Landau, Eugen, 10
Landau, Moshe, 92, 262, 287, 288, 309, 337, 346, 349, 356, 362, 385, 386
Landau, Yitzhak, 92
Landers, B. L., 222
Langsdorf, Alexander, 222
Laskov, Haim, 209
League for Friendly Relations with the U.S.S.R., 232
League of Nations, 87
Lebanon, 225
Lebeson, David, 224
Lebeson, Herman, 224

de Leeuw, Abraham, 267
Leffert, Mrs. Herman, 383
Leichter, Sinai, 383
Leigh, Michael Kennedy, 340, 374, 384
Leipzig, 92
Lenji, Ilana, 288
Lenz, Ehud, 396
Leshtziner, Yehuda, 141
Lev, Naomi, 292
Levi Shabbetai, 15, 24, 25, 65, 66, 67, 68, 73, 76, 77, 78, 187, 217, 254, 262
Levi, Yuval, 351
Leviant, Israel, 247
Levin, Nahum (Nyoma), 112, 113, 116, 141, 182
Levin, Schmaryahu, 8–10, 11, 12, 13, 21, 25, 26, 27, 29, 30, 31, 35, 38, 39, 40, 41, 42, 45, 52, 54, 55, 59, 62, 63, 87, 111, 113, 114, 124, 132, 186
Levine, Harry, 222
Levine, Laurence N., 148
Levinson, Sam M., 222
Levitan, Louis, 400
Levy, Gustav, 262
Levy, Howard, 221
Levy, Lionel F., 221
Levy, Mario Giacomo, 166
Levy, Mordecai M., 215, 243, 278, 321
Lewin, Hermann, 91
Liberia, University of, 321
Libertovsky, Israel, 209–210
library, 113, 130, 190, 195, 308, 324, 351, 368–369, 393
Libyan desert, 203
Lichtheim, Richard, 60
Lidow, Eric and Leon Building, 292
Liebman, Joshua Loth, 222
de Lieme, Nehemiah, 87
Lindner, I. J., 279
Linenthal, Mark, 222
Lipshitz, Yosef, 109
Lipsky, Louis, 218
Lithuania, Zionist Organization of, 130
Littauer, Sebastian B., 288, 300
Locker, Berl, 158, 161, 262
Lockshin, Louis, 383
Lockspeiser, Sir Ben, 364

Loew, E. M., 222
Loewy, Joseph, 98
London, 10, 54, 62, 68, 69, 70, 76, 79, 85, 88, 89, 100, 101, 102, 126, 127, 128, 155, 167, 173, 195, 246, 384, 386
London Board of Trade, 101
London School of Economics, 109
London, University of, 180, 324
Loren, Sophia, 326
Los Angeles, 382
Lovinger, Ronald T., 399
Lowdermilk, Walter C., 222, 267
Lowenherz, Arthur, 41, 52
Lowenstein, Dr., 83
Luntz, A. M., 221
Lurie, Joseph, 45, 55, 62, 136, 153
Lwow, 125, 137

Maccabi Sports Organization, 77, 174, 188
MacDonald White Paper, 199, 200
Machane Yehuda, 100
Machanik, Barnett, 149
Mack, Julian W., 54, 68, 69, 70, 71, 122, 219
Magen David Adom, 243
Magnes, Judah L., 33, 55, 63, 72, 81, 102, 119, 120, 123, 124, 125, 129, 142, 148, 152, 158, 180, 211
Majevski, G., 267
Makower, Felix, 73
Mallah, Asher, 24, 25
Management Sciences Research Center, 379
Manchester, 19, 240, 266, 288
Manchester University, 239, 240
Mandate, British, 87, 89
Mandatory Government, British, 49, 105, 111, 114, 121, 140, 141, 149, 153, 158, 174, 178, 179, 190, 193, 198, 208, 209, 211, 224, 228, 231, 234, 245, 260, 263
manpower studies, 364–366, 385
Mansfeld, Alfred, 336, 382
Maof, 239
Mapai, 327
Mapam, 327

Mardor, Munya, 208, 209
Marine Carp, 224
marine engineering, 83, 194, 227, 239, 389; *see also* coastal and ocean engineering
Mark, Jacob, 219
Markson, Yoland, 222
Marshall, George C., 249
Marshall, Louis, 12, 42, 47, 54, 56, 58, 59, 61, 72
Marx, Karl, 19
Mashav, 319
Massachusetts Institute of Technology, 219, 283, 388
master plan, campus, 252, 261, 324–325, 341, 374, 399
materials engineering, 342
Materials Engineering, Dep't of, 344, 391
mathematics, 112, 115, 242, 268, 301, 311, 351, 354, 365, 373, 394
mathematics, applied, *see* applied mathematics
Mathematics, Dep't of, 266, 310, 369, 398
Mauerberger Foundation, 400
Mauerberger, Morris, 282, 292, 384
Mayer, Armand, 247
Mayers, Rabbi Baruch, 47
Mazer, Rose, 292
mechanical engineering, 20, 83, 96, 98, 108, 176, 190, 215, 220, 227, 242, 275, 328, 350, 359, 360, 364, 365
Mechanical Engineering, Dep't of, 215, 241, 265, 289, 344, 391, 394
Mechanics, Dep't of, 266, 358, 392
Mechanics Research Center, 379
medal, Technion, 375
medical engineering, 383
Medical School, *see* Medicine, Faculty of
Medicine, Faculty of, 104, 356–359, 368, 381, 385, 386, 389, 390, 391, 394, 396, 398, 399
Medina, 77
Megiddo, battle of, 67
Mehl, David W., 246, 274
Meir, Golda, 316, 320, 375

Melbourne, 400
Melchett, Lord, 81, 90, 111, 113, 114, 194
The Melting Pot, 70
Mendelsohn, Erich, 93, 173
Mendelson, Morris, 222
Mendelssohn, Lilian, 383
Merhavia, 66, 85, 254
Merlub-Sobel, Menahem, 232, 241, 242–243, 254, 266
metallography, 190, 205
metallurgy, 289
Mevorach, Yaacov, 321
Mexican Technion Society, 246, 400
Mexico, 246, 274, 321, 384, 398
Mexico Student Hostel, 398
Meyer, Andre, 383
Meyer, Reuben Manasseh, 282
Michael, Suse, 197
Michigan, 82
Middle East Supply Centre, 202, 203
Mifal Hapayis, 357
Mikveh Israel Agricultural School, 4, 5, 52
Milan, 166
Milch, Pearl, 383, 400
Miller, H. A., 353
Miller, R. Stevenson, 192
Milwaukee, 221
mineral engineering, 115, 322, 344, 378
mining engineering, 83
Mining Engineering, Dep't of, 289, 392
Ministry of Agriculture, 267
Ministry of Commerce and Industry, 317, 356, 364
Ministry of Defense, 235, 251, 257, 266, 349, 386
Ministry of Education and Culture, 237, 255, 271, 310, 331, 348, 355, 365, 373, 386
Ministry of Finance, 237, 251, 258, 269, 293, 309, 317, 331, 345, 346, 348, 382
Ministry of Foreign Affairs, 319
Ministry of Higher Education and Science, 237, 331, 349
Ministry of the Interior, 251
Ministry of Labor, 316, 317,

331, 343, 366, 380
Ministry of Transport, 342, 391
Minsk, 26
Mishcon, Victor, 279, 384, 400
Mishkin, Eliezer, 232, 265
Mitelman, Mordecai, 268
Mitrani, Marco, 312
Mitweide, 96
Moabit Prison, 173
Mohl, E. H., 82
Moisseiff, Leon S., 125, 147, 148, 368
Mokady, Raphael, 351
Moller, Eric, 262
Monasch, Harold, 224
Mond, Sir Alfred, *see* Melchett, Lord
Montgomery, Gen. Sir Bernard L., 222
Montor, Henry, 195, 218
Montreal, 246, 383
Moscow, 3, 10, 12, 31, 104, 113, 120, 132, 274
Moscow Technical Institute, 3
Moses, 1, 92
Motzkin, Leo, 57, 59, 159
Mouchly, Joseph, 321
Mount Scopus, 117, 156
Mufti of Haifa, 48
Muggia family, 166, 169
Muggia, Giuseppe, 166, 194
Munich, 108, 120
Munich Conference, 196, 197
Munk, Nicholas M., 246, 400
Munnich, Robert, 247
music at Technion, 201, 299, 371
Mydans, Max, 222

Nachlat Yehuda, 100
Nadaf, Yaish, 27, 28
Nahalal, 116
Nahariya, 236
Naiditch, Isaac, 84
Nairobi, Royal Technical College, 318
Nairobi, University College, 321
names of the Institute, 8, 12, 42, 47, 90, 91, 103, 121, 244, 263, 355
Namir, Mordecai, 316
Naor, Pinhas, 341, 354, 355, 365
Naot, Yehuda, 234

Naphtali, Peretz (Fritz), 108, 172
Naples, 166
Napoleon's soldiers, 66
Nasser, Gamal Abdel, 349
Nathan, Lord, 246
Nathan, Paul, 5–17, 19, 20, 23, 29, 30, 35, 36, 37, 38, 39, 40, 41, 42, 43, 46, 48, 50, 53, 54, 56, 57, 58, 59, 60, 62, 65, 68, 73–74, 75, 93, 96, 110, 173, 251
National Council for Higher Education, *see* Council for Higher Education
National School for Senior Technicians, 380
Natovich Orthopedic Rehabilitation Research Center, 398
Nautical School, 192, 195, 196, 199, 201, 215–216, 243
Naveh Shaanan, 185, 258, 259, 260–261
Nazis, 203, 208, 209, 213, 220, 233, 362
Neaman Institute for Advanced Studies in Science and Technology, 383, 385, 391
Neaman, Morris, 222
Neaman, Samuel, 382, 400
Negev, 265, 276, 359
Nesher Cement Works, 85, 279
Nessyahu, Yehoshua, 109, 112, 215, 292, 309, 348
The Netherlands, 190, 220, 384, 400
Netter, 41
Neufeld, Zipporah, 113, 114, 141
Neumann, Alfred, 328, 339
Neumann, Heinrich, 120, 124, 135, 136, 147, 151, 154, 191, 202
New Delhi, 202
New Jersey, 221
The New Palestine, 274
New York, 11, 12, 40, 58, 62, 82, 89, 130, 148, 149, 176, 179, 195, 218–224, 249, 345, 383
The New York Times, 47, 56
New York University, 239
The New Yorker, 309
New Zealand, 2
Newmark, Morris, 221

434

Nigeria, 321
NILI, 67
Nir, Dov, 321
Nissan, Isaac, 348, 396
Noah, 142
Norman, Edward, 221
Norsa, Comm., 166
nuclear energy, 352, 389
nuclear engineering, 289, 343–344
nuclear science, 323, 376, 389
numerus clausus, 1, 152
Nuremberg Trials, 304
Nurick, Henry J., 82

OBE (Order of the British Empire), 283
Odessa, University of, 107
Ohel Aharon, 353
oil, "discovery" of, 40
Oliphant, Laurence, 15
Ollendorff, Franz, 173, 191, 192, 196, 197–198, 202, 206, 236, 239, 249–250, 275, 381
opening ceremony, 110–112
Oren, Eliyahu, 215
Osrin, Gus, 282, 384
Ostrow, Seniel, 291
Osunsanya, Tundeh, 319
Ottawa, 125
Ottoman Empire, *see* Turkey
Ottoman Law of Societies, 263
Oxford University, 194

Palestine Council for Scientific and Industrial Research, 238
Palestine Council for Technion, 102
Palestine Economic Corporation, 148, 149, 176
Palestine Electric Corporation, 131, 191, 192, 205
Palestine Executive, 80, 93, 101, 102, 152, 159
Palestine Management Committee, 101, 105, 106, 118, 120, 121, 122, 123, 124, 128, 132, 134, 136, 144, 159
Palestine Maritime League, 192, 216, 221, 243
Palestine Orange Growers' Loan, 71
Palestine Pavilion, World's Fair, 218

Palestine Post, 214
Palestine Railways, 206
Palestine Restoration Fund, 71
Palmah, 210
Pamm, Max, 125
Pardo-Roquez, Comm., 166
Paris, 6, 40, 59, 167, 186, 393
Parma, 166
Pasadena, 127
Pasha, Jerusalem, 46
Passfield White Paper, 200
Pat, Yaakov, 131, 181, 185
Patai, Raphael, 215
Patinkin, Don, 365, 366
Peel Royal Commission, 189, 200
Peled, Avraham, 244
Peled, Yaakov, 262
Pelleg, Frank, 315
Penner Wing, 398
Peranio, Anthony, 313
Peres, Asher, 298, 367
Perlstein, Harris, 222
Persian language, 7
Peru, 321
Petach Tikva, 15
Peter, Yehuda, 321
Les Petites Soeures de Pauvres, 357
Petrie Pavilion, 398
Pewsner, Leah, 15
Pewsner, Shmuel, 14, 15, 22, 29, 52, 71, 79, 96, 102, 106, 110, 113, 116, 123, 124, 127, 129, 133, 134, 135, 136, 137, 145, 146, 150–151
Philadelphia, 65, 221
Philadelphia Hostel, 291
Phillipsohn, Martin, 10, 41
Phoenicia, District Governate, 68
physics, 95, 112, 115, 180, 190, 227, 242, 301, 304, 311, 322, 351, 365, 373, 394
Physics, Dep't of, 266, 292, 304, 310, 390, 391, 398
Piattelli, Fidia J., 239, 241, 266
PICA, 137
Pincus, J., 83
Pincus, Sol, 220
Pinkas, David Zvi, 161, 181, 215
Pinshow, David, 325
Pinsk, 104

Piper Cubs, 206, 236
Pisa, 166, 304
Pittsburgh, 194, 221, 222, 374
Plastic Materials Laboratory, 379
Plugat Solel Boneh, 208
plywood levy, 318
Poale Zion, 27
Point Four Program, 289
Polak Michel, 279, 292, 378
Poland, 65, 92, 126, 137, 138, 171, 172, 190, 199, 200, 215
POM, 208
Port Harcourt, 321
postage stamp, Technion, 293, 375
Potash Works, 205
Potsdam, 15
Prague, University of, 304
Pratt and Whitney engines, 204
pre-academic preparatory program, 311, 373, 374
President, office of, 370
President, residence of, 397
press, academic, 242
Prime Minister's Office, 237, 238, 255, 346, 354, 366, 375
Princeton University, 219, 241, 266, 282
Pritzker, Asher, 246
Prochownik, 92
production engineering, 289, 365
Propp, Arthur, 91
Providence, R.I., 221, 222
Provident Fund, 245
Provisional Committee for Management of the Technion, 76, 79, 84, 85, 86, 87, 88, 89, 90, 100
Provisional Executive Committee for General Zionist Affairs, 63
Prussia, Government of, 21
public relations, 215, 219, 223, 243, 274, 375, 377
Public Works Department, 98, 140, 147, 149, 165
Purdue University, 220

Quat, Jacob, 288

Rabi, Isidor I., 315

Rabinowitz, John, 91
Rabinowitz, Sidney, 222
Rablinsky, Walter, 91
Raczkowsky, J., 274
radioactivity of Tiberias Hot Springs, 265, 274
Ramat Hasharon, 397
Rambam Hospital, 356, 357, 368
Raphael, Zeev H., 321
Rappaport, Baruch, 368
Rappaport Family Medical Sciences Building, 399
Ratner, "Jenka", 208–209
Ratner, Yohanan (Eugene), 120, 151, 183, 184–185, 191, 202, 207, 213, 215, 235, 261, 262, 274, 288, 294–295, 321
Ravikovitz, Shlomo, 218
Rawraway, Nathan, 91
Razsvet, 65, 128
Reading, Lady, 247
Reali School, 20, 28, 29, 52, 75, 76, 79, 80, 93, 95, 103, 107, 108, 129, 133, 152, 155, 161, 164, 174, 177, 255
Recanati, Ernesto, 166
rector, 333
Red Crescent, 65
Redstone, Louis, 222
Rehovot, 240
Reich, Simon, 97
Reichstag, 60, 171
Reiner, Markus, 98, 147–148, 232, 303, 322, 376, 382
Reiser, Jacob, 181, 215, 262, 295
Remez, David, 102, 123, 217, 255
Ernst Renan, French warship, 66
reorganization, 295, 308, 370, 384, 395
research, 195, 264, 267, 298, 322–323, 362, 378–379, 381, 392–393
Research and Development Foundation, 267, 275, 305, 324, 378
Research Regulations, 298
Resnick, William, 268, 309, 327, 333
Revisionists, 158, 169
Rheinisch-Westfalische Technische Hochschule, 393

rheology, 322, 376
von Ribbentrop, Joachim, 199
Riesman Plaza, 399
Rigbi, Zvi, 298
Rijksuniversiteit, 393
RILEM, 304
Rinkoff, Eliezer, 141
Road Safety Center, 342, 378, 391
Robinson, Nathan, 276
Robinson, Sir Robert, 246
robots, 389
Rohrbach, Paul, 60
Rom, Yael, 351
Rom, Yosef, 382
Roman catapults, 184
Romania, 6, 215
Rome, 166, 239
Rome, University of, 109
Rommel, Field Marshal Erwin, 213
Rose, David, 220, 262, 282, 374, 379
Rose, Rebecca, 324
Rosen, Abraham, 371
Rosen, Edward E., 279, 384
Rosen, M., 82
Rosen, Maurice, 346, 382
Rosen, Maurice and Ruben, Solid State Institute Building, 374, 390
Rosen, Nathan, 268, 298, 309, 333, 360
Rosen, Pinhas, 262
Rosenberg, J.E., 222
Rosenberg, Yitzhak, 27
Rosenblatt, Bernard, 181, 219, 232
Rosenblatt, David B., 274
Rosenbloom, Charles J., 222
Rosenbluth, Felix, 218
Rosenfeld, Y., 102, 110, 123, 137
Rosenwald, Julius, 35, 54, 58, 71, 135
Rosenzweig, A., 82
de Rothschild, Baron Edmond James, 11, 54, 59, 167
de Rothschild, Evelyn, 370, 384, 388
Rothschild Hospital, 368
Rothschild Memorial Group, 310
de Rothschild School, Evelina, 5

Rothschild, Y., 65, 66
Royal Air Force, 65, 204
Royal Corps of Engineers, 179, 204, 208
Royal Fusiliers, 255
Rubien, Murray, 282
Ruppin, Arthur, 37, 111, 261
Rural Building Research Center, 379
Russia, 48, 49, 59, 85, 104, 112, 113, 120, 126, 136, 138, 139, 185, 190, 213, 215; see also Soviet Union
Russian language, 43, 108, 120, 322
Rutenberg, Avraham, 191, 262
Rutenberg, Pinhas, 77, 127, 191, 206

Sabinia, 224
Sacharov, Eliahu, 318
Safed, 5, 47, 189
Sagi, Ram, 371
St. Louis, 221, 222
Salaman, Redcliffe N., 180, 246
Salenger, Oscar and Ethel, 341
Salomon, Haim, 170
Salomon, Yaakov, 233, 305
Salonica, 24
Sambursky, Samuel, 93
Samuel, Sir Herbert, 71, 179, 247
Samuel, Myron, 246
Samuel, Rudolf, 93, 195–198, 218–221, 223, 226, 247, 283
San Francisco Agricultural Engineering Building, 340
Sanhedrin, 59
sanitary engineering, 97, 176, 289, 344
Sapinsley, Milton, 222
Sapir, Eliyahu, 15
Sapir, Pinhas, 274, 317, 345, 346, 347
Sarajevo, 59
Schacht, Lawrence, 382
Schacht Park, 399
Schaudinischky, Leo, 206, 236
Schechter, Solomon, 12, 41, 47, 53, 54
Scher, Charles, 400
Schiff, Jacob, 11–13, 42, 52, 53, 54, 56, 58–60, 62, 63, 68–72, 74, 82, 132

Schiff, Mrs. Jacob, 194
Schiff, Ludwig, 13
Schiff, Mortimer, 12, 54
Schiff, "Otto", 70
Schlesinger, George, 19, 35, 41, 43, 50, 173
Schleswig, 92
Schlosser, Erika, 298
Schmidt, Eduard, 46, 48, 53
Schnepp, Otto, 268
von Schoen, Secretary of State, 24
School of Commerce, Jerusalem, 39
School for Senior Technicians, 343
Schooler, Nathan, 274
Schulman, 194
Schultz, Morris, 224
Schumacher, Gottlieb, 21, 26
Schwartz, William, 282
Schwerin, Edwin, 172, 191
Science, Faculty of, 266, 305
Scientific Editing Office, 378
Segall, Karl, 222
Seigel, Saul, 400
Seligman, Adrian, 192
Seligman, Richard, 192
Senate, 190, 191, 298, 308, 330, 333, 336, 337, 338, 342, 343, 355–356, 358, 386, 396; see also Teachers' Council
Senate Building, 280, 324
Senator, Werner, 196
Senior, Woolf, 282, 324
Sensibar, Jacob R., 382
Sephardim, 14, 310, 311, 312, 352
Septuagint, 376
Serbia, 59
Shaag, Shlomo, 141, 182
Shachefet, 243
Shafer, H. Jerome, 266
Shalon, Rahel, 131, 176, 182, 184, 242, 268, 298, 303, 309, 316, 320, 321, 327, 330, 333, 341, 365
Shalon, Uriel, 156, 254, 262
Sharef, Zeev, 348–349, 364
Sharett, Moshe, 39, 47
Sharon, Aryeh, 259
Sharp, Max, 84
Shazar, Zalman, 262
Shear-Yashuv, Aharon, 353

Shein, Max, 384, 400
Shein Student Hostel, 398
Shemen, 22
Shenburg, Bernard, 108
Shenkar, Aryeh, 217, 262
Shereshefsky, Judah L., 241
Sherman, George and Beatrice Building, 391, 398
Sherman, Harry and Abe, 280
Sherman, Leonard and Diane, Environmental Engineering Research Center, 378, 398
Sherman, Nate, Family Forum, 399
Shertok, Moshe, see Sharett, Moshe
Shevelov, Katriel, 18
Shevelov, Moshe, 377
Shevelov, Tova, 231
Shfar-Am, 258
Shiloni, Ephraim, 112
Shimoni, Emanuel, 399
Shine, Barnett, 292, 340, 370, 374, 384
Shipyards, Israel, 210
Shklarsky, Bezalel, 35, 46
Shklarsky, Elisha, 257, 268, 338, 341, 369–370, 371
Shochat, Israel and Manya, 28
Shoolman, Max, 149, 194, 222
Shukri, Hassan, 186
Sideman, Samuel, 371
Siegel, David T., 292, 378
Silkin, Lord, 246, 262, 279
Silver, Abba Hillel, 221
Silver, Brian, 371
Silver, Julius, 358, 390, 398
Silver, Julius, Biomedical Engineering Institute, 379
Simon, Ernst, 108
Simon, James, 5–7, 9, 10, 16, 22, 37, 38, 41, 48, 53, 57, 58, 59, 60, 65, 67, 70, 73, 93, 110
Simon, Julius, 87, 148
Sinai, 349, 351
Singapore, 218, 282, 321
Singer, Paul, 396
Sinyaver, A., 217
Sitte, Kurt, 288, 304–305
Sivan, Raphael, 371
Six-Day War, 334, 349–351, 360, 364
Slomovitz, Philip, 222
Smoira, Moshe, 170

Smuts, Jan, 246
Smyrna, 14
Sobell Chemical Engineering Building, 374
Sobell, Sir Michael, 279, 384
Sobeloff, Isidore, 222
Sochazewer, Eliyahu, 85, 229
Society of Zionist Engineers, *see* Zionist Society of Engineers
Soil Conservation Service, 267
soil engineering, 176, 190, 292, 344
soil mechanics, 289
Soil and Road Testing Laboratory, 378
Soils and Fertilizers Laboratory, 379
Sokolow, Nahum, 91, 102, 150, 158, 161, 163
Solan, Alexander, 298
Solel Boneh, 84, 100, 103, 200, 208, 317
Solid State Institute, 342, 379, 386, 390
Solomon, King, 142
Solomon, Lt. Col. H., 90
Soloweitchik, Max, 90, 189, 215
Sonneborn, Ferdinand, 148, 195, 219
Soroka, Itzhak, 321
Sourasky, S. D., 181
South Africa, 125, 149, 167, 274, 282, 384, 400
South African Technion Society, 247, 384
South African Zionists, 81
South Indian Railways, 143
Soviet Union, 199, 213, 232–233, 322, 374, 381, 394; *see also* Russia
Spalato Cement Co., 81
Spandau, 15
Spektor, Shlomo, 141
Spertus Auditorium, 340
Spertus, Maurice, 262
Spiers, Frederick, 122
Spinoza, Baruch, 19
Spira, Ephraim, 321
Split, 81
Sporn, Philip, 219, 220, 323, 324
sports, 109, 298–299, 371
Sprinzak, Yoseph, 158

Stander, Isaac J., 82
Stanford University, 127
Starr Chemical Laboratories, 83
State Department, 267, 285
State of Israel, 234, 235, 263, 276, 380
Stearns, Eugene, 400
Steering Committee, 396
Stein, Benjamin Walter, 224
Stein Food Engineering Building, 398
Steinberg, Saul, cartoon, 309
Sterling, Sir Louis, 279, 280
Stern, Edgar, 279
Stiefel, Otto, 274, 377
Stoch, Leslie, 321
Stock, Emilio, 81
Stone Technology Center, 378
Strauss, Nathan, 54, 130
Strauss, Samuel, 12, 42, 54
Strich, Eliezer, 108, 109, 132
strike, staff, 137, 265, 293, 332, 337, 381
strike, students, 113, 199, 231, 270, 271, 278, 300–302, 346, 373–374
Student Association, 115, 158, 209, 301, 314, 352, 370, 371–372, 373, 397
student ball, 117, 137
student dormitories, 20, 149, 216, 228, 278, 292, 294, 298, 324, 340, 374, 383, 398, 399
student enrollment, statistics of, 112, 117, 121, 147, 171–172, 190, 215, 229, 270, 295, 310, 333, 340, 373, 374, 381, 385, 394
Student Union Building, 308, 340, 370
Study of Man at Work Center, 379
Sturman, "Sabta", 28
Suchman, Yehoshua, *see* Nessyahu
Sudetenland, 196
Suez Campaign, 290
Suez Canal, 65, 208
Sugarman, M. Henry, 179, 195, 219
Sulzberger, M., 36, 42, 54
summer semester, 365, 367, 381
Supreme Court, Israel, 92, 170

Susman, Louis and Rebecca, 324
Swahili, 319
Switzerland, 2, 166, 168, 190, 197–198
Swope Fund, 307
Swope, Gerard, 219, 282–285
Swope, Henrietta, 284
Swope, Mary Hill, 283–284
Sydney, 400
Symes, Col., 111
synagogue, 130, 353
Syracuse University, 304
Syria, 140, 156, 165, 184, 203, 208, 225, 350, 386
Szold, Henrietta, 136, 144, 149
Szold, Robert, 195

Tabb, Jay, 289, 321
Tachauer, David, 52
Tanganyika, 318
Tanne, David, 317
Tantura, 27
Taub, Alex, 275–276
Taub, Ariel, 321
Taub, Henry, 400
Taub, Henry and Marilyn, Computer Science Building, 374
Taub, Morris and Sylvia, Computer Center, 340, 374
Tcherniavsky, Aharon, 95, 96, 107, 111, 112, 113, 114, 124, 133, 134, 135, 136, 145, 146, 147, 151, 170, 190, 231, 262, 270
Tcherniavsky, Mrs. Aharon, 117
teacher training, 239, 302, 322, 354, 391; *see also* Education in Science and Technology
Teachers' Association, *see* Hebrew Teachers' Association
Teachers' College, Jerusalem, 39, 45
Teachers' Conference, 3, 4
Teachers' Council, 122, 123, 124, 129, 133, 135, 136, 138, 150, 153, 155, 156, 169, 228, 232, 235, 241, 242, 264, 270; *see also* Senate
Teachers' Union, 169, 170
Technical High School, *see* Bosmat
Technion City, 45, 84, 261,

286, 292, 300, 324, 334, 335, 343, 345, 354, 358, 377, 396, 398, 399, 400
Technion Society of Great Britain, *see* British Technion Society
Technion Year Book, 285
Technische Hochschule, Charlottenburg, 19
Teicher, Yosef, 357
Tel Aviv, 7, 22, 75, 77, 85, 97, 117, 123, 136, 154, 165, 193, 217, 255, 258, 275, 355, 356, 359, 365, 377
Tel Aviv University, 343
Templar colony, 37
The Temple, 47, 142
Temple Emanuel, New York, 58
textile engineering, 356
Thalheimer, Alvin, 221
Thau, Avshalom, 268
Thon, Jacob, 7
Throop Institute, 127
Tiberias, 5, 40, 189, 259, 269
Tiberias Hot Springs, 265, 274
Tiegerman, Moshe, 141
Tietz, Oskar, 22
Timendorfer, B., 10
Times Square, 159
Tirat Hacarmel, 251–252, 255–258, 261, 268, 324
Tisch, Laurence A., 382, 400
Tobias, Lily, 384
Tolkowsky, Dan, 346, 384
Tolkowsky, Samuel, 97
Tooval, Ben Zion, 243, 276
Toren, Tuvia, 380
Toronto, 84, 246
town planning, *see* architecture and town planning
Town Planning, Research Institute for, 215
Trade School, *see* Bosmat
Transjordan, 225
transportation engineering, 227, 239
Transportation Research Institute, 391
Trieste, 81
Truesdell, C., 376
Tscherniak, Zipporah, *see* Neufeld, Zipporah
Tschlenow, Yechiel, 12, 17, 29, 40, 41, 42, 55

437

Tsimhi, Yehoshua, 267
Tu B'Shevat, 298
tuition fees, 115, 118, 122, 129, 150, 164, 178, 193, 194, 215, 230–231, 253, 270, 273, 278, 290, 300, 346, 367, 373, 374
Tulin, Abraham, 220, 257, 262, 283, 284, 307, 383, 385
Tumpeer, Joseph, 222
Tunis, 220
Tunisia, 207
Turetzky, Shimshon, *see* Ben-Tur, Shimshon
Turkey, 4, 6, 7, 8, 12, 13, 14, 15, 20, 24, 29, 43, 49, 55, 62, 65, 66, 72, 75, 79, 225, 321
Turkish army, 65, 66, 76
Turkish language, 36, 44, 55
Turkish Revolution, 24
Tzerkowitz, Guido, 173

Uganda, 5, 6
U.I.T., 249
U.J.A., 250, 344–345, 347, 361
Ullmann, Jacob W., 382
Ullmann Teaching Center, 340
UNESCO, 289
United Fruit Co., 219
United Israel Appeal, 248, 249
United Nations, 222, 231, 233, 283
United Nations Special Fund, 323
United Nations Technical Assistance, 289
United States, 8, 53–59, 60, 61, 71, 81, 82, 90, 125, 144, 147–148, 159, 167, 190, 199, 218–226, 229, 235, 246, 247–249, 282–286, 344–345, 357, 361, 393
U.S. Air Force, 289, 303, 323
U.S. Bureau of Standards, 323
U.S. Department of Agriculture, 323
U.S. International Cooperation Administration, 289
U.S. Office of Naval Research, 289
U.S. Operations Mission, 289
University College, London, 101
University Institute, *see* Haifa University

University of Israel, 238
University, Jewish, 1, 2, 3, 40, 71
University of the Negev, *see* Ben Gurion University
U.P.A., 195
Urban and Regional Studies Center, 378
Uretzky, Leib, 141
Urey, Harold C., 283
Uruguay Technion Society, 246, 384, 400
Ussishkin, Menahem M., 2, 3, 5, 16, 68, 75, 78, 82, 83, 84, 87, 88, 90, 95, 96, 97, 101, 102, 111, 123, 124, 125, 126, 127, 129, 132, 145, 150, 160, 189, 191, 201
U.S.S.R., *see* Soviet Union

Vaad Ha-Lashon Ha-Ivrit, 130, 375
Vaad Leumi, 102, 158, 160, 161, 167, 174, 186, 237, 262
Vaad Menahel, 102, 123, 133, 135, 136, 137, 147, 152, 161, 164, 166, 168, 170, 174, 180, 181, 207, 215, 228, 231, 232, 235, 238, 241, 242, 243, 249, 254, 255, 261, 262, 286, 327, 329, 333, 334, 343, 345, 358, 396
Vajda, Edward, 282
Valdora, 216
Van Arsdale Wing in Electrical Engineering, 374
Vaturi, Josef, 400
Venezuela, 384, 400
Veterans' Administration, 224
Vienna, 1, 41, 154, 158, 176, 368
Vienna Technical School, 159
Vilna, 47, 85
Vilna Jewish Technical School, 193
Vilnay, Zev, 46
Visitors' Reception Center, 377
Vizansky, Zvi, 141
Volcani, *see* Wilkensky-Elazari
Vorenberg, Frank, 222
Vulcan Foundries, 203
Vulcan, S. S., 224

Wachman, Avraham, 337, 338

Wachs, Alberto M., 289
Wadi Rushmiyah, 233–234
Wallace, Henry A., 231
War of Independence, 184, 224, 238
War Supply Board, 202
Warburg, Felix, 149, 194
Warburg, Mrs. Felix, 130, 179, 194
Warburg, Max, 12, 152
Warburg, Otto, 72, 73, 79, 80, 81
Warsaw Ghetto, 200
Warsaw, Polytechnic Institute, 82, 131
Washington, D.C., 89, 221
Wasserman, Shimon, 141
Water Resources Research Center, 378
Watson, Rivka, 274, 377
Wattenberg, Judah, 219, 282
Wauchope, Sir Arthur, 179, 186
Wavell, Gen. Archibald P., 207
Wechsler, Albert H., 222
Weinbrand, Abigail, 210
Weisgal, Meyer, 248, 249, 291
Weissbrem, 91
Weizmann, Chaim, 1, 2, 3, 19, 33, 41, 57, 59, 68, 69, 80, 81, 83, 87, 89, 90, 101, 102, 105, 106, 111, 118, 123, 125, 127, 128, 143, 149, 150, 157, 158, 167, 214, 219, 223, 240, 244, 247, 368
Weizmann Institute, 222–223, 231, 237, 238, 248, 249, 255, 273, 291, 293, 343, 360, 361
Weizmann, Rachel-Leah, 368
Weizmann, Selig, 78
well, 20, 28, 52, 85, 106, 118, 154
Werksman Building in Physics, 374, 398
Wertheimer, Joseph, 222, 274
West Bank, 351
West Indies, 319
West Nigeria, 319
Western Desert, 204
White, Lazarus, 221, 224, 225, 247, 282
White Paper, 200, 209
WHO (World Health Organization), 289

Wilbushewitz, Gedalyahu, 26–28, 31, 35, 41, 51, 97–99, 187
Wilbushewitz, Nahum, 14
Wilentchuk, Yitzhak, 262
Wilhelm II, 24, 38, 45, 61
Wilinsky, Charles F., 222
Wilk, Benjamin, 222, 274
Wilkensky-Elazari, Yitzhak (Volcani), 160
Wilmington, Martin W., 203
Winterthur, technical secondary school, 2
Wise, Stephen S., 218
Wiseman, Gdalyah, 321, 371
Wissotzky, David, 9, 10, 56, 57, 59, 81, 110
Wissotzky family, 11, 57, 63, 68
Wissotzky Foundation, 10, 29, 30, 52, 56
Wissotzky, Kalonymous Zeev, 9, 10, 11, 74, 132
Wissotzky trustees, 13, 65
Wix, Michael and Anna, 291
Wolberg, Samuel, 222
Woldenberg, Max, 222
Wolffsohn concern, 92
Wolfson, Charles, Charitable Trust, 384
Wolfson Foundation, 384
Wolfson, Sir Isaac, 279, 400
Wolman, Abel, 221, 262
Women's Division — American Technion Society, 221–222, 291, 383, 400
Women's Division — Canadian Technion Society, 383
Women's League, *see* Women's Division — American Technion Society
Woodhead Commission, 200
World Union of Jewish Students, 158
World's Fair, New York, 218
Wunsch, David, 220
Wunsch, Harry, 220
Wunsch, Joseph W., 219, 220, 224, 225–226, 247, 262, 282, 315
Wunsch, Samuel, 220

Yaarot Hacarmel, 258
Yach, Solm, 400

Yaffe, Bezalel, 288
Yale University, 127
Yasky, Abraham, 273
Yellin, Aviezer, 45, 142, 155
Yemen, 28
Yemenites, 27, 28, 296
Yiddish, language, 7
Yiftach, Shimon, 323, 343
Yitzhaki, David, 141, 268
Yom Kippur, 72, 386
Yom Kippur War, 375, 380, 387, 388, 399
Young Turks, 24

Zacks, Yaakov, 184
Zaira, Y., 276
Zakai, Moshe, 396
Zakai, Shlomo, 111, 141
Zakon, Eliyahu, 273
Zangwill, Israel, 6, 70, 71
Zaslavsky, Aharon, 268
Zaslavsky, Dan, 321

Zebulun Valley, 250
Zehavi, Menahem, 396
Zeitlin, J., 10
Zeitlin, J. G., 289
Zemurray, Samuel, 219
Zichron Yaakov, 2, 3, 54
Zimels, L., 91
Zimmermann, Arthur, 42, 53
Zini, Eliahu, 353
Ziniuk, Yaakov, 210
Zionism, 36–38, 41, 48, 53, 56, 61, 70, 71, 155
Zionist Actions Committee, 44, 45, 52, 79, 95, 127, 128, 153, 167, 171, 189
Zionist Commission in Palestine, 74, 76, 77, 78, 80, 84, 88
Zionist Congress, First, 1
Zionist Congress, Fifth, 1
Zionist Congress, Sixth, 2
Zionist Congress, Eleventh, 41
Zionist Congress, Twelfth, 80, 86, 87, 89

Zionist Congress, Thirteenth, 90, 91, 99, 103
Zionist Congress, Fifteenth, 133
Zionist Congress, Sixteenth, 145, 151, 154, 157, 159
Zionist Congress, Seventeenth, 151, 155, 157, 158–159, 163, 171, 208
Zionist Congress, Nineteenth, 172
Zionist Congress, Twentieth, 199
Zionist Congress, Twenty-first, 199
Zionist Congress, Twenty-second, 239
Zionist Executive, 74, 81, 86, 89, 90, 97, 102, 119, 121, 122, 123, 124, 126, 129, 133, 134, 135, 136, 137, 138, 140, 142, 144, 145, 147, 150, 151, 153, 157, 159, 162, 166, 178, 186, 207
Zionist General Council, 68, 70, 145
Zionist Organization, 2, 4, 16, 17, 44, 55, 56, 57, 59, 61, 62, 67, 68, 69, 70, 71, 73, 81, 84, 85, 87, 88, 89, 90, 91, 95, 100, 115, 119, 121, 123, 126, 128, 133, 137, 152, 153, 158, 159, 163, 167, 196, 215, 237
Zionist Organization of America, 68, 69, 71, 82, 274
Zionist Society of Engineers and Agriculturists, 82, 83, 148, 219
Ziskind, Jacob, 292
Ziv, Jacob, 396
Znamirowska, Rachel, *see* Shalon, Rahel
Zorea, Meir, 209
Zur, Benjamin, 321
Zurich, 2, 393